W9-CPF-882

# THE ART OF EDITING

# THE ART OF EDITING
## Sixth Edition


Floyd K. Baskette

*Late, University of Colorado*

Jack Z. Sissors

*Medill School of Journalism, Northwestern University*

Brian S. Brooks

*Missouri School of Journalism*



**Allyn and Bacon**

Boston • London • Toronto • Sydney • Tokyo • Singapore

A Viacom Company
160 Gould Street
Needham Heights, Mass. 02194

Internet: www.abacon.com
America Online: Keyword: College Outline



*Vice President, Humanities:* Joseph Opiela
*Series Editorial Assistant:* Kate Tolini
*Marketing Manager:* Karon Bowers
*Production Coordinator:* Thomas E. Dorsaneo
*Editorial Production Service:* Melanie Field, Strawberry Field Publishing
*Design and Composition:* Wendy LaChance/By Design
*Composition and Prepress Buyer:* Linda Cox
*Manufacturing Buyer:* Suzanne Lareau

**Library of Congress Cataloging-in-Publication Data**

Baskette, Floyd K.
The art of editing / Floyd K. Baskette, Jack Z. Sissors,
Brian S. Brooks —6th ed.
p. cm.
Includes index.
ISBN 0-205-26219-8
I. Journalism—Editing. 2. Copy-reading I. Sissors, Jack Zanville
II. Brooks, Brian S. III. Title
PN4778.B3 1996 96–29201
808'.06607—dc20 CIP

Printed in the United States of America

10 9 8 7 6 5 4 3 2 1 01 00 99 98 97 96

# CONTENTS

# PREFACE

No previous editon of *The Art of Editing* has undergone as many changes as this, the sixth edition. There is ample justification for that; the media marketplace is changing rapidly, and with it the news business is changing. We have altered and updated this text to reflect those changes.

No longer do newspapers dominate the scene. Nor does television. Nor do magazines. Indeed, today's media marketplace is more fractured than ever. New computer-based competitors—the so-called new media—are emerging as serious challengers for the public's attention. The proliferation of magazines targeted at well-defined audiences—and cable television channels that do the same—also are having a major impact.

Almost every week, it seems, media companies are in the news: Disney buys ABC. Westinghouse buys CBS. Microsoft and NBC collaborate to produce an all-news cable channel and an Internet-based news service. All of it is happening at a dizzying pace.

We continue to stress the importance of editing as an art while recognizing the monumental changes taking place in the corporate media arena in which the editor operates. This edition gives more attention than previous ones to the process of editing for the broadcast media, for corporate and general magazines and, yes, for the new media. Longtime users of the book will find that we have retained the depth and breadth that have made *The Art of Editing* the most successful editing text of all time. We think that both those users and new ones will be delighted with the changes and additions that make this the most up-to-date text of its kind.

In this edition, as in earlier ones, we have included numerous examples of editors' successes and failures as illustrations of how to edit and how *not* to edit. We have taken examples, both good and bad, from newspapers, magazines and broadcast stations coast to coast. Through them, we learn.

Journalism is an interesting, stimulating and exciting profession. Editing, in turn, is a vital part of journalism, both print and broadcast. Newspapers, magazines, radio, television and on-line services would not be nearly as good without editors as they are with editors. They can be superb with top-flight editors. We hope this book inspires some of you to become just that.

Many of our examples are taken from newspapers, where most of the conventions of editing evolved. But the techniques described herein apply just as readily to magazine, broadcast and on-line editing. Most of the techniques are the same, and where they are not we have highlighted the differences.

The excitement of producing the news is universal, and it is a process in which editors are full partners. Still, it is difficult for any book to capture the excitement of editing because the beginner must first master the intricacies of the editor's art. Attention to detail is of primary importance to the editor, and we believe this book attends to that detail more thoroughly than any other. We hope we have done so as interestingly as possible.

Those of you who are attentive to detail will notice a variation from AP style. We have adopted the book publisher's convention of italicizing newspaper, magazine and book titles.

We are indebted to our colleagues, students and editors who read chapters and offered many helpful suggestions during the revision process. We extend special gratitude to Sandra Davidson, an attorney and teacher whose advice was invaluable in updating the chapter on media law, and to Barbara Luebke, who wrote the information on stereotyping.

We also would like to thank the reviewers whose comments have helped greatly in the improvement of the sixth edition: Dr. Douglas J. Carr, St. Bonaventure University; John E. Newhagen, University of Maryland; Tom Donohue, Virginia Commonwealth; Don Zimmerman, Colorado State University; Bill Brody, University of Memphis; Glen Bleske, California State University–Chico; Jay Goldman, American Association of School Administrators; Bill Ferguson, Columbia College; Emil Dansker, Central State University; Janet Rohan, Pikes Peak Community College; and Karen Springen, Newsweek.

We also extend thanks to Joe Opiela, our editor at Allyn & Bacon, and his talented staff, especially Kate Tolini. The professionalism of the Allyn & Bacon staff made simple our transition from Macmillan Publishing Co., the publisher of earlier editions from whom Allyn & Bacon purchased rights to the book.

We hope this edition has been prepared with the same high standards set by our colleague Floyd K. Baskette, who died in 1979. His name remains on this edition because his work was of enduring quality, and some of it remains from the first two editions.

We have changed and updated, but one axiom holds true: Editing is an art no matter where or by whom it is practiced. To those who will accept the challenge of careful and thoughtful editing, this volume is dedicated.

*Brian S. Brooks*
*Jack Z. Sissors*

# THE ART OF EDITING

# PART 1

# EDITING IN THE INFORMATION AGE

# CHAPTER 1

# EDITING FOR TODAY'S CHANGING MEDIA

## THE MEDIA ARE CHALLENGED

For generations, news has been mass-produced for public consumption in assembly-line fashion. Much of it still is. Reporters gather and write it, editors edit it, and others produce and distribute it in print or broadcast form to mass audiences. That model, born in the Industrial Revolution of the 19th century, remains dominant today. But the old model is beginning to yield to the communications model of the future, one born in the Information Age. It is a model dramatically different from that of the past.

Explains Peter Leyden, a staff writer for the *Star Tribune,* published in Minneapolis:

> The media business is, by definition, one of the core industries of the Information Age. As the Digital Revolution fundamentally changes the way information is produced and disseminated, expect the media to be traumatized first.
>
> All sectors of the media, from broadcast television to the movie industry to the book-publishing world, are already experiencing varying degrees of trauma that will only worsen in the coming decades.
>
> Cheaper digital tools, such as desktop publishing and multimedia equipment and software, will allow much smaller competitors to emulate the feats that once took multimillion-dollar budgets. And the new information infrastructure will open up a distribution channel that will undermine the monopolistic channels used now.
>
> The handful of television networks, already hit by cable TV, will really suffer when 500 channels or more can flow into your home. The elite book-publishing houses, now begged by authors to publish their manuscripts, will find those authors can distribute their ideas in other electronic forms.

Leyden predicts that one industry will suffer even more:

> Newspapers were born at the dawn of the Industrial Age and matured through every stage in its evolution, and they've ended up as reflections of that era.

> Newspapers are elaborate factories for mass-producing news on an assembly line.... This method of delivering news has worked fine for generations. But it's on a collision course with the Digital Age....
>
> A key concept of the Digital Age is the difference between "atoms," which make up physical products, and "bits," which make up the intangible digital language of computers....
>
> Newspapers are now in the atom business, producing a physical product. But what they're really selling is pure information, which is easily converted into intangible bits. And once the new information infrastructure reaches into the home and information "appliances" become more sophisticated, which is generally expected to happen in the next 10 years or so, the newspaper's product will rapidly shift from atoms to bits.

Most pundits envision those information appliances to be computer-based systems capable of receiving text, audio and video in digital format over high-speed lines into the home, probably provided by telephone or cable television companies. This vision of an *information superhighway,* a term used by Vice President Al Gore and others, is far from a mere abstraction. The Internet, a worldwide computer network, already offers much of that capability. It also has the advantage of already being used by an estimated 30 million people, a figure that will almost certainly be hopelessly outdated by the time this book is published. The world is rushing to the Internet in a way that is likely to make it the *infomedium,* or information medium, of choice in the years ahead. Further, the Internet is gaining the capability to support on-line financial transactions, which will make it an attractive and profitable place for business. Already, companies are setting up shop on the Internet with its user-friendly front-end, the World Wide Web. There they envision not only providing information about their products but selling them as well.

Business already is transacted on the *public information utilities*—CompuServe, America Online, Prodigy, Delphi, GEnie, MSNBC and others. Together with the Internet, to which all of them provide links, they are known collectively as the *new media.* And they are doing nothing less than changing the way news is consumed.

Just how rapidly the climate is changing is well-illustrated by news coverage of the Federal Building explosion in Oklahoma City in 1995. Writes Wade Rowland in the *Toronto Star:*

> Three hours after the explosion, the Internet and the big American on-line computer service providers found themselves in the news business in a big way. America Online, Prodigy and CompuServe all worked rapidly to provide subscribers with the latest information on the blast, forums for discussion and links into Oklahoma City computer bulletin boards.
>
> Internet providers in Oklahoma City went a step further, presenting live video from local television and photos from the city's newspapers. Lists of survivors and their whereabouts were posted long before they were available from other sources.
>
> It was all a bit primitive, a little amateurish, but it was there.

## *What Makes It News?*

Since newspapers first were published, editors have been grappling with the question of how to define news. One definition is that to be news a story should be *relevant* to people's lives, *useful* to them or *interesting.* More traditional definitions of news take these factors into consideration:

- **Audience.** No two audiences are alike, so it is reasonable to assume that readers' tastes differ from city to city. Audiences may differ even within a city. Readers of *The New York Times,* for example, may have tastes that differ significantly from those of readers of the New York *Daily News.* Good editors have a feel for the interests of their audiences, and in many cases readership research has helped to clarify those interests.
- **Impact.** The number of people affected by an event is often critical in determining how extensively an account of it will be read. If garbage rates are to increase throughout the city, the story has more impact than it would if the garbage rates of only 15 families were affected.
- **Proximity.** If the event happened nearby, it may be more interesting to a newspaper's readers than it would be if it happened in another country.
- **Timeliness.** News is important when it happens, and old news is of little value to readers.
- **Prominence.** Prominent people are of more interest than those who are not. If the president changes his hairstyle, that may be news. If the local butcher does so, chances are that few people care.
- **Novelty.** A 30-pound tomato may be interesting, and therefore newsworthy, because such a tomato is unusual. Events that are firsts or lasts, and therefore historic, also may be unusual enough to merit attention in the news.
- **Conflict.** Sports events, crime, political races and disputes often are newsworthy because conflict, unfortunately, plays such an important role in modern society.

Seldom does a news event qualify for inclusion on all these counts. The editor weighs each story to determine if it has one or more of these values. If the story does, there is a good chance it will be printed or broadcast. If it includes most or all of these criteria, it may well be worthy of Page One or the lead story on a newscast. Judging the news, then, is at best an inexact science.

Ann Brill, a journalism professor and new-media expert, told Rowland, "Watching what happened on-line in this case convinces me we're looking at a [new] medium, a new kind of newsroom." That new medium is digital, on-line and multimedia-capable. It poses significant threats to the existing media. It also presents them with opportunities.

## THREATS AND OPPORTUNITIES

The existing media are taking note of the changing media marketplace. As this was written, more than 1,700 newspapers, magazines and broadcast stations, most in the United States, already had a presence on the Internet, and the number was growing daily. Few, if any, were making money on such services. Most were convinced, however, that they must learn to do so and that the threat to their advertising base is real.

Almost no one believes that newspapers, magazines, radio and television are doomed by all this. Instead, what we are witnessing is the emergence of a new competitor in the media marketplace. Indeed, as Brill suggests, it is a new medium—one that clearly has the potential to change the nature of the media industry. That transformation will not take place overnight, which means that the existing media will survive, even thrive, for years to come. Still, any new competition constitutes a threat.

Leyden is correct in predicting that newspapers are threatened most. The newspaper business is troubled by the shrinking number of U.S. dailies (down about 200 in recent years to slightly more than 1,500) and weeklies (down even more) and a stagnant, if not eroding, advertising and circulation base (see Figures 1–1 and 1–2). Indeed, most U.S. cities now have only one daily newspaper; with rare exception, that's the only way big dailies remain profitable.

Newspapers, like their magazine cousins, face some other formidable challenges: the rapidly escalating price of paper, not to mention the environmentally unpopular need to harvest trees to produce it; the high cost of ink; and, perhaps most significantly, the high cost of labor in the media industry's most labor-intensive sector. Newspapers, a manufactured product in the truest sense of the

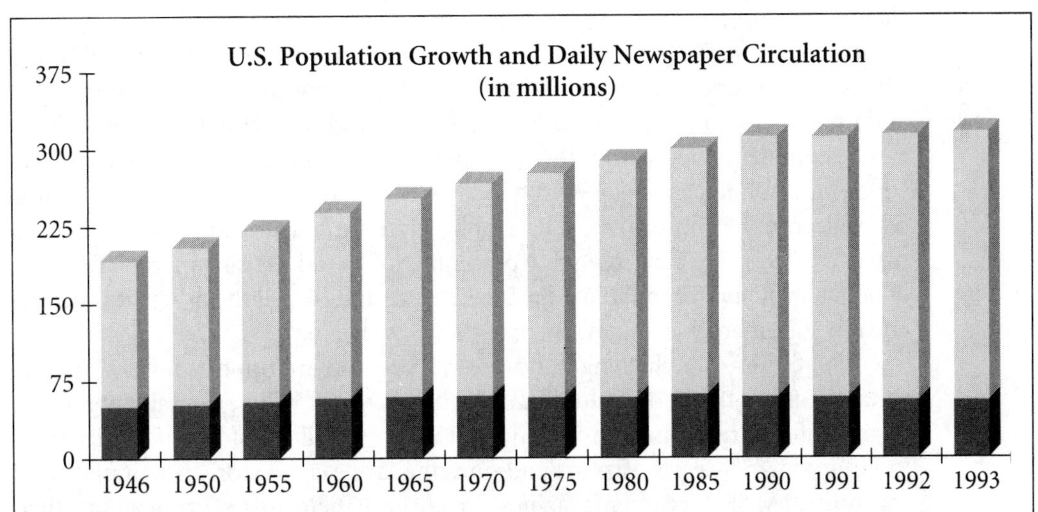


***Figure 1–1** U.S. daily newspaper circulation has remained relatively flat during the past 50 years, while the population has increased dramatically. Circulation peaked at 62.8 million in 1987 and has declined each year since. The result is declining market penetration, or the percentage of households receiving a daily newspaper.*

*Sources:* U.S. Census Bureau and the Newspaper Association of America.

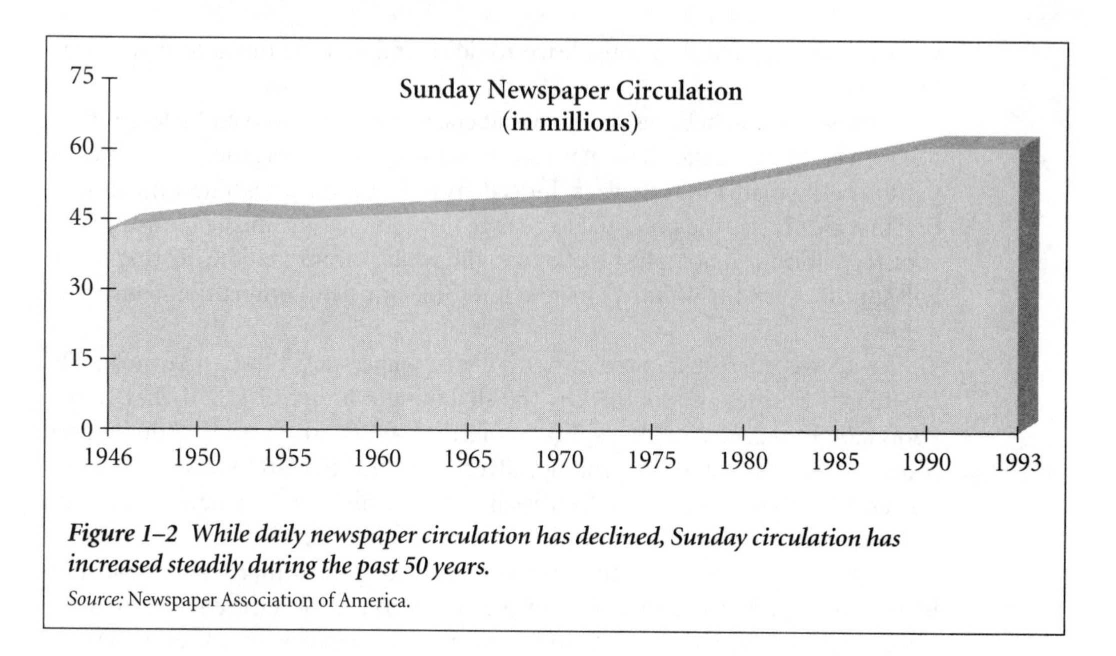


***Figure 1–2** While daily newspaper circulation has declined, Sunday circulation has increased steadily during the past 50 years.*

*Source:* Newspaper Association of America.

Industrial Revolution, remain one of the nation's largest employers. With the high cost of labor, that's not good. Magazines have the additional burden of coping with rapidly increasing postal costs, the delivery mechanism used for most of their circulation.

It's easy to see why newspapers and magazines are seriously threatened by the prospect of competitors delivering the same information, with even more depth and immediacy, through electronic means. With today's technology, that delivery can be complemented with full-color photos, audio and even full-motion video. Video on demand challenges even television, which offers information on *its* time schedule, not the consumer's. Television's electronic cousin, radio, isn't much of a player even today; it captures a tiny percentage of the advertising market. Radio is, at best, a niche player in the media marketplace, but like all the others it is threatened by the emergence of new media.

Where there are challenges, however, there also are opportunities. The one certainty about the infomedium of the future is that it will be digital, not analog. Digital information can be stored in computers as millions of bits of information in binary form—series of zeros and ones that are easy to store, manipulate and transmit. Newspapers and magazines already have their stories written and edited in digital form, and they have vast libraries of valuable information stored that way. Further, almost all newspaper photos are now processed digitally. Newspapers also have the largest news-gathering staffs in virtually every city. Those are significant advantages as newspapers seek to redefine their role in information delivery.

Television has problems beyond the proliferation of cable competition. Almost all television (and radio) is transmitted using the time-tested analog format. But if the future is in digital services, which it is, then analog television is about to become an anachronism. If demands of the marketplace don't drive television to digital format, the government probably will. The Federal Communications Commission wants television to convert to digital transmission, which can best be accomplished by moving it to delivery over those high-speed lines into the home. That would free valuable broadcast frequencies for redistribution to the fast-growing cellular-telephone industry, electronic beeper services and the like. So, for television to become a real player in the information arena of the future, it must make the transition from analog to digital. That will be a costly and difficult process because today's transmitters and television sets are not capable of handling digital information and must be replaced. Even worse, during the transition period television must be broadcast in both analog and digital form, a costly proposition.

Television has other problems, too. Its local news staffs are small and not well-designed to produce the massive amounts of information that consumers will demand in an infomedium that must provide depth as well as eye appeal. Still, television is the foremost provider of the sights and sounds that customers demand. Because of that, it is almost certain to be a major player in the Information Revolution.

The fact is that no existing medium is ideally positioned to take advantage of the changes about to occur. That reality has led to a proliferation of media-industry mergers as corporate media moguls try to position their companies for the Information Revolution. News, unfortunately for journalists, is an afterthought in the process. That's because much of the change will be driven by the fact that this new infomedium is even better-suited for delivering entertainment. Like television, the new infomedium will be primarily an entertainment medium; its role as a news medium will be secondary, though important to society.

Imagine the possibilities. With video on demand, a consumer can watch a movie in the comfort of home without a trip to the video rental store. That consumer also will be able to see the movie when he or she desires. Waiting for the show to begin will be a thing of the past. Shopping on-line will be easy, too. With a plethora of on-line stores to visit, the medium will be a comparison shopper's paradise. Banking services also are likely.

News, then, will be of secondary importance to most consumers. But imagine being able to sit at home and, on one appliance, watch and listen to the news supplied by hundreds of radio and television stations. And imagine being able to watch and listen on your schedule, not the provider's. Or imagine having instant access to thousands of newspapers and magazines each day. This new medium can do all those things. And, because it is based on the computer, this new medium can be programmed to present that news in ways that most interest the consumer. In addition to the top news stories, as ranked by journalists, consumers will be able to

## *Media Merger Mania*

Media companies are hot properties in the 1990s, fueled by the expectation of huge profits resulting from the Information Revolution's sweep into the home. Mergers and strategic alliances are the result, and some blue-chip corporations are involved. Some examples:

- The Walt Disney Co., operator of film studios, amusement parks and retail stores, created the world's largest media company with the purchase of ABC/Capital Cities, owner of the ABC television network and newspapers, including the *Fort Worth Star-Telegram* and *The Kansas City Star.* The company was valued at more than $43 billion.
- Westinghouse Corp., known for its electric appliances and electrical infrastructure products, including transformers, purchased CBS-TV.
- Microsoft Corp., the world leader in personal computer software sales, launched a joint venture with NBC-TV to create an all-news cable channel to compete with CNN. The two also operate a news-oriented Web site on the Internet.
- IBM Corp., the world's largest computer company, and Sears, one of the nation's top retailers, until recently co-owned Prodigy, the public information utility.
- Time-Warner, publisher of *Time* and other magazines and owner of the Warner Bros. film studio, merged with Ted Turner's media empire, which includes CNN and some of cable television's main attractions, including the Atlanta Braves baseball team.

program their information appliances to sort through thousands of news and information sources to provide all they want to know about soccer, fine art or any other topic of interest.

If all this sounds like futuristic mumbo jumbo, think again. All of those things are being done today on the Internet or on the public information utilities. Suppliers of those services are limited only by the speed of transmission into the home. Once that problem is solved, transmitting audio and video will no longer be as difficult as it is today. Then, the new media will really take off, providing a serious threat to the existing media.

While the new media loom as a significant threat to the existing media, the existing media also are ideally positioned to be key players in the change. The media already mentioned, along with book publishers, movie producers and other entertainment conglomerates, have the content the public wants. That's why media properties are hot properties in the current merger craze, and that's why *media companies* are trying to become *multimedia companies.* The road won't be an easy one. Competition will come from all corners, and it will come from megacompanies with solid grounding in entertainment, computers and communications. Companies as large and powerful as Sony and Microsoft already are players.

Telephone and cable television companies also are destined to play key roles; so is government—local, state and federal—and whether, or how, it decides to regulate. Clearly, the trend is toward minimal regulation in the United States, although questions remain about such things as who controls the wiring of neighborhoods and the extent of controls on pornographic and obscene material. Remember, though, that the Internet is a worldwide network. Experts question whether it is possible for any one government to regulate such a behemoth.

The pace of change is almost breathtaking, but there is little hope of slowing the momentum. The Information Revolution is well under way.

## THE NATURE OF NEWS

Not only are the news media changing, but so is the nature of news itself. Indeed, the Information Revolution is changing even the way news is defined. In the latest edition of its popular text, *News Reporting and Writing*, the Missouri Group writes:

> Webster's Unabridged Dictionary defines news as:
>
> "1. New information about anything . . . .
> "2. recent happenings . . . .
> "3. reports of such events, collectively.
> "4. a newspaper."
>
> That definition raises more questions than it answers. Read your newspaper, listen to a radio newscast, watch the evening news on any network. Clearly, not just "anything" is reported. Nor are all "recent happenings." And there are differences among what you see, hear and read. So how do journalists decide which pieces of new information, which recent happenings, are worth reporting?
>
> The criteria used by professional reporters and editors can be summarized in three words: *relevance, usefulness* and *interest.*

That's a dramatically new way of looking at news, even when compared to the way it was defined by the same authors in earlier editions. Those changes reflect new ways in which journalists look at news. But the nature of news is being changed by external forces, too. Traditionally, news wasn't news until journalists reported it. With today's new media, anyone can be a reporter, and anyone can be a publisher or broadcaster. Therefore, almost anyone can produce news. The public is left to decide whether the source of that news is credible.

This reality diminishes the role of the journalist as *gatekeeper* and *agenda setter.* For many years, journalism research has focused on these critical roles of the media. In the traditional model of news distribution, journalists have served as gatekeepers by deciding what information reached the public. Any newspaper or newscast has a limited amount of space or time for news. Journalists decided how that space or time was used and therefore functioned as gatekeepers controlling the flow of information to the public. With the emerging new media, however, space and time are virtually unlimited. It's possible to put *all* the news into the system and let the consumer decide what to see, hear or read.

In the traditional model, the journalist also helped to set the agenda for a given audience, usually local but sometimes regional or national. By deciding to put the city council's decision to raise taxes at the top of Page One, the journalist made that a topic of public debate, thus setting the agenda for the community. With new media, the *consumer* may decide what's most important, thus diminishing the journalist's role.

Most media critics revel in this turn of events. Some see an opportunity to erode the media's perceived "leftist bias," diminish the role of media in our society and empower consumers. But the fact remains that media oversight of government, in particular, has ferreted out many wrongs against society. Few could argue persuasively that the media have not been of great service to American society, dating back at least as far as the muckraking investigative reporting of Ida Tarbell and others in the early 20th century.

If the new media in fact erode the influence of editors, who, then, performs that role? Without someone or some institution doing so, does news become what anyone wants it to be? And who sets the agenda for societal discourse? Politicians? The government? Special-interest groups? Those are troubling questions to be answered as the definition of news—and the nature of news—continues to change.

## THE ROLE OF THE EDITOR

In today's busy world, many consumers *want* someone to help them sort through the news and tell them what is important. And they *want* someone to help set the public agenda. Few have the time to sort through all the day's news to determine what's important. They want someone to do that for them. They are accustomed to having editors perform that service. And, while they may not agree with the editor's priorities, they go back for more day after day. It helps to have someone suggest priorities, even if one disagrees.

That reality is the reason the role of the editor will remain an important one as society embraces the new media. Because almost anyone can be a reporter or publisher in the new media, there will be an increasing need to separate fact from fiction, to know the source of information and to determine its credibility. Editors are trained to do just that.

Editors will, however, have more competition. Consumers will look to other sources of information to help guide their decision making. One such source will be the on-line chat or discussion areas found on most new-media services. Chat areas serve much the same purpose as town-hall meetings or call-in talk shows on radio and television. They provide a forum for a variety of views, which in turn enables the consumer to form an opinion about a pending issue.

Information provided in town-hall meetings or on talk shows is not always accurate, however. That's where editors can help. With their excellent training in fact checking and their ability to separate truth from fiction, they can continue to play a major role in agenda setting if not in gatekeeping. The importance of that role will be no greater than warranted by the quality and accuracy of the services they provide. As it has been for decades, the foundation of journalistic practice is accuracy. Without it, credibility suffers and influence diminishes.

## THE ART OF EDITING

In the first edition of this book, published in 1971, we asserted that editing is an art, no matter where or by whom it is practiced. That axiom remains true today. Although the medium may change, the role of the editor remains clear: Provide timely and accurate information in the best form possible.

This edition was rewritten extensively to update and to recognize the realities of the changing media marketplace and to discuss the skills needed for the new media. In this period of transition, however, there is still a need to address the peculiarities of the traditional mass media—newspapers, magazines, radio and television—and the parallel industries that service them—corporate communications (or public relations) and advertising. Anyone entering any of those fields will edit. Anyone entering any of those fields can benefit from reading this text and acquiring the skills it teaches.

It's worth remembering that despite the unsettled nature of the media industry, there is absolutely no chance that newspapers, magazines, radio stations and television stations will disappear in our lifetime. There may be fewer of them, but they will continue to exist for many years. Therefore, some who learn from this text may well spend their entire careers in what we refer to as the traditional media. The emergence of the new media will take time.

One thing is certain: The skills that can be acquired from this text will help, no matter which medium you enter. Not everyone is an artist, and not everyone can be an editor. Those who learn here can be both.

# CHAPTER 2

# THE EDITOR AND THE AUDIENCE

## THE DISCONNECTED AUDIENCE

Editors and their audiences are too often disconnected. Consider these indications of that reality:

- In 1970, 98 percent of U.S. households subscribed to a daily newspaper, but by 1993 that number had dropped to 62 percent.
- Weekday circulation of newspapers remains flat while the population continues to increase dramatically (see Chapter 1).
- According to the Newspaper Association of America, in 1970 about 78 percent of the nation's adult population read newspapers daily, but by 1995 that percentage had declined to 64.2.

Those are damning figures for newspapers, and the situation may be even worse than it appears. Further examination of newspaper circulation trends indicates that declines are most prevalent among people 35 and younger. That's bad news because young people are on the advertiser's most-wanted list. People in their late 20s and early 30s are in the process of setting up households and making major expenditures. Advertisers covet their attention.

Recent statistics from the Newspaper Association of America show that only 57 percent of Americans aged 25 to 34 read newspapers daily; the percentage drops to 54 percent for those aged 18 to 24. A study by the National Opinion Research Corp. indicates the situation may be even worse; it reports that everyday readers of newspapers among those aged 30 to 34 dropped from 76 percent to 46 percent from 1967 to 1990 and from 60 percent to 25 percent among those 18 to 22.

Perhaps the biggest problem for newspapers is that there's little hope of reversing that trend. Study after study shows that if the newspaper reading habit is not acquired early in life, it probably will never be acquired. As a result, newspaper readers are increasingly older. Worse, as older readers die, younger readers are not taking their place.

One anecdotal incident illustrates the problem vividly: After being exposed to an "electronic newspaper" used in a University of Missouri research project, one junior high student was asked what he thought of the new medium.

"Cool," he responded.

"Cool or way cool?" asked the researcher, using a popular idiom in the age group.

"It can't be way cool," responded the student. "It's a newspaper."

For those fond of newspapers and the significant role they have played in winning and maintaining democracy in the United States, that comment hurts deeply. For the editor or publisher trying to ensure a newspaper's future, it is equally devastating. But it is a view shared by many school-age children. Newspapers, they believe, are boring, written for older people and simply not relevant to their lives.

That same study, however, showed that the so-called new media have a chance to turn that attitude around. After a two-year study of elementary and junior-high school students, researchers concluded that computer-based media could convert young people into news consumers if not into newspaper readers.

"Children would stay in from recess to read the [digital newspaper]," one teacher reported. "I've never seen anything like it." And researchers confirmed that the students weren't just looking at comics and reading horoscopes. News ranked high on the list of items consumed. Clearly, young people are more willing—and more eager—to consume information from a computer screen than from the printed page.

Perhaps that's because the younger generation was weaned on television and computers. But television has never managed to attract young viewers to its news; they are entertainment consumers almost exclusively. Computer-based media may well be the last great hope for reaching this generation, which needs news information if it is to make well-informed decisions in the voting booths of America.

If young people are disconnected from the media, so, too, are many adults. Study after study shows that the public views journalists with disgust. Journalists, as a whole, are ranked among the least popular of professional groups—often below politicians and used-car vendors. Those dismal and declining newspaper readership figures are further confirmation of disgust with the media.

Television also has problems. The decline of network and local news operations is an often-discussed topic among broadcast journalists. The consensus seems to be that budget cutbacks and pandering to ratings has led to a marked decline in the quality of broadcast news. Today's television news, it is said, revolves around "talking heads" who look good on the screen but have limited journalistic talent and ability. Gone are respected figures like Walter Cronkite, the former CBS news anchorman who once ranked as the most trusted person in America.

The fragmentation of audiences also has hurt. The three big networks—ABC, CBS and NBC—once dominated network news, and their affiliates dominated the local news scene. Today, the networks compete with CNN and others, and cable has made local stations' audience shares increasingly smaller. No longer are the local stations the only thing to watch. The networks are fighting back by starting their own cable news operations, further fragmenting audiences and compounding the problem.

## ADVERTISING PROMPTS CHANGE

Newspaper editors and broadcast news executives have done plenty of self-examination in an effort to arrest some of these alarming trends. In reality, they can do little about most of the problems by changing editorial content. That's because the big mass-media outlets of the past are giving way to smaller media designed to reach specific segments of the audience. All that is led by advertising, not news.

Audience fragmentation is fed by advertisers' desires to reach targeted segments of the reading, listening or viewing audience—teen-agers, young adults, even cyclists or exercise addicts. A few products (soap, for example) can and should be mass-marketed items. Other products are marketed most efficiently and at the lowest cost by reaching those who are more likely to be interested in buying. Why advertise Maseratis to a mass audience when only the elite have the money to buy them? Target marketing goes after those who can afford to buy or who need the product.

Radio lures certain audiences with station formats (country and western, oldies and rock, for example). Magazines increasingly are aimed at targeted audiences with titles like *Skiing, Boating* and *Popular Mechanics,* while the number of general-interest magazines (*Life, Look* and *The Saturday Evening Post*) has dwindled dramatically. Radio and magazines deliver target audiences efficiently, and television (with targeted cable channels on subjects as diverse as sports, health and fitness and cooking) isn't far behind. Newspapers, on the other hand, are poor vehicles for targeting audiences because they are, by nature, a mass medium. It's difficult enough to print special editions for one part of the city and almost impossible to target specific socio-economic groups. Why place an ad in a newspaper (with a high cost per thousand consumers reached) when radio (with a much more cost-effective pricing plan) will get the same results?

As if these shifts in the existing media aren't enough food for thought, just consider the likely impact of advertising in the new media. Computer-based media will be able to tell advertisers exactly how many people read an ad for how long. Even more significantly, they will be able to tell that advertiser the demographic profile of the potential customer or customers and—with the consumer's permission—generate a firm sales lead. Advertisers will be willing to pay a lot for that capability.

## *Measuring Audience Reaction*

Media companies know that gaining and maintaining an audience of acceptable size is important to their bottom line. For that reason, they go to great lengths to measure audience reaction.

Newspapers and magazines do that with *readership studies,* which measure the relative popularity of articles, photos, graphics and other features. Such studies often lead to changes in content, as when an editor drops an unpopular column.

*Focus groups* also are increasingly popular. Representative groups of readers are brought together to discuss what they like and don't like about the publication. Editors often are present or watch videotapes of the sessions. It's one more way for them to keep in touch with readers.

Radio and television depend on *Neilsen ratings* to determine the relative popularity of programs, including news shows. Neilsen uses sample families and measures what they are watching at certain hours. Low ratings often lead to shake-ups in the newscast, including the replacement of lead newscasters.

Certain "sweeps" periods, when critical ratings measurements are done, prompt broadcasters to prepare special series with high audience appeal for the evening newscasts. They do so to wring out every possible point in the ratings.

The new media have a distinct advantage in this area. Because they are based on computers, it is possible to measure the exact number of times an item is accessed and the duration of that access. As the new media refine that technique, they should have a competitive advantage over the existing media, which depend on audience estimates, not actual data. New media data should be more reliable.

Marty Levin, advertising director of the Microsoft Network, asked executives of one large automobile company what they would be willing to pay for such information. The answer? "Name your price." That means the existing media face formidable competition from new media in the years ahead.

## AN INDUSTRY IN TURMOIL

If all this makes it seem as if the media industry is in turmoil, it is. Not only are great shifts taking place in the impact of one medium versus another, but the arrival of new media forms has complicated the situation. All this has led to an increased emphasis on packaging and marketing of media products.

Newspapers are a great example. In the 1980s, many newspapers launched massive redesign projects to improve their appearance. Many also created marketing committees made up of executives from various departments—circulation, advertising and news among them—who tried to settle on joint approaches to improving and selling the product. In prior years, such cooperation had been frowned on for fear the news department would be influenced by market considerations to cover stories that advertisers wanted and ignore those they wanted to hide. Most of those barriers broke down as editors and publishers decided that something had to be done. Said Gregory Fauvre of *The Sacramento Bee,* "I think we have to quit putting out newspapers for 54-year-old editors like me."

Newspapers' eagerness to reconnect with youthful readers, in particular, led to some rather questionable pandering. *The Syracuse* (N.Y.) *Herald-Journal* tried youth-page articles that addressed readers as, "Hey you, yeah you"; or, on occasion, "You knuckleheads"; or even, "Yo, buttheads." Explained the youth-page editor, Larry Richardson, "We decided we would give [our readers] things they wanted to read, not only things we think they should read."

Critics charged that newspapers were guilty of doing what they perceived television had done: pandering to the market concept of simply giving people what they want. The media, critics argued, are a public trust and should be above the pressures of marketing. But the reality is that media operations in free societies produce products that must be marketed. There is no government subsidy to keep them running, and as competition increases, marketing pressures grow accordingly. The reality is that broadcast programs and newspapers are products that must be marketed and sold if they are to survive.

That reality led to the formation in the late 1980s of a journalistic think tank called New Directions for News. The organization, based at the University of Missouri School of Journalism, is funded by almost all the major newspaper groups in the United States. NDN has been a leading force in convincing editors to rethink the way they edit newspapers. A hallmark of its efforts has been the organization of brainstorming sessions at which editors have come up with suggestions as diverse as printing newspapers that smell like baked bread to wrapping a news digest around the outside of the regular newspaper (see Figure 2–1). While not all the ideas are good, or even plausible, such sessions force editors to contemplate change, something that all newspapers need to consider.

NDN also has helped to spark the creation of prototype sections for newspapers. Many of these have attempted to appeal to those disappearing younger readers.

The consensus among editors is that newspapers must change if they are to survive. Unfortunately, there is no consensus at all about what that change must encompass.

## THE CONTRARY VIEW

Despite a consensus that newspapers must change, some argue that concerns about the industry's health are overstated. They point to a powerfully persuasive statistic to support their view: Newspapers still get the largest slice of the advertising pie (see Figure 2–2). According to McCann Erickson Inc. and the Newspaper Association of America, newspapers get 22.8 percent of the U.S. advertising dollars spent each year, while broadcast television gets 20.8 percent and cable television only 2.0 percent.

***Figure 2–1 Editors of Hearst Newspapers participate in a brainstorming session led by executives of New Directions for News.***

Photo courtesy of New Directions for News.

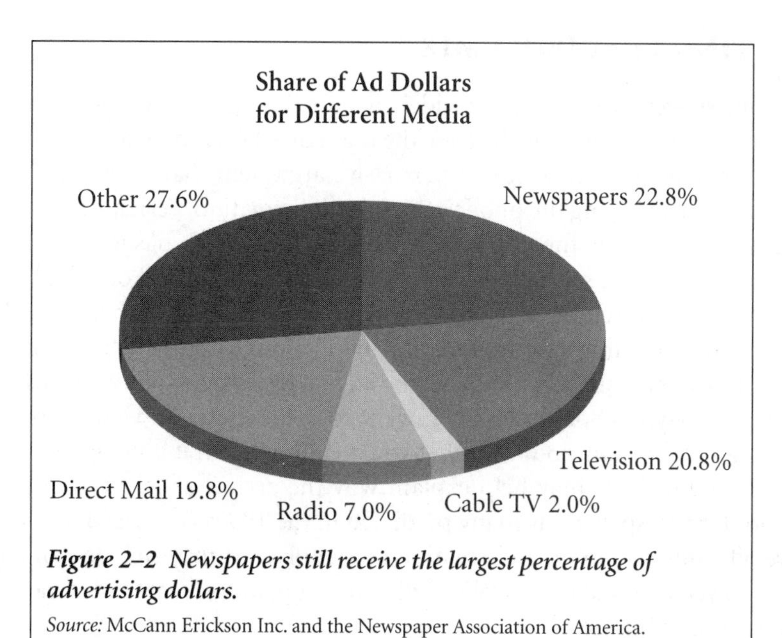


***Figure 2–2 Newspapers still receive the largest percentage of advertising dollars.***

*Source:* McCann Erickson Inc. and the Newspaper Association of America.

One who thinks newspapers are relatively healthy is Benjamin Bradlee, former executive editor of *The Washington Post.* "I don't see any point in shutting down now and saying it's too hard [to publish a newspaper]," Bradlee says. "If we could lick the ink [rub-off] problem, I think we could do very well."

Indeed, the newspaper industry continues to be one of the most profitable around. Late 1994 figures suggested that the industry as a whole averaged a profit of about 15 cents on the dollar, and only one of the major public media companies reported profit margins in the single-digit range. Fifteen percent is an extremely high profit margin compared to other industries, although it's down from the newspaper industry's historic profit margin of more than 20 percent. The grocery industry, in comparison, typically earns a profit of about 2 percent of every dollar handled.

The reality of newspapers' profitability is sometimes lost among all the talk about declining readership, flat circulation, the disappearance of afternoon dailies and the decline in the number of newspapers overall. Most industries would be thrilled with profit margins of 15 percent, and newspapers still enjoy such a rate.

The broadcast industry is doing quite well, too. Local stations are doing extremely well, and the networks are making a comeback after some lean times in the early 1990s. The magazine industry thrives, and even local radio is in most cases profitable. And new media still haven't made so much as a dent in the profitability of traditional media.

## EVALUATING THE MEDIA MIX

No one suggests that in an age widely heralded as the Information Age, the desire for information will wane. Instead, the main question is how that information will be consumed. One can make a persuasive argument that no medium is doing a good job of satisfying the public's thirst for information. Perhaps that's because no one medium has all the technological advantages. Radio has immediacy of delivery on its side. But so does television, which adds the appeal of color and moving pictures. Conversely, newspapers are far better suited than either of the broadcast media to carry large amounts of information and to provide the space necessary for analysis. Newspapers also have the advantage of portability and selective consumption—you choose what you want to read when you want to read it. With radio and television, you take what you are given when it is being offered.

Those realities may help explain why the arrival of television did not spell doom for newspapers, as many predicted in the 1950s. They also may explain the overall confusion in the information marketplace. It may well be that technology has not yet delivered the so-called killer medium, the one that is destined to put the others out of business.

It's not too difficult to envision what kind of medium that might be. Imagine a television set (or home computer) that allowed subscribers not only to watch the evening news but to call up *The New York Times'* text version of stories that piqued their interest. Such a device also would allow subscribers to print out the classified ads or grocery coupons and consume the news in the form they wanted, when they wanted.

Even better, such a device might be programmed especially for the individual. While giving the latest world news, it also might inform a subscriber that a doctor had canceled a medical appointment. Or it might help that subscriber balance a checkbook or order groceries for home delivery. It might even alert the stamp collector to the latest news in the world of philately.

One who envisions such a system is Nicholas Negroponte, director of the Media Lab at the Massachusetts Institute of Technology. "I would pay $10 a day for that newspaper," he says. "That would be a really important newspaper for me to read every morning." Negroponte's goal is to design just such a device at an affordable price. The key, of course, is making such a device affordable. The quality of screen images also must improve; it's clear that reading text on current screens is not something many consumers choose to do in their spare time. Various experiments with video delivery of written information have proven that.

As the technology to develop the ultimate information medium evolves, editors are experimenting with pieces of that technology to bring new delivery mechanisms to consumers. Starting in the 1970s, editors began experimenting with *videotex* and *teletext,* which are means of delivering information electronically to computer or television screens in the home. More recently, newspapers have been marketing fax newspapers—summaries of the news delivered to businesses or individuals—or audio newspapers (also known as *audiotext*), which offer recorded up-to-the-minute news on dial-in telephone lines.

Many believe the forerunner of Negroponte's vision is the Internet, which emerged from relative obscurity in 1994 and 1995 to become the hottest new medium around. More than 1,700 newspapers, magazines and television stations now have sites on the Internet, and companies are scrambling to figure out how to make a profit with it. Even the more established public information utilities, including CompuServe, Prodigy, America Online, MSNBC and others, are now required by market demands to provide connections to the Internet from their once-proprietary systems. A sure sign that the Internet will play a major role in the media mix of the future is the rush of media companies to hire students well-versed in Internet site creation, including the programming language of the Internet's World Wide Web, HTML (HyperText Markup Language).

## THE INFOMEDIUM

If Nicholas Negroponte's view of the future is close at all to what we can expect, a process of evolution will bring the various media closer together. What is a medium that has moving pictures as well as text? Is it a newspaper, a television or neither? Is it merely a news medium or an information medium, complete with a wealth of information about how we can improve our lives? How such a medium evolves will be determined as much in the laboratories of engineers as in the corporate boardrooms of the media industry. Driving it all will be the desires of the public, which will discover what it wants only when it knows what is possible and how much it will cost.

Ultimately, newspapers, television and perhaps the other media are likely to evolve into a new infomedium that combines the best attributes of them all. The wise young journalist of today should therefore view himself or herself not as a print journalist or a broadcaster but as an information provider. After all, what is more important, the message or the way it is delivered? Says Susan Miller, director of editorial development for Scripps Howard Newspapers, "[Newspapers] aren't dying, but I think we have to change." She told executives of her group that for the year 2000 "we have to stop worrying about whether it will be a fax newspaper or computers. If we get the topics right... then we'll survive. If we get the topics wrong, nobody's going to want us in any form."

Clearly, the selection of the proper marketing approach is critical to the future of all media, not just newspapers.

## A RESPONSE TO CHANGE

Newspapers make an excellent study on the inevitability of change. As newspapers increasingly moved to a marketing approach in the 1980s, newsrooms were forced to change their relationships with other departments. There was a time not so long ago when newspaper editors had the luxury of working almost independently of the other managers of their newspapers. Newsrooms were sacred, and no one in advertising or circulation departments presumed to tell editors how to edit their newspapers.

Now, editors meet regularly with top management executives and the heads of other departments to develop marketing strategies. There is a realization that news departments must work with other departments to develop marketing plans for their products. Sometimes that results in decisions to change the content of newspapers. Usually the final decisions about such changes are left to editors, but today the heads of other departments are more likely to make suggestions. Most editors are inclined to listen. Advertising and circulation people have extensive contact with the public, and often they hear compliments and complaints about their newspapers that editors never hear. Good editors appreciate hearing those things.

The danger, of course, is that the editor may go too far in an attempt to accommodate others. A newspaper full of nothing but what the public *wants* to read would be a poor one in most editors' judgment. Editors must balance what the public *wants* to read with what it *needs* to read. Only then will the newspaper fulfill its role as guardian of the public's welfare. Today's editors know that the key to a newspaper's integrity—not to mention its existence—lies in its ability to remain financially sound. A marginally solvent newspaper may be more susceptible to advertiser influence on editorial decisions. So attention to the demands of the marketplace is important. But it also is important to remember that the press is the only private institution mentioned in the U.S. Constitution. That is tacit recognition that newspapers, while competing in the marketplace, also have a special mission in U.S. society.

The fact that editors are now active participants in marketing the newspaper should not be viewed negatively. The fact is that a newspaper is a manufactured product, a commodity, and like other products it must be marketed aggressively. A newspaper's marketing strategy probably will work best if all departments—including the news department—are working to accomplish the same goals.

Editors of the 1990s know that successful market strategies are a plus: Their newspapers will be stronger, and more people will read the work of their staffs. Magazines, radio stations and television stations have been forced to make similar changes as an awareness of increased competition became widespread. Almost invariably, the self-examination that resulted has improved the media's responsiveness to their audiences. Understanding those audiences is the key to success in the marketplace.

## UNDERSTANDING U.S. AUDIENCES

Newspapers, magazines and broadcast stations frequently conduct audience studies to determine not only the demographics of their audiences but also the attitudes of consumers about the job they are doing. Each publication or station should conduct its own study. What works in New York City may not work in Fort Scott, Kan. Extensive audience research has been conducted at the national level, however, and these studies yield some important clues about how the public perceives the media.

One of the more important studies of this type was conducted by Ruth Clark, president of Clark, Martire & Bartolomeo Inc., a market research firm, for the American Society of Newspaper Editors. The survey, conducted in the 1980s, remains one of the landmark studies of audience perception of newspapers. Major findings:

- Readers expect *news* in their newspapers, whether it's national, state, regional or local. What editors call *hard news,* or late-breaking developments, is most important.
- Readers want information about health, science, technology, diet, nutrition and similar subjects but will figure out for themselves how to cope with these problems. Newspapers don't need to tell them.

- Overall, readers like their newspapers. Most think they are indispensable, although younger readers aren't so sure. Most agree that newspapers are here to stay, regardless of the potential that television and computer screens may have for disseminating information.
- On the negative side, readers sometimes feel manipulated by editors and question whether they are fair and unbiased in covering and allocating space to various constituencies.

People believe their own newspaper tends to be better than most, and that should be encouraging to editors. A minority of readers, although a substantial one at 39 percent, believes the local newspaper is biased. Most (57 percent) believe that newspaper stories in general are usually unfair. More than half (53 percent) believe that newspaper stories are usually accurate, but 84 percent describe their own paper as accurate.

Editors need to deal with the issue of bias, whether real or perceived, because it has enormous implications. They need to find ways to make their newspapers more believable. The problem is greater at the national level, however. While most people believe that newspapers are unfair, 88 percent believe their local paper really cares about the community. Translating that feeling of concern to the national level would lead to major improvement in the public's perception of newspapers.

Clark's study found that readers want complete newspapers, even in small communities. Readers of smaller newspapers (under 75,000 circulation) place more emphasis on local and regional news, while readers of larger newspapers tilt toward national and international news. But when it comes to performance, readers of smaller papers give them poorer marks on national and international news, while readers of larger papers are not satisfied with the local coverage they receive.

If newspapers are to win the battle for the reader's time, Clark believes they should follow these suggestions from readers:

- Give us a complete, balanced paper with solid reporting of national and international news but equal quality in coverage of local news.
- Strike a balance between bad news and other important news we need to know. Don't sensationalize or attempt to manipulate public opinion.
- Give us the important details we don't get from television news, but remember we're short of time.
- Do a better job of covering the new subjects of major interest—business news, health, consumerism, science, technology, schools and education, family, children and religion.

## *Public Journalism: Connecting with the Audience*

Editors trying to find ways to connect better with the public have engaged in a new form of journalism called *public journalism* (sometimes also called *civic journalism*). Public journalism seeks to find new and better ways to listen to the public, to focus attention on key public issues and to help citizens think through major decisions on public policy.

One public journalism project involved collaborative efforts by a local newspaper, radio station and television station. Together, they focused the public's attention on key issues—such as the quality of trade education in the local school system for noncollege-bound students—by timing high-profile stories on the topic to run in all three media at the same time. Surveys showed that, compared to citizens who consumed other local news media, those who gained their information from the participating media knew more, grew more interested in the issue and felt more positive toward the media.

This activist brand of journalism is not without its critics. Traditionalists argue that the media should merely *measure* public opinion, not *manage* it. In fact, the media have long tried to manage public opinion for the good of the community. Public journalism merely formalizes the process.

The strong pro-community motivation of public journalism clearly places it on a higher plane than other attempts to improve communication between the media and the public. Television's infamous "happy talk," in which newscasters engage in mindless chitchat on the air, is one example of pandering to the public for the purpose of merely improving ratings, not content.

- Realize that women today are interested in the sports and business pages you used to put out for your male readers. Of course, women are still interested in food, fashion and other traditional subjects, and you should keep those features going. But recognize they are not the attractions they once were.
- Make us feel we belong. You need to widen your focus if you are to win more regular readers among young people, working women and members of minority groups. We look at your newspaper to see if we belong, and too often we feel we do not. It's a sore point and a source of grievance.
- Tell us more about yourselves—your editors and reporters.
- Count us in on the fight to preserve the First Amendment. We're ready to support *our* right to know.
- Get to know us better. We are a far more serious, concerned, interested and demanding audience than you have served in the past.

That advice, while intended for newspapers, applies equally to magazines and broadcast stations. What it amounts to is this: Get to know and understand your audience. Only then can you produce a publication or broadcast of interest and appeal.

## CREDIBILITY AND THE MEDIA

A major factor in the newspaper-reader relationship is credibility. If readers believe that newspapers are biased and unfair, as Ruth Clark found, it is only reasonable to assume that they will read them with skepticism.

The media reinforce such doubts about their credibility when they obstinately refuse to admit their errors, when the names of people and places are consistently misspelled or inaccurate, and when hoaxes of one sort or another are uncovered. One such event that shocked editors occurred in 1981 when the Janet Cooke affair was revealed. Cooke, a reporter for *The Washington Post,* won a Pulitzer Prize for a story about a child named Jimmy who became a heroin addict. Subsequently, it was learned that Jimmy didn't exist and that Cooke had invented the fictional youngster as a composite of situations involving children she had learned about while doing research for her story. The resulting publicity damaged the credibility of the media nationwide.

*The Washington Post* also was involved in one of the most celebrated scandals in U.S. history, the Watergate affair, which eventually led to the resignation of President Richard Nixon. Through a series of stories featuring anonymous sources, the *Post* and reporters Carl Bernstein and Bob Woodward unraveled the involvement of Nixon and his aides in the burglary of the Democratic Party headquarters in the Watergate apartment complex in Washington, D.C. The service the newspaper performed in that investigation probably is unparalleled in U.S. newspaper history, but along the way the many stories with anonymous sources raised serious questions about the credibility of the media. Editors today are reluctant to use anonymous sources without compelling reasons to do so.

Editors concerned about their credibility have tried to find ways to convince the public that newspapers are in fact reliable. These range from simple steps, such as the attempt to reduce annoying typographical errors, to elaborate schemes designed to check the accuracy of reporters' work.

To enhance the newspaper's image, today's editors readily admit errors. Some papers run a daily notice, prominently displayed, inviting and encouraging readers to call attention to errors in the paper. Another editor regularly conducts an accuracy check of his newspaper's locally written news stories. A clipping of the story is mailed to the source along with a brief query on the accuracy of facts in the story and headline. Another editor invites persons involved in controversy to present amplifying statements when they feel their positions have not been fully or fairly represented.

More corrections are being printed, even though this practice is distasteful to editors. When the old *Minneapolis Star* had to print four corrections on one day, the editor warned the staff, "Let's hope it is a record that is never equaled—or something besides the sky will fall." The *Boca Raton* (Fla.) *News* candidly tells its readers of its corrections under the heading, "Dumb Things We Did."

More balance in opinion is evident in the use of syndicated columnists whose opinions differ from those of the newspaper and in expanded letters-to-the-editor columns. Some newspapers are using ombudsmen to hear readers' complaints. More are providing reader-service columns to identify newspapers with readers' personal concerns. More attention is also being given to internal criticism in employee publications or at staff conferences.

Broadcast stations are making similar efforts to increase their credibility. Many stations now have electronic mail addresses for their news directors. Listeners and viewers are encouraged to write with questions or complaints. Still, studies show that the public considers broadcast news more credible than print news. The reason? It's easier to believe what you can see and hear.

## MEASURES OF READABILITY

Broadcast news is written to be spoken, so sentences are short and to the point. Newspaper and magazine writing is more literary, so the print media constantly run the risk of making stories too difficult to read. Difficult reading results in lost or confused readers. As a result, newspapers and magazines periodically test their stories for readability. Researchers have devised formulas that measure the ease with which an item can be read. Or, more accurately, they try to gauge some of the factors that make reading difficult.

Most readability formulas are based on concepts long familiar to newspaper editors. Short sentences generally are easier to read than long ones, and short words generally are more comprehensible than long ones. Two of the better-known formulas developed by readability experts use sentence and word lengths as key ingredients. The Flesch formula, devised by Rudolph Flesch, uses 100-word samples to measure average sentence length and number of syllables. The formula multiplies the average number of words in the sentence by 1.015 and the total syllable count by 0.846. The two factors are added, then subtracted from 206.835 to arrive at a readability score.

Robert Gunning uses a similar procedure to determine the *fog index.* He adds the average sentence length in words and the number of words of three syllables or more (omitting capitalized words; combinations of short, easy words like *butterfly;* and verb forms made into three syllables by adding *-ed, -es* or *-ing*). The sum is multiplied by 0.4 to get the fog index.

Suppose the sample contains an average of 16 words to the sentence and a total of 150 syllables. By the Flesch formula, the sample would have a readability score of 64, which Flesch rates as standard or fitting the style of the *Reader's Digest.* In the same sample, assuming hard words make up 10 percent of the text, the fog index of the Gunning scale would be 10, or at the reading level of high school sophomores and fitting *Time* magazine's style.

Neither Flesch nor Gunning tests content or word familiarity. All they suggest is that if passages from a story or the whole story average more than 20 words to the sentence and the number of hard words in a sample of 100 words exceeds 10 percent, a majority of readers will find the passages difficult to understand.

The formula designers would not recommend that editors pare all long sentences to 20 words or fewer and all long words to monosyllables. Long sentences, if they are graceful and meaningful, should be kept intact. Mixed with shorter sentences, they give variety to style and provide the pacing necessary in good writing. A long word may still be a plain word.

An editorial executive of *The New York Times* preferred to measure density of ideas in sentences rather than sentence length itself and came up with a pattern of "one idea, one sentence." A special issue of the newsroom publication *Winners & Sinners* was devoted to this pattern and reports of reading tests on two versions of the same articles. One test was done on the comprehensibility of the articles as written originally; another on the articles when rewritten to lower the density of ideas in the sentence. The "one-idea, one-sentence" dictum is not taken literally even at the *Times,* but the editors insist, "Generally it speeds reading if there is only one idea to a sentence."

The number of unfamiliar words in passages also has been found to affect readability. Edgar Dale and Jeanne S. Chall at The Ohio State University prepared a list of 3,000 words known to 80 percent of fourth-graders. The word-load factor in the Dale-Chall formula consists of a count of words outside the list. Only 4 percent of the words on the Dale-Chall list have three or more syllables.

Editing stories to reduce the number of words outside the word-familiarity list would be time-consuming and impractical. The lists would have to be revised periodically to take out words that no longer are familiar and to add new words that have become part of everyday language—even to fourth-graders.

Most readability formulas use a few fundamental elements but neglect context or story structure. Thus, a passage in gibberish could rate as highly readable on the Flesch, Gunning and Dale-Chall scales. This was demonstrated by Wilson L. Taylor at the University of Illinois Institute of Communications Research. Taylor developed the *cloze* procedure (from "close" or "closure" in Gestalt psychology) to test context. In this procedure he omitted certain words—usually every fifth word—and asked respondents to fill in the missing words. He then graded them on the number of correct words they could fill in. Of passages from eight writers, the Taylor method ranked samples of Gertrude Stein's semi-intelligible prose as the second most difficult. The most difficult was a passage from James Joyce. Both the Dale-Chall and the Flesch scales rated the Stein passage as the easiest to read and the Joyce passage in a tie for fourth with a passage from Erskine Caldwell. To test for human interest, Flesch measures personal words and sentences. Sentences that mention persons and have them saying and doing things increase readability. The cloze procedure suggests that unfamiliar words may be used and understood if they are placed in a context in which the reader can guess their meaning.

For years, editors were quick to point out that in the fast-paced world of newspapers, magazines, radio and television, there was no time to worry about the academic exercise of measuring readability. Now that news stories are written on computers, that no longer is a valid argument. Many word processors have utilities for measuring readability of an article. Some even have the capability built in. If editors truly want to connect with readers, they will make regular use of those tools.

## CHANGING NEEDS OF CHANGING CONSUMERS

If the media industry is changing, so are the lives of media consumers. U.S. Census and other data reveal much about changing lifestyles:

- Both spouses are working in more and more families, and there are more and more single-parent households.
- The number of people choosing to remain single also is increasing, contributing to a rapid increase in households. (That, by the way, makes those newspaper circulation numbers even worse. Market penetration, a key measure of media effectiveness, is measured by dividing the number of households in a given market area by circulation.)
- Leisure time is shrinking, and media use is a leisure-time activity.
- More and more competitors—direct mail, special-interest magazines and an increasing number of television channels—are competing for that decreasing amount of time.

All that puts pressure on editors to know and understand their audiences. They do so by reading everything they can find about their communities. They also keep an open ear to topics of discussion at the local health club, at parties and anywhere else people gather. An editor who knows what people are talking about has a good idea of what stories would interest them.

Some editors have tried other means of taking the pulse of the community, including formal readership studies, open forums or focus groups—representative samples of media consumers assembled into small discussion groups. Such sessions often are recorded for the benefit of editors. Whatever the technique employed, there is no substitute for learning your community, even if that is an inexact science. The editor who is part of the community—who participates actively in it—is in the best position to be connected to the audience.

# PART 2

# THE FUNDAMENTALS OF EDITING

# CHAPTER 3

# THE EDITING PROCESS

## THE EDITOR'S ROLE

Every editor edits. That is, every editor determines to some extent what will and will not be published or broadcast. Usually, those decisions are based on that editor's perception of the mission and philosophy of the publication or broadcast station.

This book emphasizes the skills of editing, but learning those skills without a thorough understanding of the philosophy of editing would be like learning to hit a baseball without knowing why hitting is important. Why bother to hit if you don't know to run to first base? In editing, it is important to know not only *when* a change in copy should be made but also *why* that change should be made.

Good editing depends on the exercise of good judgment. For that reason it is an art, not a science. To be sure, in some aspects of editing—accuracy, grammar and spelling, for example—there are right and wrong answers, as often is the case in science. But editing also involves discretion: knowing when to use which word, when to change a word or two for clarity and when to leave a passage as the writer has written it. Often, the best editing decisions are those in which no change is made. Making the right decisions in such cases is clearly an art.

The editing skills taught herein will be those used at newspapers in general and at newspaper copy desks in particular. Those same skills, however, apply directly to magazine and broadcast editing. Editing for those media differs slightly from newspaper editing because of special requirements, so separate chapters to highlight those differences are included in this book. Still, the skills required of all editors are much the same as those required of newspaper copy editors, the valuable members of a newspaper's staff who have the final crack at copy before it appears in print. Copy editors, it has been said, are the last line of defense before a newspaper goes to press. As such, they are considered indispensable by top editors but remain anonymous to the public. Unlike the names of reporters, who frequently receive bylines, copy editors' names seldom appear in print.

Some believe that absence of recognition accounts for the scarcity of journalism graduates who profess interest in copy desk work. Editing, it is said, isn't as glamorous or as exciting as reporting. But those who view desk work as boring clearly have never experienced it. To the desk come the major news stories of the day—the spacewalk, the eruption of a volcano, the election of a president, the rescue of a lost child. The desk is the heart of the newspaper, and it throbs with all the news from near and far. Someone must shape that news, size it, display it and send it to the reader.

The copy editor is a diamond cutter who refines and polishes, removes the flaws and shapes the stone into a gem. The editor searches for errors and inaccuracies, and prunes the useless, the unnecessary qualifiers and the redundancies. The editor adds movement to the story by substituting active verbs for passive ones, specifics for generalities. The editor keeps sentences short so that readers can grasp one idea at a time and still not suffer from writing that reads like a first-grade text.

Ah, but editing isn't as much fun as writing, some say. Why learn editing skills if you want to write? Columnist James J. Kilpatrick knows. Although considered one of the great writers of our time, he bemoans the demise of *editing* skills:

> To read almost any American daily today is to conclude that copy editors have vanished as completely from our city rooms as the ivory-billed woodpecker has vanished from the southern woodlands. We appear to have reared a generation of young reporters whose mastery of spelling, to put the matter mildly, is something less than nil. . . . Once there was a white-haired geezer in an eyeshade to intercept a reporter's copy, and to explain gently but firmly to the author that phase and faze are different words, and that *affect* and *effect* ought not to be confused. The old geezer has gone, and literacy with him.

Kilpatrick's fond memories of the good old days probably are enhanced by the passage of time. The fact is that newspapers always have made errors, and the newspapers edited by crotchety old copy editors wearing green eyeshades were no exception. Still, few would disagree with Kilpatrick that language skills in general have deteriorated. Newspapers, without a doubt, have been affected.

Too many reporters and editors at today's newspapers are products of an educational system with misguided priorities. There was a time not so long ago when it was fashionable to consider phonetic spelling adequate. Rote memorization of spelling words was a waste of time, educational trendsetters told us. Grammar was viewed as an exercise in nit-picking. That abandonment of the basics is commonly acknowledged today as one of the great tragedies of modern education. Now, a back-to-basics movement has swept the country, and there is evidence that teachers in today's elementary and secondary schools—some of whom were victims of the errors of the past—are at least attempting to emphasize language skills. Unfortunately, that won't help those who failed to learn, including many reporters and editors now on the job.

## THE VALUE OF THE COPY EDITOR

No position on the newspaper offers greater opportunity for growth than that of copy editor. Work as a copy editor provides the chance to continue an education and an incentive to climb to the top of the newspaper's hierarchy (see Figure 3–1). Copy editors must of necessity accumulate a warehouse full of facts they have gleaned from the thousands of stories they have been compelled to read and edit or from the references they have had to consult to verify information.

Copy editors are super detectives who incessantly search stories for clues about how to transform mediocre articles into epics. The legendary Carr Van Anda of *The New York Times* studied ocean charts and astronomical formulas to find missing links in a story. Few editors today would correct an Einstein formula, as Van Anda did, but if they are willing, they can probe, question, authenticate and exercise their powers of deduction.

Historically, a stint on the copy desk has been considered important to professional advancement at newspapers. The desk serves as an important spawning ground for administrative editors because those who serve there have a more complete picture of how the newspaper operates than those who do not. The reporter has little feel for copy flow and production requirements; copy editors develop that in the normal course of their duties (see Figure 3–2). Thus, if two equally talented individuals are contending for promotion, the one with copy desk experience probably will have the inside track.

There are encouraging signs that the lot of the copy editor is improving. Many newspapers pay copy editors more than reporters as an incentive for the best and brightest to work at the desk. Journalism schools and departments have awakened to the reality that copy editors are more difficult to find than reporters and have responded by improving course offerings in editing. This, in turn, encourages more good students to pursue careers in editing. No longer do editors require that newcomers work as reporters before becoming copy editors; many now realize that wanting to be a copy editor is more important than having experience as a reporter. Many of the best newspapers in the United States and Canada now have editors who never worked as reporters.

Furthermore, the artificial barriers that once prevented women from becoming copy editors have been torn down. No longer is the copy desk a man's domain. Women with editing skills have risen to important positions at many newspapers, large and small.

All this indicates that new life may yet be breathed into the profession of editing. If, as Kilpatrick and others suggest, there is a serious problem with the quality of editing, many journalists have hope that things will improve. They have hope that the art of editing is not a lost art.

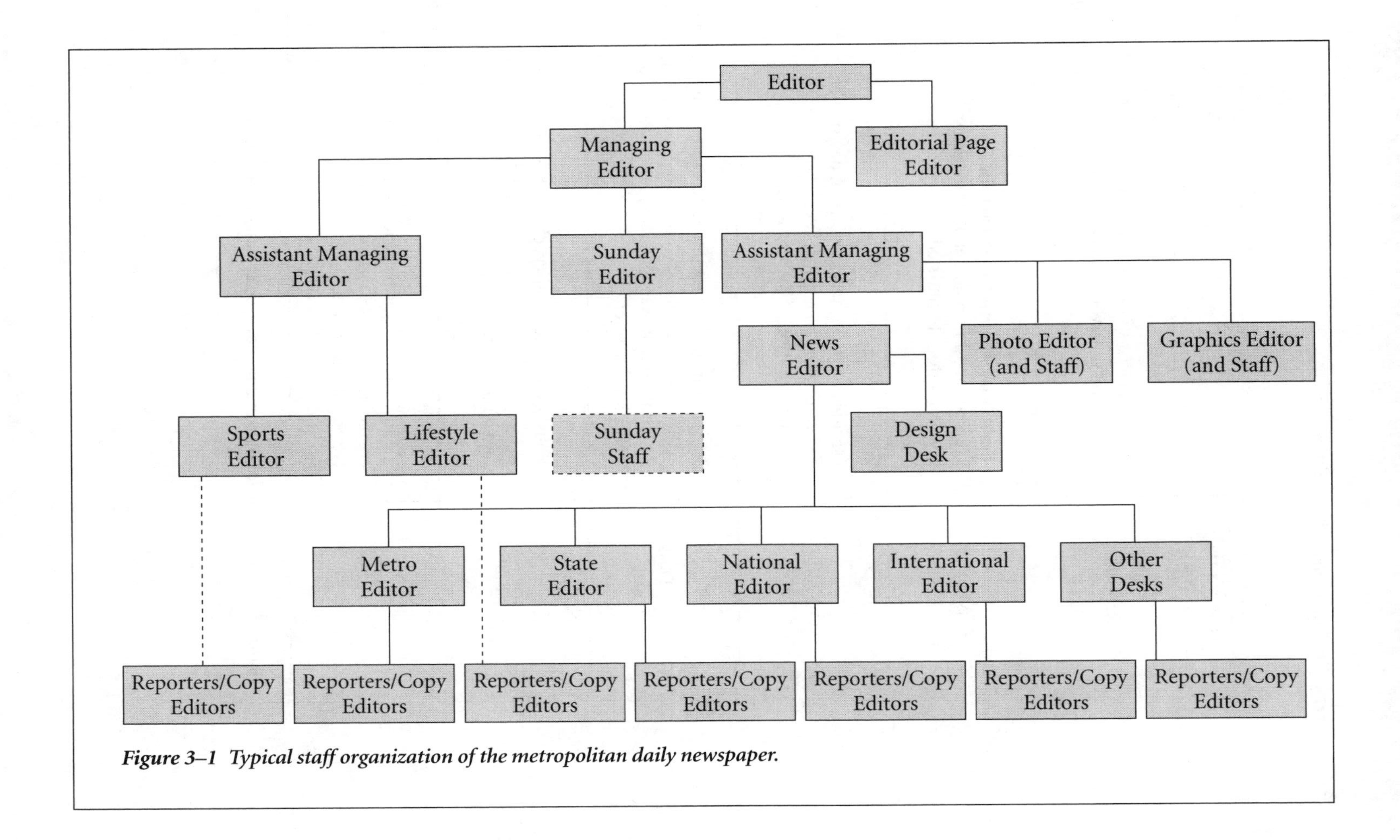


***Figure 3–1*** *Typical staff organization of the metropolitan daily newspaper.*

| Individual | Action |
|---|---|
| Reporter | Gathers facts, writes story, verifies its accuracy, forwards to city editor. |
| City Editor (Or assistant) | Edits story, returns to reporter for changes or additional detail (if necessary), forwards story to news editor. |
| News Editor (Or assistant) | Decides on placement of story in newspaper, forwards story to copy desk chief for implementation of instructions. |
| Copy Desk Chief | Prepares page dummy that determines story's length, setting and headline size, forwards to copy editor. At some large newspapers, a separate design desk may play this role. |
| Copy Editor | Polishes writing of story, checks for missing or inaccurate detail, writes headline, returns to copy desk chief for final check. |
| Copy Desk Chief | Verifies that story is trimmed as necessary and that correct headline is written, transmits story to typesetting machine. |

***Figure 3–2*** *How copy flows through a newspaper news department.*

## *Organization of the Newspaper Newsroom*

It's not easy to describe the typical organizational structure of a newspaper news department. That's because no two are organized alike. Still, most newsrooms have editors with fairly common job descriptions and titles, as summarized here:

- **Editor.** The editor, or editor-in-chief, tops the organizational ladder and is responsible for all editorial content of the newspaper. This includes everything from local to international news in categories ranging from sports to business to entertainment. The editor's responsibilities even include the comics and the editorial page. Today, those responsibilities may extend to editorial content in the newspaper's on-line service.
- **Editorial page editor.** Midsized and larger newspapers may have a separate editorial page editor. This editor, as the name implies, is in charge of production of the editorial page and often chairs the newspaper's committee that determines editorial positions. The editorial page editor is the immediate supervisor of the newspaper's editorial writers.
- **Managing editor.** The managing editor is responsible for all day-to-day news-gathering operations. Often, he or she prepares the newsroom budget. At larger newspapers, there may be one or more *assistant managing editors,* who often are responsible for specific areas of coverage—AME-News, AME-Graphics, AME-Features and so on.
- **City editor or metropolitan editor.** The city editor supervises the staff of local reporters and may also direct reporters located in remote bureaus. At all but the smallest papers, there may be two or more *assistant city editors.* The city editor plays an important role in setting the newspaper's agenda because he or she decides which stories are covered.
- **News editor.** This title means different things at different newspapers, but it usually refers to the individual in charge of the copy desk. There, *copy editors* edit stories into polished form and write headlines. At some papers, the copy desk may also do page design.
- **Graphics editor.** The graphics editor is in charge of the staff of *graphic designers* who produce charts, maps and other nonphotographic illustrations for the newspaper. The importance of information graphics in telling the news has made this an increasingly common position at U.S. newspapers. At small papers, one person may handle all these chores.
- **Photo editor.** The photo editor supervises the *photographers* who take pictures for the newspaper. In addition to news photos, the photo department produces photo illustrations and studio shots to complement food and fashion stories. At some small papers, the photo editor also is in charge of producing graphics.
- **Section editors.** Many newspapers have editors who supervise staffs of reporters assigned to special categories of news—sports editors, lifestyle editors and business editors supervise reporters who specialize in covering news in those areas. Often, a section also will have its own staff of copy editors.

Figure 3–1, on Page 33, shows the newsroom organization of a typical metropolitan newspaper. Smaller papers have simplified versions of the same basic organization.

## THE EDITOR'S RESPONSIBILITIES

Of all the editor's duties, editing for accuracy is probably the most important. A publication that is inaccurate soon loses its credibility and, ultimately, its readers.

Good reporting, of course, is the key ingredient in ensuring accuracy. But all who edit the story share that responsibility. Editors ensure accuracy by questioning the reporter about the information obtained and the means of obtaining it, and by checking verifiable facts. Usually the city editor questions the reporter, and the copy editor checks verifiable facts. The spelling of names and the accuracy of addresses can be checked in telephone books, city directories and similar sources. The publication's library, almanacs, stylebooks, dictionaries, databases and other references are used to check other facts. The good publication provides its staff with these source materials, and the good editor uses them frequently.

Ensuring accuracy is important, but the editor's job entails much more than that. This chapter and those that follow outline what is expected of the copy editor. In general, those responsibilities include:

- Ensuring accuracy.
- Trimming unnecessary words.
- Protecting and polishing the language.
- Correcting inconsistencies.
- Making the story conform to style.
- Eliminating libelous statements.
- Eliminating passages in poor taste.
- Making certain the story is readable and complete.
- Ensuring fairness.

The editor who does each of these things in every story will be a valuable addition to the staff. The importance of accuracy already has been emphasized, but a brief review of the other responsibilities may be useful.

When unnecessary words are being trimmed, adjectives and adverbs often are suspect. If an event is *very interesting,* it is sufficient to say it is *interesting.* Very is an overworked word that has become almost meaningless. Similarly, *totally destroyed* is worse, not better, than *destroyed.* Meaningless phrases also should be eliminated. *At the corner of Ninth and Elm streets* is no better than *at Ninth and Elm streets.*

To polish the language, the editor must know the difference between a conjunction and a preposition and be aware of the perils of misplacing modifiers. The editor also must be able to spell and to use words properly. All editors, and certainly students in an editing course, would benefit by getting into the habit of using a dic-

## *The Journalist and the Three Schools of Ethics*

Unlike lawyers and physicians, journalists have no written code of ethics. Nor is there anyone to license journalists based on adherence or non-adherence to such a code. In the United States, journalists instead must develop their own sense of ethics, which often is rooted in personal experience or religious belief.

Most journalists are *situation ethicists.* That is, they believe each situation must be evaluated to determine what is right and what is wrong. An editor who is a situation ethicist may well break a newspaper's rule never to publish a juvenile's name. When that editor encounters a situation in which it seems to make sense to do so—when the public good is served—the name goes into the paper or newscast.

Two other schools of ethics are common. In *deontological ethics,* it is the person's duty to do what is right. Some actions are always right, some are always wrong. There is no deviation from right and wrong, and the end never justifies the means. Those who adhere to this theory of ethics are often called *absolutists.* A reporter who is an absolutist would never agree to pose as a shop owner, for example. Doing so would be lying, and lying is always wrong.

In *teleological ethics,* the consequences of the act, not the act itself, make an act ethical. So, it may be virtuous for a reporter to pose as a shop owner and offer bribes to health inspectors for the purpose of catching unethical inspectors.

Editors face ethical dilemmas almost daily as they judge whether to run a quotation that is sure to embarrass or humiliate, a fact that will damage a person's reputation or a revelation that may end a public official's career. Usually, they weigh the pros and cons and make a decision based on the facts at hand.

Those decisions are tough ones, even for experienced editors. Junior editors are expected to defer to their superiors for decisions about whether to use such information.

tionary and learning to use words properly. Chapter 5 shows how the copy editor deals with the language.

The top-flight editor also recognizes inconsistencies within a story. Here is a good example:

> Mrs. John E. Simpson, 81, of 914 E. Texas Ave., died Saturday at Boone County Hospital after an illness of several months.
>
> Mrs. Simpson was born in Audrain County Dec. 21, 1914, to the late John and Mary Simpson.

The inconsistency is that Mrs. Simpson likely did not have the same maiden and married names. Therefore, her parents' surname must be checked. The alert copy editor would have detected that possible error and questioned the city desk about it. And, depending on the year in which the story is written, the age and date of birth may not agree.

Annoying inconsistencies also result from failure to conform to the newspaper's style. The word *style,* as used here, refers to the consistency provided by rules of usage and in no way limits a reporter's individual writing style. Style, for example, dictates that the correct spelling is *employee,* not *employe,* which can be found in some dictionaries. It may be annoying to the reader to find the word spelled one way in one story and another way in the next. The importance of style is emphasized in Chapter 6.

The editor faces other problems, many of which cannot be solved by consulting dictionaries and stylebooks. One of those problems is recognizing and eliminating libelous statements. Libel laws vary from state to state, so one of the first duties of a newly employed editor is to check the laws of the state. A thorough discussion of legal limitations on the press can be found in Chapter 7. By reviewing the law and interpretations of it, the editor can usually determine whether a statement is libelous.

More difficult to resolve are questions of good taste. Some newspapers have policies banning profanity of any type; others permit the use of some words but not others. Most editors would object to a story detailing what occurred during a rape; some would not. These are matters of judgment. When in doubt, the wise editor consults a higher-ranking editor for a ruling.

The editor makes certain the story is readable. Writing that is good, and therefore readable, has five characteristics:

- It is precise.
- It is clear.
- It has a pace appropriate to the content.
- It uses transitional devices that lead the reader from one thought to the next.
- It appeals to the reader's senses.

Writing is precise when words are used in the way they should be used, and it is clear when simple sentences and correct grammar are employed. Pacing is a matter of using varied sentence length. The use of short, choppy sentences conveys action, tension or movement. Conversely, long sentences slow the reader. Transitions help the reader tackle the story, and sharp, descriptive words appeal to the reader's senses to bring him or her close to the action.

Good writing employs all these principles. There is no reason why they should be used in novels but not in newspaper stories. The argument that reporters don't have time to produce high-quality writing is a poor excuse, understandable as it may be, for mediocrity. It can be done, and the best newspapers in this country are proving it. The quality of writing depends upon the reporter, but the editor can do much to help. Good editing invariably complements good writing. Occasionally, good editing can save mediocre writing. Poor editing can make writing worse or destroy it.

Editors have no business changing a writer's style. But editors have every obligato insist that the story be correct in spelling, grammar and syntax: It is their duty to make copy compact and readable.

Most copy can be tightened. Even if only a few words in a paragraph are removed, the total saving in space will be considerable. Some stories, notably material from the wires and syndicates, can be trimmed sharply. But the editor should not overedit. If a story is so poorly organized it has to be rewritten, the story should be returned to the originating editor. Rewriting is not an editor's job. Nor should the editor make unnecessary minor changes. Indiscreet butchering of local copy is a sure way to damage morale in the newsroom.

## EDITING THE STORY

Now that we have discussed the editor's role and what that editor is expected to do, let's examine the editing process itself. Most experienced editors suggest that the process be divided into three distinct steps:

- Read the story.
- Edit it thoroughly.
- Reread the story.

Editors frequently skip the first step or abbreviate it by scanning the story for the gist of the news. To do so may be a mistake because intelligent editing decisions cannot be made unless the editor understands the purpose of the story and the style in which it is written. That understanding is developed with a quick but thorough reading.

Some editors try to skip the third step, too. They do so at the risk of missing errors they should have detected the first time or those they introduced during editing. Few sins are greater than to introduce an error during editing. The more times a story is read, the more likely errors will be detected.

Unfortunately, deadline pressures sometimes dictate that step one or three, or both, be skipped. When this is done, it becomes increasingly important for the editor to do a thorough job the first time through the story.

To illustrate the editing process, let's see how one copy editor edited the story shown in Figure 3–3 on the defeat of a bill in the Missouri General Assembly. This alert copy editor recognized immediately that, while the lead was well-written, it also was inaccurate. The state's streets and stores are not deserted on Sundays. The reporter, reaching for a bright lead, overstated the case.

Because the story is from the newspaper's state capital bureau, it needs a dateline, or place of origin, which the editor inserted. The second paragraph became

BLUE LAW

State capitol bureau

~~Missouri's city streets and stores will continue to be deserted on Sundays, at least for another year~~.

The House Tuesday afternoon rejected a bill that would have sub~~mitted the~~ repeal of the states blue laws to the voters in November.

The measure was ~~soundly~~ defeated 97-53 ~~with two members present but not voting~~.

The ~~state's~~ blue laws, or Sunday closing laws, prohibit ~~the~~ sale of nonessential ~~chosen merchandise~~ including autosmobiles, clothing, jewelry and hardware ~~items~~.

During a ~~lengthy~~ debate ~~that lasted four hours~~, the original bill was amended ~~to place~~ the issue on th~~is~~ November ballot.

~~The bill also was amended to require~~ that counties and cities ~~exercise local option to approve~~ repeal ~~of blue~~ laws ~~locally if the measure was approved by voters in Nov~~.

Several amendments were defeated ~~that~~ would have required employers to ~~pay as much as twice the wages per hour~~ to employes work~~ing~~ on Sundays

~~Rep. James Russell, a Florissant Democrat, included in one amendment the proviso that employers be fined~~ $1,000 for each violation of the wage requirements. ~~His amendment lost.~~

Another amendment ~~introduced by Rep. Jerry McBride, St. Louis, would~~ have levied a ~~1¢~~ excise tax on retail businesses open on Sunday. Proc ~~eeds~~ from this were to have gone to city and county treasuries.

~~The McBride amendment also was defeated.~~

~~During debate~~ much of the support for blue law repeal cane from urban representatives whose ~~consitituents~~ include large retail chain stores and businesses that supported repeal ~~of the existing law~~ during committee hearings.

~~In the end, the majority of the house agreed with Rep. Walter Meyer D. Bellefontaine Neighbors, that "this bill is all screwed up."~~

~~Local~~ Democrat~~ic Reps~~.Ray Hamlett of Ladonia, Joe Holt of Fulton and John Rollins, of Columbia, and Republican ~~Reps~~ Larry Mead and Harold Reisch all voted in favor of putting repeal on the ballot.

*Figure 3–3 Good editing helps readers make sense of a story.*

the lead, and the third paragraph was tightened and combined with the second. Note that the editor struck the phrase "with two members present but not voting." In many states there is a difference between present but not voting and abstaining. An abstention allows the legislator to change his or her vote later; present but not voting does not. The editor decided that in this case it wasn't important, so the phrase was deleted.

The story is confusing as written, probably because the reporter was too close to the subject to realize that readers would have difficulty following its meaning. The editor clarified fuzzy passages, tightened the story and distinguished between the amendments that were adopted and defeated before the bill itself was defeated.

The editor deleted the next-to-last paragraph because the legislator quoted used a colloquialism that may have been offensive to some readers. It added nothing to the story, so it was deleted.

The last paragraph illustrates how important it is to use common sense when editing. Conformity to the stylebook would have called for the identification of each legislator by party affiliation and district number. Because there are five of them, the paragraph would have been difficult to read. The editor decided that the legislators' hometowns gave more information to the reader than the district number, and grouping by party affiliation made the paragraph less awkward than it would have been if the editor had followed stylebook practice. Such editing helps the reader make sense of the story. As a result, it is good editing.

That story needed plenty of attention. More typical are stories that need only a bit of polish and clarification. Here are unedited and edited versions of the same story. Look carefully to see the changes the editor made:

| UNEDITED | EDITED |
|---|---|
| LOS ANGELES (AP)—A little girl born with a perpetually grumpy look underwent surgery Friday that could literally bring a smile to her face. | LOS ANGELES (AP)—A little girl born with a perpetually grumpy look underwent surgery Friday that could literally bring a smile to her face. |
| "I'm fine, and excited," 7-year-old Chelsey Thomas, clutching a favorite doll, said as she walked with her parents into Kaiser Permanente Medical Center in Woodland Hills. | "I'm fine, and excited," 7-year-old Chelsey Thomas said as she walked with her parents into Kaiser Permanente Medical Center in suburban Woodland Hills. |
| The corners of the blond, blue-eyed girl's mouth sag because she was born without a key nerve in her face, a condition called Moebius Syndrome that afflicts up to 1,000 people nationwide. The nerve transmits commands to facial muscles that control smiling, frowning and pouting. | The corners of the blond, blue-eyed girl's mouth sag because she was born without a key nerve in her face. It is a condition called Moebius Syndrome that afflicts up to 1,000 people nationwide. The nerve transmits commands to facial muscles that control smiling, frowning and pouting. |
| In an eight-hour operation that began at midmorning, doctors planned to remove | In an eight-hour operation, doctors planned to remove muscle and nerve from |

**UNEDITED (continued)**

muscle and nerve from Chelsey's leg and transplant it to one side of her face. If the transplant succeeds–something that won't be known for at least three months–the other side will be done in about six months.

"She'll be able to smile for the first time, and that's something every parent waits for. Usually it happens in the first few weeks of their life, but we've had to wait a little bit longer," said Lori Thomas, her mother.

The $70,000 two-step operation is covered by the family's health insurer, Kaiser Permanente, because it is not considered cosmetic.

"Just being unable to smile has numerous problems associated with it, social and psychological," said Dr. Avron Daniller, one of Chelsey's surgeons.

Her mother said: "It's been hard for her because people think she's unfriendly or ignoring them or bored. It's been hard. Kids stare at her. Adults are pretty understanding, but she has a worse time with kids."

Kaiser flew renowned Canadian surgeon Dr. Ronald Zuker, who pioneered the procedure, to Los Angeles to lead the surgery. Chelsey was expected to be hospitalized five days.

Last month, Chelsey said she was eager to have the operation: "After the surgery, I'm going to smile at all my friends and have a party."

**EDITED (continued)**

Chelsey's leg and transplant it to one side of her face. If the transplant succeeds–something that won't be known for at least three months–doctors will do the other side in about six months.

"She'll be able to smile for the first time, and that's something every parent waits for," said Lori Thomas, her mother. "Usually it happens in the first few weeks of their life, but we've had to wait a little bit longer."

The $70,000 two-step operation is covered by the family's health insurer, Kaiser Permanente, because it is not considered cosmetic.

Kaiser flew renowned Canadian surgeon Dr. Ronald Zuker, who pioneered the procedure, to Los Angeles to lead the surgery. Chelsey was expected to be hospitalized five days.

Chelsey's mother said: "It's been hard for her because people think she's unfriendly or ignoring them or bored. It's been hard. Kids stare at her. Adults are pretty understanding, but she has a worse time with kids."

The editor first sought to trim the article's length under instructions from a higher-ranking editor. The lead, or opening paragraph, was both punchy and to the point, so the editor left it alone. In the second paragraph, the editor chose to eliminate "clutching a favorite doll" because the doll was never again mentioned in the story and because the sentence was awkward and cluttered. Removing that phrase seemed to help. The editor inserted "suburban" in front of Woodland Hills to show that the city is a suburb of Los Angeles, where the story is datelined.

In the third paragraph, the editor broke a long sentence in half. In the fourth, the editor eliminated "that began at midmorning" on grounds that the story would be dated with its inclusion.

The mother's quote in the fifth paragraph of the original was not identified until too late in the paragraph. The editor chose to move the attribution to the end of the first sentence. That way, the editor explained, the reader would not have to guess who was speaking. Another option would be to introduce the two-sentence quote with something like this:

Lori Thomas, Chelsey's mother, said:

The editor also chose to move together two paragraphs about Kaiser Permanente's role in the surgery. That enabled the editor to avoid excessive shifting back and forth among sources within the story. The editor also deleted a dated quote from the child and ended with a quotation from the mother.

Were all those decisions the right ones? That, of course, is a matter of judgment, and each editor's judgment is different. Arguably, though, this editor's decisions were good ones, and the story is better as a result.

## RECOGNIZING WRITING STYLES

One of the keys to being a good editor is understanding the process of writing. That begins with a strong understanding of the styles in which articles are written and recognition of why the writer chose that style. Gone are the days in which almost all newspaper stories were written in the time-honored inverted pyramid format. Today, newspaper and magazine writers employ all the tricks of the writing trade, including use of story forms more often associated with literary writing (see Figure 3–4).

A good editor takes the time to read a story first before beginning to make changes. One reason for that is to recognize what the writer is trying to accomplish with the story form he or she has chosen. Learning to respect that choice is a critical first step toward becoming a good editor and an important part of the editing process.

Let's take a look at the most common story forms used in journalistic writing, beginning with what's still the most important.

### The Inverted Pyramid Approach

This classic style, with its crammed leads and recitation of facts in descending order of importance, should not be dismissed. It remains the staple of news writing. Fast-breaking news stories lend themselves to this treatment. Indeed, the form was developed by newspaper writers, who found it worked well in meeting the needs of their fast-paced business.

*Figure 3–4 The newspaper newsroom is a busy place as reporters approach deadline.*

Photo by Philip Holman.

The inverted pyramid story has three main parts:

- The lead.
- Support and development of the lead.
- Details in descending order of importance.

Typically, stories written in this form have the traditional lead that tells the reader who, what, when, where, why and perhaps how. Here is an example of an inverted pyramid story:

---

SPRINGFIELD, Mass. (UPI)–A man screaming "I'm the king" slashed his way through a crowded hospital emergency room with a long-bladed paring knife Monday, killed a 6-year-old boy and injured five other people.

Willie Robinson, 42, infuriated when he and his young granddaughter were burned with cleaning fluid during a domestic quarrel at his apartment, stabbed his wife and ran amok at Bay State Medical Center's Springfield Unit a short time later, police said.

"He had a really wild look in his eyes," said Sgt. George Bishop, a hospital security guard. "There was an awful lot of blood."

Robinson had been taken to the hospital by his nephew for treatment of burns, then "went berserk." He wounded five persons in the emergency room and killed Anthony Lombardi, 6, of Agawam, police said.

Police said Robinson stabbed several women and a guard and then pinned the boy down at an ambulance entrance, stabbed him repeatedly and screamed, "I'm the king, I'm the king. Now I've done it. I'm done."

The boy died about two hours later of wounds in the neck, throat and arms.

The boy's mother, Rose Lombardi, 30, was listed in serious condition with multiple stab wounds.

Police said Robinson would be charged with murder and seven counts of assault and battery.

---

The inverted pyramid was developed to make it easier to trim the story in the composing room as the deadline neared. Because facts were in descending order of importance, trimming from the bottom was simple. Traditional copy-processing procedures made story length estimates imprecise, so trimming was frequently necessary. Computer technology used in modern newspaper plants permits editors to determine the exact length of a story before it is typeset, so page layouts can be more precise. This makes it more feasible for newspapers to use alternative writing styles, which, unlike the inverted pyramid, often employ formal closings.

## The Narrative Approach

Some newspaper editors believe there is an increasing need to use the narrative, or magazine, style in newspaper stories. A narrative may have no breaking news at all, but rather a wealth of new and interesting information that will make it news nevertheless.

This concept already has been tried successfully. For example, some news events heavily covered by television have been presented in the narrative rather than the traditional news style. Here is an example of the narrative approach, which is little more than good storytelling:

---

MADRID (UPI)—"You will write about me and my photo will be published on every front page," boasted 16-year-old Mariano Garcia. "I will be a great bullfighter one day."

This was his dream and the theme he returned to time and again during our 200-mile drive from Saragossa to Madrid.

It was on Feb. 23 in the outskirts of Saragossa that Mariano had asked me for a ride.

"I want to go to Madrid and start a career as a bullfighter," the boy said in the accents of his native Mancha.

Mariano's was a classic story of Spain: The eager youth deserting the misery of his sunbaked village with its whitewashed walls for the danger and glory of the bullring.

## *The Importance of Leads*

The lead of a story is its most important element because the lead's quality may well determine whether the reader is hooked or skips the story in favor of another. Editors hesitate to impose formulas for writing by limiting sentence length, but on one point they are almost unanimous: The lead of a story must be short. How short? The *Chicago Tribune* applauded this three-word lead: "Money and race." An even better one read: "Are nudes prudes?"

Ralph McGill, late editor of *The Atlanta Constitution*, liked what he called a flawless lead in the Bible: "There was a man in the land of Uz whose name was Job." That all-time best seller probably has another of the best leads ever written: "In the beginning God created the heaven and the earth."

Columnist Tom Fesperman uses short leads and short sentences to move his readers along:

> There was a quail whose name was Hercules.
>
> Don't ask me how he got that name. I have spotted a lot of quail in my time, and I wouldn't call any of them Hercules.
>
> Even so. Let us get on with the story.
>
> It is not exactly accurate to say that Hercules was born in Arizona. It's more exact to say he was hatched.

The editor's eye brightens when he or she reads leads that rank with these classics:

> Only in Russia could Peter and the Wolf die on the same night. (Stalin's death)
>
> Today the Japanese fleet submitted itself to the destinies of war—and lost.
>
> They're burying a generation today. (Texas school explosion)
>
> The moon still shines on the moonshine stills in the hills of Pennsylvania.
>
> Fifty thousand Irishmen—by birth, by adoption and by profession—marched up Fifth Avenue today.

Most lead problems arise when reporters try to see how much they can pack into the lead. More often, they should try to see how much they can leave out:

| ORIGINAL | EDITED |
|---|---|
| The former girlfriend of a man charged with killing a local bartender almost four years ago testified Monday that she has never seen another man who claims he killed the victim and that she was with him that night. | A man who says he actually committed the murder for which another man is on trial was contradicted in court Monday. |
| **ORIGINAL** | **EDITED** |
| Columbia Gas Transmission Corp. has started discussions with its subsidiaries and major customers that could result in increasing the supply of natural gas to its Kentucky subsidiary—and thus to industrial users in the central and northeastern parts of the state who now face a total cutoff of supplies this winter. | Columbia Gas Transmission Corp. has started a search for ways to save industrial users in central and northeastern Kentucky from a threatened total cutoff of gas this winter. |

One way to avoid long, cluttered leads is to substitute simple sentences for compound ones.

| | |
|---|---|
| LONDON—Forty years ago today, a 27-year-old princess was crowned Queen Elizabeth II, and Britain, in an outpouring of emotional fervor unmatched since, hailed her coronation as the beginning of a new Elizabethan era of splendor and achievement. | LONDON—Forty years ago today, a 27-year-old princess was crowned Queen Elizabeth II. Britain, in an outpouring of emotional fervor unmatched since, hailed her coronation as the beginning of a new Elizabethan era of splendor and achievement. |

The lead should be pruned of minor details that could come later in the story if they are needed:

| | |
|---|---|
| Donald E. Brodie, son of William Brodie, for more than three decades a member of the display advertising department of the News, and Mrs. Brodie of 106 W 41st St. was graduated from Jefferson Medical College last week. | Donald E. Brodie, son of Mr. and Mrs. William Brodie of 106 W. 41st St. was graduated from Jefferson Medical College last week. For more than three decades, William Brodie was a member of the display advertising department of the News. |

The lead need not be long to be cluttered and unreadable:

| | |
|---|---|
| First National is one of four American banks and the American Express which issues travelers checks throughout the nation. | First National, three other American banks and the American Express Co. issue travelers checks throughout the nation. |

## Leads That Mislead

A good lead contains qualities other than brevity. It must inform and summarize. It must be straightforward; it cannot back into the action. It sets the mood, the pace and the flavor of the story. It accomplishes what the term implies: It guides, directs, points to and induces. If it is a suspended-interest lead, it must be so tantalizing and intriguing that the reader cannot help but continue.

In an effort to get the maximum punch in the lead, the overzealous reporter may overstate. When this happens, the lead is unsupported by the story or ignores facts contained in it. It stretches and therefore distorts. It is the type of lead that reads, "All hell broke loose in city hall last night." Then the final sentence reads, "When the dispute subsided, members of the council shook hands, and the mayor adjourned the session."

No matter how appealingly you wrap up the lead, it is no good if it gives a wrong impression or tells a lie. The "souped-up" lead invites a sensational headline. If the editor lets the overextended lead stand, then tries to top the lead with a calm headline, the editor is likely to have the headline tossed back with the suggestion that more punch be put into it.

Akin to the sensationalized lead is the opinion lead. This type of lead offers a judgment rather than fact. Often it fails to distinguish between mere puffery and news: "Construction features described as newer in concept than space travel will be part of the easy-to-shop-in Almart store to open soon on the Kirkwood Highway."

*(Continued)*

# The Importance of Leads (continued)

## Don't Back into It

John Smith returns home from a meeting and his roommate asks, "What happened at the meeting?"

"Well," replies John, "Jim Jones opened the meeting at 8 and Bill Prentice read the minutes of the previous meeting, and . . ."

"Good night, John."

Too many news leads read like secretarial reports, and that is especially true of leads on speech stories. Few care that a nursing consultant spoke at a group's "regular monthly meeting" or care to know the title of the talk. Readers want to know *what* that person said. That's the news. They won't wade through three paragraphs to learn in the fourth paragraph that "the epileptic child should be treated as a normal person." The lead should get to the point of the story immediately, like this:

| ORIGINAL | EDITED |
|---|---|
| Dean F. Snodgrass of the University of California's Hastings College of Law says the American Bar Association's longtime ban on news photographs in courtrooms is archaic and unrealistic. | The rule against news photographs in courtrooms is archaic and unrealistic, a law school dean says. |
| **INDIRECT** | **DIRECT** |
| An Atlanta businessman who joined two anti-black, anti-Jewish groups and turned over information to the FBI today associated a man on trial for dynamiting the Jewish temple with race-hating John Kasper. | A man on trial for dynamiting the Jewish temple was linked today with race-hating John Kasper by an FBI undercover agent. |
| **INACTIVE** | **ACTIVE, TIGHT** |
| A top-ranking rocket AFD space weapons expert coupled a disclosure of his resignation from the Air Force today with a blast at the senior scientists upon whom the services rely for technological advice. | A top-level Air Force space weapons expert blasted civilian scientists today and said he has resigned. |

## Cliché Leads

Clichés, because of their familiarity, often produce leads that sound like many the reader has read before. The result is a "so-what?" attitude:

Quick action by two alert police officers was credited with saving the life of...

Police and volunteers staged a massive search today for a man who...

## Say-Nothing Leads

If a say-nothing lead causes the editor to ask, "So what else is new?" chances are the readers will have the same reaction.

DETROIT—Somber was the word for the memorial services to American dead of the War of 1812 Sunday.

Fire so hot it burned the mud guards off and melted a small section of an aluminum trailer damaged the trailer.
(The latter is almost like writing, "Fire so hot that it burned the roof and walls and destroyed all the furnishings damaged the house of John Doe.")

## Illogical Leads

Frequently, illogical leads occur when the writer presents the idea backward or uses a non sequitur, a phrase that is nonsensical or defies logic.

State police attributed an auto collision and the alertness of witnesses to the rapid apprehension of Benjamin Petrucci.
(Either the apprehension was attributed to the collision and the alert witnesses or the collision and the witnesses were credited with the apprehension.)

Other examples:

Hoping to encourage transient parking at its facilities, the city parking authority voted Monday to increase rates at two lots.
(Charging more for parking hardly seems to be the way to encourage more of it.)

Three small brothers died last night in a fire that burned out two rooms of their home while their father was at work and their mother was visiting a neighbor.
(Note how much clearer the revision is: "Left unattended, three small brothers perished in a fire last night that destroyed two rooms of their home. Their father was at work and their mother was visiting a neighbor.")

## Excessive Identification

Another problem is excessive identification, sometimes even preceding the name:

Former Assistant Secretary of State for Latin American Affairs Lincoln Gordon said today...
(This results in unnecessary clutter in the lead. Long titles should follow names.)

## Too Many Statistics

Burdensome statistics or numbers also may clutter the lead. Including those statistics later in the story may help alleviate problems such as these:

At 7 p.m. Monday 60 persons fled a three-story apartment building at 2523 E. 38th St. when a carelessly discarded cigarette sent smoke billowing through the building.

Louis Ezzo, 29, of Plainville, a school bus driver, was charged by state police with speeding and violation of a statute limiting school bus speeds to 40 miles an hour at 3:30 p.m. Tuesday on I-95 Groton.

A 14-year-old boy fired three shots into a third-floor apartment at 91 Monmouth St. Friday to climax an argument with a 39-year-old mother who had defended her 9-year-old daughter against an attack by the boy.

*(Continued)*

### Overattribution

Sometimes attribution isn't needed in the lead. Including it there merely creates clutter:

> Mimi La Belle, a Springfield exotic dancer, was arrested on a charge of indecent exposure last night, officers George Smith and Henry Brown said Monday.

### Underattribution

On other occasions the attribution must be included in the lead. An Associated Press reference book comments, "Don't be afraid to begin a story by naming the source. It is awkward sometimes but also sometimes is the best and most direct way to put the story in proper perspective and balance when the source must be established clearly in the reader's mind if he [or she] is properly to understand the story." An example:

> All Delawareans over 45 should be vaccinated now against Asian flu.

The attribution should have been in the lead because this is opinion, the consensus of a number of health officials.

### Second-Day Lead on a First-Day Story

Every veteran wire editor knows that frequently a wire story's first lead is better than its second, third or fourth. A lead telling the reader an airliner crashed today, killing 50 passengers, is better than a later lead reporting that an investigation is under way to determine the cause of an airline crash that killed 50 passengers. If the first lead tells the story adequately, why replace it with a second and often weaker lead?

The first lead:

> HOLYOKE, Mass.–At least six persons–four of them children–perished early today when a fire, reportedly set by an arsonist, swept a five-story tenement.

The second lead (with a second-day angle):

> HOLYOKE, Mass.–The body of a little boy about 2 years old was recovered today, raising the death toll in a tenement house fire to seven.

And the third lead (back on the beam):

> HOLYOKE, Mass.–Seven persons–five of them children–perished when a general alarm midnight blaze, believed set, destroyed a five-story tenement.

Conversely, a first-day lead sounds like old news when placed on a second-day story. Copy editors allow such mistakes to appear in print when they fail to keep up with the news. It is important that copy editors read the newspaper each day. Those who don't fail to recognize whether the story is the first on the subject or a follow-up. That may make a big difference in the way the lead and headline are written.

When he learned that I was a reporter, Mariano pulled a pencil stub and sort of calling card out of his pocket. The card bore an amateurish sketch of the Virgin Mary and Jesus. Across it he scrawled his name in bold letters.

"The Virgin and the Christ are my protectors," he said.

Then he talked with wide-eyed dreaminess of how the bull paws the ground with his left forefoot when excited, of his sudden charges, of the secrets of the matador's capework and of the "pata, rabo y orejas"–the hoofs, tail and ears of the bull–the highest honors the fickle Spanish crowd can pay a matador.

For two years, Mariano tried to get a start on the road to fame and fortune. It came not in Madrid but at San Martin de Lavega.

The bull he met Thursday was six years old, fat, limping, and the sharp tips of his horns had been clipped. But the old bull could be dangerous. He had survived the ring once and remembered well how the matador evaded his charges and the hooking of his horns.

Mariano made three or four passes with his cape. But the bull was old, and the boy was young, and the fickle crowd was bored.

"Just a meletilla," some shouted, "just a beginner."

Another matador stepped in to divert the bull. But the animal had his eyes fixed on Mariano. Suddenly, he charged the boy and knocked him down.

Then the old bull drove his blunted left horn into Mariano's skull.

It was a quick kill, as the good matador's sword should be quick.

Mariano Garcia's mother and father carried his body in a bloodstained sheet to the local cemetery. And today I wrote the story Mariano promised I would write.

---

This type of storytelling has the quality once described by Earl J. Johnson of UPI: "to hold the reader's interest and stimulate some imagination to see, feel and understand the news."

## The Personalized Approach

Business and financial news may not be the most glamorous of subjects for most readers, but *The Wall Street Journal* knows how to make such stories compelling. Here's an example of the *Journal*'s personalized approach:

---

**HOME ON THE RANGE IS A PART-TIME DEAL FOR MANY COWBOYS**

**They Keep Cows Just for Fun, Helping Keep Prices Low; Job After Banker's Hours**

By Marj Charlier
Staff Reporter of *The Wall Street Journal*

HUNTLEY, Mont.–C.T. Ripley has a redwood ranch house with cowhides adorning the walls and floors. He also has a red pickup truck, with a color-matched stock trailer and six cowboy hats, for whatever the occasion demands.

"I am a cowboy," he says. "That's all I've ever wanted to be."

In truth, he is an oil-refinery employee. But on the side for the past couple of years, he has been putting together his 200-acre "Almosta Ranch." Now he is ready to buy some cattle. He will probably lose money on them, but there is the income from the job sustaining him and the romance of the open range spurring him. He will do it anyway.

Such part-time ranchers are a big problem for full-time cattlemen, many of whom are losing their shirts. The would-be cowboys add up to too many cows, standing the law of supply and demand on its head.

Demand for red meat has tumbled the past decade, depressed by health concerns and recession-induced penny-pinching. As a result, the price ranchers command for cattle has fallen 25 percent, adjusted for inflation. Cattle-herd liquidations would seem to be in order, to bring supply in line with demand and strengthen prices. Some cows have been culled, but analysts suggest the herds must be cut much more to ensure long-term profits.

Small herds compound the price problem. More than 80 percent of the nation's herds have fewer than 100 head of cattle, kept mostly by part-time farmers and those who make their living raising grain. Many of these maintain their livestock, despite the low prices, because cattle aren't their main source of income. Some cattle also are kept at a loss because they provide a tax shelter against other income.

The bottom line: Cow-calf operators (the cowboys who keep beef cows that produce calves for sale to feed lots) have sustained losses eight out of the last 10 years....

---

The *Journal*'s personalized approach is part of a well-designed formula that newspaper has popularized. It is a four-step process:

- Focus on the individual.
- Transition to the larger issue.
- Report on the larger issue.
- Return to the opening character for the close.

To tell the part-time ranchers' story in the inverted pyramid formula probably would have been less effective.

## The Chronological Approach

American newspapers of the colonial period often used the chronological approach in their news accounts. This approach should be used sparingly because it can be boring, but occasionally it can be effective. A detailed account of a gunman's siege may lend itself to this treatment; so may a reconstruction of a pilot's frantic final moments before an airplane crash or a step-by-step account of how a legislative bill was altered to appease lobbyists.

When the chronological approach is used, the formula works this way:

- A summary lead.
- Transition to the chronological account.
- The chronological account.
- A closing summary or reversion to the inverted pyramid.

One good time to use th… …hronological approach is when the subject of an interview relates … …ries of events. There's nothing wrong with allowing the subj… …ith a minimum of interference from the writer. Here's an exc… …United Press International:

---

COLUMBIA, … …aves sloshed over her face, the chilly water of Lake Marion numb… …moment Lynne Heath thought about giving up and drowning. But … …ok over and she kept struggling toward shore.

"I fussed at … …that way," Ms. Heath, 20, said in an interview Sunday.

"My father is … I told him, 'Hey Dad, put in a good word for me with God. I'm too … …bborn."

The interview … …ath has talked with the news media about the boating accident M… …ce and two other people died. Ms. Heath managed to survive by s… …the 52-degree water of Lake Marion until she reached shore.

Ms. Heath, her fian… …vin Morris, 20, and assistant county solicitor Harrison Heller III, 29, … …on the huge lake about a mile from shore when the wind picked … …e rough.

The gas tank in the 16 … …fted, Ms. Heath said, and the boat took on water and quickly sank.

Heller—who could not … …wearing a life preserver. The other preservers went down with th…

Ms. Heath said they look… …ng—flotsam or a stump or anything to hold onto—but there was not… …ipping across the 20-foot-deep water.

"I didn't realize how muc… …she recalled. "Kevin (Brown) looked back at Kevin Morris. He could… …self if he had not gone back to help. Please tell everyone that he we… …ers.

"I looked at him and I said, 'I… …he look on his face is one I'll never forget.

"I started swimming toward so… …n shore. I would swim a little way and the waves would push me ba… …come over my head, and you would have to hold your breath and then … keep going.

"I have never been a strong swimmer, but I was determined to make it. I swam with the back stroke awhile, the breast stroke, the side stroke. I saw white birds flying over, and it made me mad because I wondered why I couldn't just fly out there.

"The sun was shining, and it made me feel better. I knew I had to get out of my clothes and shoes. I took off my tennis shoes, and my jeans were waterlogged and heavy. I had trouble getting them off.

"I had Girl Scout training and so I tried to make my jeans into a life preserver, but the zipper wouldn't work.

"Soon I had stripped down to my purple tube top and panties. I am 5-foot-2 ½, and I weighed 122 then. I weigh 112 now. Maybe the baby fat helped with hypothermia.

"At one point I was fixing to give up, and I said no, 'My mom will have a nervous breakdown.'

"I'm very stubborn, and I have a very hot temper. I think it was my temper. I was mad.

"Finally, after 90 minutes, I got to this beach," she said. "I sat there for about 10 seconds and got my legs to stop shaking so badly. Then I started half crawling and running to the houses about a mile down the beach.

"Briars ripped at my legs, but I didn't care. When I got to the house I found these six steep steps. I ran up and just fell in through a screen door."

Mr. and Mrs. Gilbert Minor, a retired couple, wrapped her in a blanket and gave her hot black coffee with lots of sugar while she blurted out the story and asked that authorities be called to help the others.

"We all know it was a miracle," said her mother, Rebecca.

Heller's body has been recovered, and Brown and Morris are presumed drowned.

---

The writer provides a strong introduction for three paragraphs and then uses the fourth as a transition. The chronological account, much of it direct quotes, begins with the fifth paragraph. Finally, the author provides a summary paragraph. The success of this story is striking evidence that the chronological approach should not be dismissed.

## The First-Person Approach

The first-person approach also should be used sparingly, but on the right occasion it is effective to have the writer become a part of the story. The first person can be used when the reporter is an eyewitness to an event. It also can be used in participatory journalism, when reporters become ambulance attendants or police officers for a day.

# CHAPTER 4

# THE FOUR C'S: BE CORRECT, CONCISE, CONSISTENT AND COMPLETE

The most important aspects of writing and editing can be summarized easily in four short imperative sentences: Be correct. Be concise. Be consistent. Be complete. Let's examine each of these qualities as it pertains to editing.

## BE CORRECT

As we explained in Chapter 2, the editor's first priority is to make certain the story is accurate. Editors do this in a variety of ways. They make certain that:

- Proper names of people and places are properly spelled.
- Writing is clear and ambiguities are eliminated.
- Verifiable facts are verified.
- Numbers are accurate.

Journalism's history is littered with examples of reporters making stupid mistakes because they assumed *Smith* was spelled this way rather than the more unusual *Smyth,* or they assumed *Bryan* when *Brian* was correct, or they didn't know *Aransas,* not *Arkansas,* is the name of a national wildlife refuge. Other mistakes result from not being aware that the smallest change in wording can be significant. For example, in some states, *driving while intoxicated* legally differs from *driving while under the influence* of alcohol. The first is more serious. If the first is correct, use it. But if it is not, a libel suit may result. Fuzzy writing and the use of imprecise language cause many legal problems for newspapers. Equally damaging is the fact that readers may be confused.

Clarity is the byword of correctness. Editors achieve clarity by making certain that stories use common words, employ the active voice (more about that in Chapter 5) and avoid technical jargon. Government stories are a particular source of unclear writing. Too often, they use terms with which readers aren't familiar—*habeas corpus, change of venue, writ of mandamus* and *mill rate* to name a few. It's the editor's job to translate those items into ordinary English.

Each newsroom has a collection of reference works in which many facts can be double-checked. If the story says the war began in 1965, check it out in the almanac. If the story says that Mary Miller was mayor from 1965 to 1967, a check of your publication's library, or morgue, should confirm it. John Bremner, a highly respected editing professor at the University of Kansas, used to tell his students, "If your mother says she loves you, check it out." Nothing more succinctly drives home the point.

Large national magazines have fact checkers who do nothing but verify facts. Unfortunately, newspapers and broadcast stations, with their much tighter deadlines, seldom have the luxury of being able to check every tidbit of information. They depend on reporters to get it right, and they depend on editors to have a healthy dose of suspicion. Editors who are sharp enough to spot potential errors are valued members of the staff. Not everyone, of course, can be as sharp as legendary editor Carr Van Anda, who spotted an error in an Einstein formula. But a well-educated editor with a suspicious nature will detect plenty of mistakes before they appear in print or are broadcast.

Reference books and the publication's library are great places to check facts. But don't overlook the on-line services, which, like the morgue, can be tapped in a matter of seconds. Among them are the Internet; the public information utilities, such as CompuServe, Prodigy, America Online and MSNBC; and the commercial database services, such as Dialog, VuText, Dow Jones News Retrieval and others. The ready availability of such resources makes it more inexcusable than ever not to have checked a verifiable fact.

There's no secret formula for ensuring correctness, and sooner or later every publication and broadcast station will be guilty of making a mistake. When it happens, admit it. As mentioned in Chapter 3, newspapers, in particular, are increasingly willing to admit their mistakes in print. Doing so helps restore a bit of lost credibility. Readers and listeners appreciate candid admissions of errors and resent failure to acknowledge them. While there is no easy way to ensure correctness, being aware of some areas that often cause writers and editors to stumble may help.

## The Hoax

Readers of *USA Today*'s LIFE section probably were intrigued to read this item, picked up from The Associated Press, about the latest romance of actress Elizabeth Taylor:

## *Another Approach: The Seven C's Plus One*

Don Ranly, a journalism professor and widely sought speaker in professional editing circles, takes a slightly different approach to writing and editing than we take in this chapter. He refers to the Seven C's Plus One. (That's eight, but Ranly argues that most people cannot remember more than seven things in a list.) To the four C's we have listed, Ranly adds:

- Be clear.
- Be creative.
- Be coherent.
- Be concrete.

In fact, we mention those very items, but we do so within the context of the others. Clarity is a key ingredient of conciseness, and it contributes to consistency as well. Creativeness is inherent in almost all forms of writing, even in the oft-maligned inverted pyramid. But news writers and editors must be creative within the boundaries of correctness, a limitation that fiction writers don't face.

Writers and editors achieve coherency by creating and maintaining a logical story structure, which they are likely to have if they are both concise and consistent. That logical structure must allow the progression of thoughts from one to the next, which writers achieve with good *transitions.* As Ranly says, crafting good transitions is more than a matter of using conjunctive adverbs such as *however, nevertheless* and *moreover* to introduce new thoughts. Good transitions tie the current paragraph to the previous one. One way to do that, Ranly says, is to use the demonstrative adjectives *this* or *these* in the first, or topic, sentence of the new paragraph.

Concrete writing is also likely to be correct, concise, consistent and complete. Ranly urges writers and editors to make certain a story uses:

- **Nouns.** Nouns give a story a concrete quality by naming people, places and things. Use adverbs and adjectives sparingly, on the other hand, because of their inherent abstract quality.
- **Transitive verbs in the active voice.** As Chapter 3 pointed out, active verbs are direct verbs and help the writer achieve conciseness.
- **Examples.** The personalized approach to writing (see Chapter 3) works well because it tells stories from the eyes of an individual. Readers can relate to people much better than to abstract thoughts. The person becomes an example of the problem or explains the problem.
- **Make comparisons.** Comparisons help readers relate things to other things with which they already are familiar. Familiarity makes for easier reading.
- **Appeal to the senses.** Nothing, Ranly says, enters the mind except through the senses. He urges writers and editors to be the eyes, ears and nose of their readers.

NEW LIZ LOVE?: Elizabeth Taylor may leave a California hospital in two weeks, but maybe not alone. The Associated Press says a biopsy performed on the star Sunday showed no new signs of lung infection. Taylor spokeswoman Lisa Flowers told the AP that the actress, 58, is fighting severe depression, and has decided to make public her romance with 23-year-old Julian Lee Hobbs. "Julian was out of the picture for so long.... She is blooming again, like a rose," Flowers told the AP, adding the couple met in 1988 and plan to travel to Switzerland.

The next day, readers learned the truth:

> CORRECTION: An Associated Press story saying that Elizabeth Taylor is in love with a 23-year-old, quoted Monday in *USA Today,* was a hoax perpetrated by an unknown woman posing as a Taylor publicist, the star's spokeswoman says.

The Associated Press and *USA Today* weren't the first victims of such hoaxes. Virtually every newspaper, magazine and broadcast station has been victimized at one time or another. Such incidents emphasize the importance of good journalistic technique, particularly the practice of knowing to whom one is talking. Hoaxes usually begin when an individual calls or stops by claiming to be someone he or she is not. Good reporters verify the identity of anyone they do not know; failing to do so is the surest way to fall victim to a hoax.

The importance of following that rule is clear in the case of a hoax perpetrated on the *Star Tribune, Newspaper of the Twin Cities* in Minneapolis-St. Paul. The *Star Tribune* ran an eight-paragraph story quoting Richard L. Thomas, president of First Chicago Corp., a large bank holding company, as saying his company was interested in purchasing troubled First Bank Systems of Minneapolis for $20 to $22 a share. That day First Bank stock jumped sharply higher before First Chicago issued a statement denying that Thomas had talked to the Minneapolis paper. The whole story, it turned out, was a hoax.

"We have absolutely no idea" who the impostor was, said Larry Werner, *Star Tribune* assistant managing editor for business. The problem began when the impostor called a real-estate reporter to talk about an earlier story written by a reporter with a similar name. The real-estate reporter tried to transfer the call, but the caller refused, saying he had already been transferred too often. The real-estate reporter gave his notes to the other reporter, who wrote the story. Neither bothered to call First Chicago to confirm that Thomas had actually made the call. A lesson was learned—the hard way.

A newspaper, magazine or broadcast station relies on its reputation for credibility, and incidents such as these tend to erode the confidence of the audience. Editors must be alert to the possibility of a hoax and raise questions if a story seems suspicious.

Longtime editors know that some hoaxes have a way of reappearing from time to time in various parts of the country. These include:

- The story of a 16-year-old baby-sitter who was stuck to a freshly painted toilet seat for hours. A doctor administered to her, tripped and knocked himself out. Both were carried off in an ambulance for repairs and both sued the family who engaged the sitter.
- A driver who was flagged by a stalled motorist needing a push. Told she would have to get up to 35 mph to get the stalled car started, she backed off, gunned the motor and rammed his car at 35 mph.

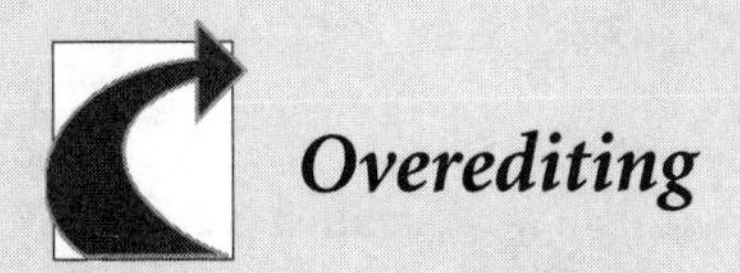

## Overediting

One of the greatest dangers facing the editor is that of overediting. Writes one editor who became a journalism professor after 40 years in the business, the last part of it as a high-ranking editor with United Press International:

> During this time, it became clear that the biggest problem we had with our editors scattered around the world was their inability to keep their blue pencils off a well-written story.
>
> Too many editors think they are better writers than those submitting copy to the desk. They often make unnecessary changes in clear, accurate copy just to put it in a form they believe is superior. Most of the time they disrupt the rhythm and continuity of the copy. Frequently, these changes cloud and distort the copy as well.

Today, that editor urges his students to avoid changing copy for the sake of change. Indeed, each change in copy must be well-justified. "I would have done it another way" is no such justification.

Here's a good rule to use when you are editing: Unless you can demonstrate that a change improves the accuracy or clarity of a story, leave it alone. In other words, let the writer write the story. The editor's job is to edit, not rewrite. If the story is so badly written that it should be rewritten, send it back to the reporter.

- The sheriff who was called to a farm to investigate the theft of 2,025 pigs and discovered that only two sows and 25 pigs were missing. The farmer who reported the loss lisped.
- A farmer armed with a shotgun who went to a chicken house to rout a suspected thief. The farmer stumbled, and the gun went off, killing all his hens.
- The story, usually from some obscure hill hamlet in the east of Europe or in Asia, of an eagle carrying off a 3-year-old child.
- A Sunday driver who called police to report that someone stole the steering wheel and all the pedals from his car. A squad car was sent to the scene, but before police arrived, the man called back and said, "Everything is all right. I was looking in the back seat."
- Someone reports he has found a copy, in near perfect condition, of the Jan. 4, 1800, issue of the Ulster County *Gazette.* The paper is prized not only for its age but because it contains a statement made to the U.S. Senate by President John Adams following the death of George Washington 21 days earlier. It refers to Washington as "Father of our country." Few copies of the original exist, but there are many reproductions.
- A story from Harrisburg, Pa., told about six students permanently blinded by looking at the sun after taking LSD. It was not until after the story had received wide play and had been the subject of editorials and columns that the hoax was discovered.

- A group of young stockbrokers got credit for plotting a hoax against New York newspapers during the depression days of the 1930s. They created a fictitious football team at a fictitious college, and every Saturday during the fall they phoned in the results of the fictitious football game. The hoax was uncovered near the end of the football season when the fictitious college team began appearing in the ranks of the untied and undefeated teams.
- The bricklayer story makes the rounds periodically, usually with a change in locale. The story may have been reworked from a vaudeville gag of earlier days. It is recorded by a comedian with a British accent as a monologue under the title of "Hoffnung at the Oxford Club." Fred Allen used it as a skit on one of his radio shows in the 1930s. In 1945 the story was retold in an anthology of humor edited by H. Allen Smith. Three versions had their setting in Korea, Barbados and Vietnam. In World War II the "bricklayer" was a sailor on the USS Saratoga requesting a five-day leave extension. Here is the Barbados version, courtesy of UPI:

---

LONDON (UPI)—The Manchester Guardian today quoted as "an example of stoicism" the following unsigned letter—ostensibly from a bricklayer in the Barbados to his contracting firm:

"Respected Sir,

"When I got to the building, I found that the hurricane had knocked some bricks off the top. So I rigged up a beam with a pulley at the top of the building and hoisted up a couple of barrels full of bricks. When I had fixed up the building, there was a lot of bricks left over.

"I hoisted the barrel back up again and secured the line at the bottom, and then went up and filled the barrel with the extra bricks. Then I went to the bottom and cast off the line.

"Unfortunately, the barrel of bricks was heavier than I was, and before I knew what was happening the barrel started down, jerking me off the ground. I decided to hang on, and halfway up I met the barrel coming down and received a severe blow on the shoulder.

"I then continued to the top, banging my head against the beam and getting my fingers jammed in the pulley. When the barrel hit the ground it bursted its bottom, allowing all the bricks to spill out.

"I was now heavier than the barrel and so started down again at high speed. Halfway down, I met the barrel coming up and received severe injuries to my shins. When I hit the ground I landed on the bricks and got several painful cuts from the sharp edges.

"At this point I must have lost my presence of mind, because I let go the line. The barrel then came down, giving me another heavy blow on the head and putting me in the hospital.

"I respectfully request sick leave."

---

Perhaps, as one editor has suggested, the original version dealt with the building of the Cheops pyramid or the Parthenon. The story seems to have magic appeal for reporters and editors.

- A man received a series of summonses to pay a tax bill. The notices said he owed $0.00 in taxes and $0.00 in penalties. He was warned that his personal belongings would be attached if he didn't pay. He sent the tax office a check for $0.00 and got a receipt for that amount. Sometimes the yarn is applied to the nonpayment of a noncharge from an electric company and a threat to cut off service unless the bill is paid—or to a tuition demand on a student studying at a college on a tax-free scholarship.
- A fake story may be hard to detect. This fraud got by the desk of a New York newspaper: "The thallus or ruling monarch of the principality of Marchantia will arrive here today on a two-day visit as part of a State Department tour."
- Newspapers have worn out the gag about the person who answered "twice a week" opposite "sex" on a census questionnaire.

## The Unwarranted Superlative

Writing that something is the "first," "only," "biggest," "best" or "a record" seldom adds to a story. Often it backfires. When Lyndon Johnson rode in a Canadian government plane, one wire service said he was the first American president to travel aboard an airplane of a foreign government. He wasn't. President Dwight Eisenhower flew in a Royal Air Force Comet from London to Scotland in 1959 to visit Queen Elizabeth.

Another wire service characterized Gouverneur Morris as "the penman of the Constitution" and Lewis Morris as the "only New York signer of the Declaration of Independence." The man who penned the Constitution was Jacob Shallus, and there were four New York signers of the Declaration of Independence.

A California obituary identified a woman as the "first postmistress" in the nation. A Missouri story reported the closing of America's "shortest commercial railroad." Both statements were disproved.

The Associated Press described Herbert Lehman of New York as the "first person of the Jewish faith ever to hold a Senate seat." The AP had to acknowledge that it was wrong by at least six men and more than 100 years. Jewish senators who preceded Lehman were David Levy Yulee of Florida, Judah P. Benjamin of Louisiana, Benjamin F. Jonas of Louisiana, Joseph Simon of Oregon, Isidore Raynor of Maryland and Simon Guggenheim of Colorado.

When United Press International described the Flying Scotsman, a famous British locomotive, as the first steam locomotive to exceed 100 mph, railroad buffs hurried to set the record straight. Records show that on May 10, 1893, the New York Central No. 999 was timed unofficially at 112.5 mph on a one-mile stretch between Batavia and Buffalo. On March 1, 1901, the Savannah, Florida and Western (later part of Atlantic Coast Line, later Seaboard Coast Line) No. 1901 was timed at 120 mph. On June 12, 1905, the Pennsylvania Special traveled three miles near Elida, Ohio, in 85 seconds for an average of 127 mph.

## *Should Quotations Be Changed?*

Editors often have to make tough calls about whether to change a direct quotation to clean up a speaker's grammar or syntax. If the goof is within quotation marks, should it remain? Did the speaker use poor English or did the writer write poor English? Do the people quoted get friendly or unfriendly treatment from the editor?

Most editors seem to agree that minor grammatical corrections within direct quotes are permissible. Some, however, insist that when the quotation is changed—even the slightest bit—the formal quotation marks should be dropped to make the passage an indirect quotation.

The problem is that we are more forgiving of grammatical errors in spoken English than in written. A live audience will forgive the school superintendent who, with a slip of the tongue, creates a subject-predicate disagreement at a school board meeting. But quoting that same thing in print is somehow more embarrassing.

Even editors who advocate cleaning up direct quotations aren't always consistent in doing so. They often clean up the superintendent's quotes but not those of a high school football player. Does that mean it is acceptable to make one person appear dumb but not the other?

In the absence of a direct policy from your superiors, we suggest it is almost always best to clean up minor errors within quotes but paraphrase when major corrections are in order.

A story from Louisville described a conviction as the first under a new law barring interstate shipment of gambling material. Two months earlier two men had been convicted under the same law. A Billy Graham rally was described as having the largest audience for a single meeting. But a Rosary Crusade in San Francisco had been attended by 500,000, a bigger crowd than Graham's.

All these are examples of abused statistics that invade the news report. One story indicated that New York City has 8 million rats. Another quoted the American Medical Association as saying that only 5 percent of Americans dream in color. How can anyone know such exact figures? Highway deaths may increase from one year to the next, but one possible explanation may be that there are more vehicles on streets and highways. Highway deaths on holiday weekends are higher than normal because such weekends often are longer. "The toll has dropped so that last year there were only 81 traffic deaths here, an all-time low." Since when? 1942? 1776? 1900? "The ships were built in record time." What was the previous record?

All superlatives should be checked. If they cannot be verified, at least they can be softened: "One of the most despicable crimes in the world . . ."; "One of the hardest-working actresses in Germany . . ."

## The Misquotation

A careful editor would be wise to keep a quotation reference at hand while handling copy containing references to often-repeated quotations or attribution of such quotations. Such a reference will stop reporters from attributing "Go west, young man" to Horace Greeley. The advice was actually given by John Babson Lane in 1851. Greeley used the expression in an editorial in the *New York Tribune* but amplified it: "Go west, young man, and grow up with the country."

Charles Dudley Warner, not Mark Twain, should get credit for "Everybody talks about the weather, but nobody does anything about it." Bill Nye, the humorist, originated the saying, "There are just two people entitled to refer to themselves as we—one is the editor, and the other is the fellow with a tapeworm." Mark Twain later revised the statement: "Only presidents, editors and people with tapeworms have the right to use the editorial 'we.' "

Voltaire is wrongly credited with the quotation, "I may not agree with what you say, but I will defend to the death your right to say it." Most likely it is a paraphrase of Voltaire's "Think for yourselves, and let others enjoy the privilege to do so too." Gen. John J. Pershing did not exclaim, "Lafayette, we are here!" It was uttered by Charles E. Stanton, chief disbursing officer of the American Expeditionary Forces in World War I.

Careless writers attribute the "gilded lily" business to Shakespeare. But what Shakespeare wrote was, "To gild refined gold, to paint the lily." Similarly, the Bible does not say that money is the root of all evil. It says, "Love of money is the root of all evil." Music doesn't have charms to soothe the savage beast. Congreve said, "Music hath charms to soothe the savage breast." And Thomas Hardy did not refer to "the maddening crowd" but to "the madding crowd."

Up to the time of his death, a South African dentist, Philip Blaiberg, had survived with an implanted heart. In an account of Blaiberg's death, a UPI reporter wrote that Blaiberg's last act was to scribble a quote from the Persian poet Omar Khayyám: ". . . for I shall not pass this way again." A Connecticut editor questioned the attribution, causing UPI to send out a correction. The probable author is Stephen Grellet and the usually accepted full quotation is, "I shall pass through this world but once. If, therefore, there be any kindness I can show or any good thing I can do, let me do it now. Let me not defer or neglect it, for I shall not pass this way again."

A sports column tribute concluded, "In the words of the late Grantland Rice: 'When the great scorer comes to write beside your name,/It's not whether you won or lost but how you played the game.' " What Rice really said was, "When the One Great Scorer comes to write against your name—/He marks—not that you won or lost—but how you played the game." Reporters should never quote poetry from memory. When poetry shows up in a piece of copy, the editor should assume it's wrong and look it up.

One story quoted a structural linguist's feelings about people who object to ending sentences with prepositions: "You remember what Winston Churchill said when an aide corrected a line in one of Churchill's speeches because it ended in a preposition? Churchill told the aide: 'This is an outrage up with which I will not put.' " Churchill was misquoted. What he said (and even this may be apocryphal) was, "This is the type of arrant pedantry up with which I shall not put."

A column criticizing overuse of the word *gourmet* said, "I am reminded of the line of poetry which told of the moth flitting its wings signifying nothing." Was it perchance not a poem but a Shakespeare play, and not a moth flitting its wings but "a tale told by an idiot, full of sound and fury, signifying nothing"?

"We've come a long way since Commodore Vanderbilt said, 'The public be damned.'" But the Commodore never said it. It was William H. Vanderbilt, son of Cornelius, the so-called Commodore, who made the remark.

"Robert Burns, the old Scotchman, said: 'Oh that we would see ourselves as others see us.'" Burns was a distinguished poet, hardly the "old Scotchman." Careful writers prefer *Scot* or *Scotsman* to *Scotchman,* as do the Scots themselves. The actual quotation: "Oh wad some power the giftie gie us/To see oursels as others see us!"

## The Imprecise Phrase or Word

A common source of incorrect writing is the imprecise phrase or word. For example, a school board (or board of education) is a group of individuals elected by the citizens to direct the operation of the school system. It is not a place, not an office, not a building, not the school system. It is incorrect to write, "He studied French under E. B. DeSauze, the retired supervisor of the school board's language department." DeSauze was supervisor of foreign languages for the public schools. The school board has no language department.

The U. S. Supreme Court did not ban prayers in school. The court banned the requirement that children pray any particular prayer, or the writing by public authorities of a required prayer. The decision had to do with public schools. It did not interfere with required prayers in church-operated schools.

*Gas* and *gasoline* are not synonymous. Gas is either natural or manufactured. Some explosions are caused by gas, some by gasoline. The story and headline should contain the precise term. Similarly, in stories of food poisoning, the copy should specify whether the story is referring to canned or bottled foodstuffs.

Reporters and headline writers are fond of writing that taxes will *eat up* a will or a fortune or an estate. Taxes may deplete the bank account, but they can't eat up anything.

"A defective 20mm cannon . . . suddenly fired and the shell killed one airman and injured another." The writer should have used *unexpectedly* rather than "suddenly," *a shell* rather than "the shell," and *bullet, slug* or *projectile* rather than "shell."

*Pistol* is a general term for a small firearm. It can be single-loading, or a revolver or an automatic. Clip-loading pistols are sometimes called automatics, but they are usually semi-automatics or self-loaders. The barrel diameter of rifles and pistols is expressed in calibers (.22). A shotgun bore is expressed by its gauge (12-gauge) except for the .410.

"A 20-year-old robber was dead as the result of a gun battle in which 14 shots were fired at point-blank range." "Point-blank range" is an archaic expression based on the firing of cannon. Because the expression is meaningless to today's readers, why use it?

### The Mislabeling of People

"It is perhaps the most cosmopolitan area in the city, stronghold of the Poles and densely populated with other ethnic groups including Czechs, Bohemians, Slovaks and some Italians," one story reported. But Czechs and Bohemians are one and the same people. The Czech lands include Bohemia and Moravia. Some Bohemians prefer to be called Czechs. Slovaks are a separate people, although there is a strong language affinity. There is a difference between a Slovak and a Slovenian, as any editor would soon realize should the two be confused.

Wire service copy sometimes fails to explain terms common in one section of the nation but not in another. For instance, readers or viewers may deduce that *bracero* is a Mexican laborer. If the word can't be explained, it should be eliminated so as not to puzzle readers who don't know Spanish.

An executive city editor gave an editor trouble for failure to catch the idiocy of an "anti-Soviet" play written in the czarist days. Even though *Soviet* technically refers to an organizational system within the Communist structure, it is now generally accepted as a reference to the U.S.S.R., which no longer exists. Russia, Bulgaria, Latvia and the other former Soviet republics are now independent nations.

*Britain* or *Great Britain* refers to the largest of the British Isles and consists of England, Scotland and Wales. *United Kingdom* should be used to mean England, Scotland, Wales and Northern Ireland. A *Briton* is a native or subject of Britain. Despite the fact that other nationals of the United Kingdom may be annoyed when *England* is used as the equivalent of *Britain* or the *United Kingdom*, the use of *England* in the wider sense is acceptable.

These are just some of the ways that print and broadcast writing can be incorrect. There are many others. Good editors learn to be suspicious of everything that is even slightly questionable.

## BE CONCISE

To be concise, start by eliminating unnecessary words. In the following sentences, the italicized words can be deleted:

> She shopped *very* late each afternoon at the store at *the corner of* 10th and Elm streets.
>
> From one week to the next, Esther went from *being the* goat to hero.
>
> Enrique shot from the top of the box, and *in an instant* the ball ripped into the upper-right corner of the net.

# *Stereotyping by Race, Sex and Age*

The public has become increasingly sensitive to demeaning and derogatory language in the news report. The editor must guard against both conscious and unconscious bias. To do that requires an awareness of stereotypes.

People live in two worlds: the one they experience directly and the one they know about from other sources. There is no doubt that the first world is smaller and that the second is created largely by the mass media. Thus, what Walter Lippmann called "the pictures in our heads"—our view of reality—are formed largely by information channeled to us through the media. For that reason alone, the process of stereotyping is insidious because it tends to distort our view of reality.

A *stereotype* is a standardized mental picture representing an oversimplified opinion, emotion, attitude or uncritical judgment of a person, group, race, issue or event. It often results in demeaning or ridiculing the subject.

We stereotype out of necessity; stereotypes provide us with a shorthand way of looking at the world. For example, we might form a stereotype based on personal experience: Whenever certain weather conditions have appeared, it has rained. So when we feel high humidity and see dark clouds, we carry umbrellas. This oversimplified reaction to an event makes our lives easier. We can react to the situation without thinking. In this example, what we believe to be true is grounded in fact, or at least in personal experience; it usually has rained on us under these conditions. So not all stereotypes are without basis in fact. Nor are they all dangerous. In this example, we might be inconvenienced if it did not rain and we had to carry umbrellas all day, but that is all. Most stereotypes, however, are not this innocuous.

Language reflects the prejudices of society. Casey Miller and Kate Swift, in *The Handbook of Nonsexist Writing,* make the point that English, through most of its history, evolved in a white, Anglo-Saxon, patriarchal society. Its vocabulary and grammar therefore frequently reflect attitudes that exclude or demean minorities and women.

Words are so much more than they appear to be. As semanticists tell us, words represent attitudes and beliefs. The words chosen by an individual say as much about the individual as anything else.

When we stereotype, we label (people especially), and once we do that, the label tends to attract more attributes than it should. Like a snowball rolling down a hill, the label becomes all-encompassing. The words and images that are caught up in it classify people indiscriminately: Women . . . are fragile. Mexicans . . . are lazy. Blacks are on welfare. Indians . . . are drunks. Professors . . . are absent-minded. Old people . . . are senile.

When people are stereotyped according to race, that is racism. In the last decade or so, the general public has become aware that language also stereotypes according to sex, which is sexism. Even newer in the public consciousness is ageism, stereotyping according to age.

While there is little excuse for reporters to let their stereotypes spill over into their writing, there is less excuse for editors to pass copy with racist, sexist or ageist assumptions and statements. For the editor, these "isms" ought to be added to the list of "red flags"—those things in copy, headlines, cutlines, even the premise of a story itself that cause the editor to step back, to wonder why.

As Miller and Swift write, "The need today . . . is to be in command of language, not used by it, and so the challenge is to find clear, convincing, graceful ways to say accurately what we want to say."

## Racism

There was a time when racism in newspapers was blatant; racial identifications were mandatory. Certain stories were not written because of the race of the participants; others were written for the same reason. Generally, that no longer is the case. Racism in the media, as in much of society, has become subtle. That is not to say that it is impossible to find examples of blatant racism.

A story about a bomb threat to a northern Wisconsin high school included this: "According to the complaint filed at the Minocqua Police Department, (the secretary) reported that the threat came from a male Indian's voice." Just how does a male Indian sound?

The *Chicago Tribune*, not too many years ago, described the subject of one of its stories as the "well-dressed, articulate black man." The implication was that it was out of the ordinary for a black man to be well-dressed and articulate.

Subtle racism is more difficult to identify. A story from *The New York Times* News Service about an explosion at a day-care center in a black section of Atlanta included this: "'I heard a real loud boom and I ran out the door and saw the smoke and dust and ran down here,'" said Eugene Drewery, 30, an unemployed construction worker."

It might be suggested that it was necessary to describe the man as unemployed to explain why he was home in the middle of the day. It is, however, sexist to assume a young man should not be home during the day and irrelevant to the story to include his job status. Instead, the reader is helped to make an unjustified connection between the man's race and the fact of his unemployment.

A writer in the *Star Tribune, Newspaper of the Twin Cities*, used the following anecdote to begin his story about a high school celebrating its state basketball championship:

> One of the posters done up Sunday for the celebration honoring North High School's boys' basketball team was misspelled.
>
> The one that said "Team Minneapolis" was all right and so were the ones that said "State Champs!" and "Polar Power," but over on the front wall of the gymnasium, right underneath the big polar bear, was a banner proclaiming *"Congradulations."*
>
> One of the women helping with the decorations spotted the error. "Oh, no," she said, "that's what they'll show on TV."
>
> Before long a new poster, with the blue tempera paint that spelled "Congratulations" still dripping, was taped up over the old one.
>
> Image is important to the people of North High School, because this school in the heart of one of the city's poorest neighborhoods has had such a bad image for so long. But on this afternoon of celebration after an evening of achievement, the people of North High were talking of pride in their community, their school and, of course, their basketball team.

A copy editor headlined the story: **NORTH HIGH SPIRIT TRIUMPHS OVER SPELLING.**

A week later, the paper's reader representative explained in his column what an eye-opener the story proved to be for the *Star Tribune*, which received numerous complaints about the headline and the story from members of the North High community, teachers and students. The Minneapolis Urban League requested and received a meeting with *Star Tribune* staff members involved in the story. As the newspaper's ombudsman explained to his readers:

> The meeting . . . forced some of us to decide that the headline and, by extension, the story did indeed display an insensitivity both to the feelings of the North High community and to the danger of reinforcing stereotypes.
>
> [The] assistant managing editor of the *Star Tribune* reflected after the meeting that [the reporter] "wrote in good faith a story in which he made choices of material that he thought told the story of a school with image problems trying to cope with those problems on a day of pride and triumph. But he produced a story that did almost precisely the opposite in the eyes of lots and lots of readers. The headline exacerbated the problem. . . ."
>
> But the *Tribune*'s real failure was not so much in the writing of the story as in not taking a second step, said [the assistant managing editor]. Some editor should have looked at the story with an awareness of the stereotypes and the danger of reinforcing them. That editor should have anticipated the effect of the story and ordered changes.

## Sexism

Like racism, sexism appears in both blatant and subtle forms. Unlike racism, however, blatant sexism is common in many American newspapers. It has been and remains common to describe women in news stories in ways that men making news seldom have been described, even though, for example, the stylebooks of The Associated Press and United Press International clearly state

*(Continued)*

that "women should receive the same treatment as men in all areas of coverage."

Irrelevant descriptions of women's appearances remain. Note the incongruity of these two consecutive paragraphs:

> At 8:58 p.m. Nancy Reagan–wearing a red dress and large gold earrings–was greeted by an ovation as she entered the gallery.
>
> Two minutes later, Reagan entered the chamber to an explosion of applause.

Apparently it was unnecessary to describe the president's attire.

UPI carried the practice to extremes when Joan Kennedy received her master's degree. Wrote UPI, "Mrs. Kennedy, dressed in black cap and gown, marched under the sunny skies with some 400 other graduates."

Editors must recognize slanted adjectives and descriptive phrases that serve only to reinforce sexual stereotypes:

> Patricia Brennan, an auburn-haired 24-year-old with the look of a mischievous tomboy, claims she never stood in front of the post office and yelled, "Nah, nah! Anything you can do we can do better."

> Alcott shared the opening day lead with Barbara Moxness. . . . The personable 24-year-old brunette from Santa Monica, Calif., started the day four strokes ahead.

> Nancy Freitas, 32, a blonde with green eyes and a preference for strictly feminine, tailored attire, is the new general manager of the San Diego Breakers of the International Volleyball Association.

One of the most common problems faced by the media is how to refer to women in stories. Many newspapers have dropped the traditional practice of using courtesy titles for women, but the use of such titles remains common. One exception is on sports pages, where it is more common to treat men and women similarly—without courtesy titles. Whatever a newspaper's policy, the editor should ensure consistent treatment.

It is incorrect to write Mrs. Mary Smith because she is not married to Mary; she is either Mary Smith or Mrs. John Smith. It is unacceptable to treat men and women of equal position differently, as in the story that made these identifications: "Ms. Long, 29, is a physician . . ." and "Dr. Stanley Mohler . . ."

Equivalent terms should be applied to men and women in a story or headline: lady/gentleman; woman/man; girl/boy. These headlines, then, should have been unacceptable:

> **LADY DOCTORS**
> **AND MALE NURSES**

> **MALE NURSE WALKS TIGHTROPE**
> **OF COMPASSION AND TRADITION**

> **BRATTLEBORO MAN FOUND DEAD**

> **MONTPELIER GIRL FOUND DEAD**

In these last two cases, he was 18; she was 26.

There are times when an editor might question the entire approach a reporter has taken in a story on the basis that it only reinforces sexual stereotypes, much as the Minneapolis reporter was cited for reinforcing racial stereotypes.

For example, a reporter began her story this way: "Mothers aren't the only ones who can tell us how to eat properly. A computer can, too." And so could a father. But this reporter reinforces the prevailing stereotype that it is the mother's role to look out for the needs of children. Further, there is a more subtle implication that a computer can replace a mother. The story easily could have been made more acceptable; the editor could have changed the lead to read, "Parents aren't the only ones . . ."

A similar sexist role is prescribed by a writer whose lead read: "The women were not the only ones who enjoyed themselves at the . . . Homemakers School last night . . . . a good time, too, sampling the good food."

How many stereotyped images can you identify in this lead, which shows that men also can be victims of reportorial sexism?

> It's your weekly trip to the hairdresser. Your hair has been shampooed, cut, colored and curled, and

you're ready to bake under the dryer. As you sit there, reading your latest issue of *Vogue* magazine, you turn to your neighbor to discuss the newest trends in skirt length.

Suddenly, you realize you're face to face with a man–and he's in curlers.

You can react three ways.

You can blush and wish you were dead.

You can quickly end the conversation and wish you'd worn makeup.

You can keep talking and wonder if he's dating anyone.

## Ageism

Ageist stereotypes have been around as long as those based on race and sex. One that comes immediately to mind is the "Confucius say..." variety—the wisdom of old age. Only now, however, are people beginning to explore the implications of stereotypes about aging and the aged, and the complex set of problems they have helped to create.

Common stereotypes portray the old as poor, isolated, sick, unhappy, rigid, reactionary, unproductive and grandparently. The Gray Panthers, activists who serve as advocates for the old, designed their Media Watch Project to allow viewers to identify and publicize ageist stereotypes in television programming and commercials. They urge alertness to several stereotypes:

- Physical appearance: face always blank or expressionless; body always bent over and infirm.
- Clothing: men's baggy and unpressed; women's frumpy and ill-fitting.
- Speech: halting and high-pitched; stubborn, rigid, forgetful.

The Gray Panthers also warn of distortions, in which old age is depicted either as an idyllic or a moribund stage of life, and omissions, in which the concerns and positive aspects of aging are omitted.

The sensitive editor will be alert for ageism in copy and headlines. In this example, from a story about a visit to central Missouri by Clare Booth Luce, both ageist and sexist stereotypes are evident. An alert editor deleted the paragraph from the story.

> She remained standing during the reception, even though she is frail since cataract surgery a few years ago. Her eyes are still bright blue, and her beauty shines despite her years. The attitude of the young college men attested her attractiveness–they all wanted to kiss her.

Apparently, age alone sometimes is enough to qualify one as grandmotherly or grandfatherly. A headline on a sports page announced: **LOCAL GRANDMOTHER WINS BOWLING TOURNEY.** In the story, no mention was made of grandchildren. The woman was, however, identified as being 58. Another reporter wrote that "Cora is a grandmotherly sort" without elaborating on just what that said about Cora. And the copy editor who asked in a headline, **CAN ONE LIVE HAPPILY EVER AFTER RETIREMENT?** implied that retirement rarely is a time of happiness.

## Unacceptable Labels

Editors increasingly are being called on to be more sensitive to the labels they place on people or groups of people. The terms *Negro* and *colored,* once common in U.S. newspapers, have disappeared in favor of *black.* Now the term *African American* is gaining favor. Similarly, *Indian* is being abandoned for the more precise term *Native American.*

Our language is full of derisive terms for people. Such terms often, but not always, are directed at minorities *(Jap, wetback, wop).* Examples of such derogatory terms aimed at the majority are *WASP* (for white Anglo-Saxon Protestant) and *honky* (a derisive term directed toward whites by blacks).

Some words *(codger, gaffer, deaf and dumb)* are offensive to some groups but commonly appear in print. Other negatives are aimed at a person's sexual preference *(dyke, queer, faggot)* or background *(redneck, hillbilly).*

A few terms may not seem at first glance to be offensive but are if one is aware of their backgrounds. An example is *basket case,* a term that began as British army slang for a quadruple amputee but has come to be used indiscriminately for those who are incapacitated.

Sensitivity to others is essential if the media are to retain their credibility. Such terms have no place in the news report.

That last sentence could be tightened further. But if a story is written in a style other than the inverted pyramid, colorful detail sometimes is necessary. Consider this sample from writer Leola Floren's description of a Tammy Wynette concert:

> On stage, she is surrounded by musicians in green suits and cowboy boots. Stuck there in the middle, Tammy looks like one smooth pearl in a bucket of green peas.

Such vivid description relies on adjectives and adverbs, the words that are usually deleted from stories written in the inverted pyramid format. But in the more elaborate writing styles now so common in newspapers and magazines, and in the descriptions so common on radio and television, the time or space occupied by those adjectives is worth every column inch and worth every second of air time. The trick for the editor is to learn when to leave such detail and when to delete it.

In most cases, however, conciseness is a virtue. Writing that is concise is also understandable. Experienced editors identify the following areas as those where conciseness often disappears.

## Excessive Attribution

Accounts of police news, and worries about possible libel problems, often cause writers to fall into the trap of excessive attribution. In too many crime stories, writers attribute every sentence to police or prosecutors. Often, there are better ways to handle it. For example:

> A Memphis man was arrested early Monday morning after a high-speed chase through South Memphis that resulted in serious injury to a pedestrian.
>
> Police arrested Jerome Caldwell, 22, of 303 S. Third St., and prosecutors charged him with resisting arrest, aggravated assault and burglary.
>
> Police gave this account of the events:
>
> Officers Jill Southerland and Carlos Rodriguez responded to a silent alarm at Southside Hardware, 2028 E. Miller St., about 4:30 a.m. They entered the building and surprised a burglar, who fled through a rear door.
>
> The man jumped into a car and led the officers on a three-mile chase through South Memphis, which ended when another squad car was summoned to help block the intersection at Hernando and Main streets. The car came to a stop there but not until it had struck Antonio D'Amato, 29, who was crossing Main Street when the car and police arrived.

Starting the description of the sequence of events with "Police gave this account of the events" made it unnecessary to attribute each sentence with "police said."

Editors can save reporters (see Figure 4–1) from attribution logjams if they remember this question: Is it clear who is talking? If the reporter shifts to a second speaker, the story should identify the new speaker immediately, not at the end of the quotation. The reader assumes the first speaker is still talking.

***Figure 4–1** A reporter takes notes as he interviews the subject of a news story.*
Photo by Philip Holman.

Another potential difficulty arises in the choice of synonyms for *said*. Do *added, pointed out, offered, admitted, disclosed, noted, revealed, indicated, conceded, explained* or *cited the fact that* give the quotation an editorial tone?

Do the synonyms for *said* convey a hint of doubt as to the veracity of the credited source (*according to, said he believes*)? *According to* actually refers to the content, not to the speaker: "According to the mayor's letter," not "According to the mayor."

Does the writer use gestures for words?

"We're gonna put on a show, too," grinned Wags.

"Now I can invite my friends to play on the grass," Donna beamed.

"The bill will be paid," the official smiled.

"I heard something pop in my shoulder," he winced on his way to the dressing room.

No matter how good a grinner or wincer, how bright a beamer or how broad a smiler, you just can't grin, wince, beam or smile a quote. If it's a quip, that should be obvious from the context. If it isn't, saying so won't make it so. An exclamation mark may be used after a brief expletive, but it looks silly after a long sentence.

*Said* is a simple verb that usually is preferable to others used in an effort to convey determination, skepticism and wit. Usually the quoted matter can speak for itself. Many of the best writers use *said* almost exclusively. *Said* used repeatedly can give emphasis; it is not weakened by repetition. It has the added virtue of being concise.

## Extraneous Facts

News presents the pertinent facts. Every story should answer all the questions the reader or viewer expects answered. If a big story returns after having been out of the news, it should contain a short background or reminder. Readers don't carry clips to check background.

Robert J. Casey, an author and former reporter for the defunct *Chicago Daily News*, once observed, "Too many facts can louse up a good story." If a fact isn't vital in telling the news, it should be omitted. Stray bits have a way of bringing trouble. A buried reference to a 30-year-old hanging "from an apple tree on Joe Smith's farm" brought a libel suit. Joe Smith was still living; the hanging wasn't on his farm. The reference added nothing to the story but taught the editor a lesson.

This could be held to a paragraph or two; it is not worth five column inches of type:

> Robert F. Kelley today was named chairman of this year's Democratic Jefferson–Jackson Day dinner.
>
> The appointment was announced jointly by Democratic State Chairman John M. C– and National Committeeman William S. P–.
>
> Kelley, administrative assistant for 12 years to ex-Sen. J. Allen F– Jr. in Washington, said he will name a dinner committee, site, date and speaker in a "few days."
>
> The Jefferson-Jackson Day dinner, traditionally held in late April or early May, is the largest meeting of its kind held by the Democrats each year.
>
> Kelley said he already is trying to line up a "nationally known" speaker for the occasion.
>
> Kelley, now associated with the legal department of the D– Co., has been a member of the dinner committee for several years. This is his first assignment as chairman of the affair.
>
> Kelley was a vice chairman of last year's Community Fund drive and has a wide background in party and civic affairs.
>
> He is a past president of the Delaware State Society and the Administrative Assistants and Secretaries Club in the nation's capital.

Deleting extraneous facts, editing out unnecessary words and eliminating excessive attribution help make a story concise and understandable. Editors who do those things routinely help their readers, listeners and viewers comprehend the news report more readily.

## BE CONSISTENT

One of the most common problems found in newspapers is the failure of copy editors (see Figure 4–2) to make stories consistent from top to bottom. The most common example is the story with numbers that don't add correctly:

> The United Way on Monday awarded $23,000 in supplemental funds to three agencies.
>
> ... The additional funds will go to the Boy Scouts, $11,000; the Girl Scouts, $9,000; and the Salvation Army, $4,000.

Here are some other classic examples:

> Tanglewood Barn Theater ended its regular season with a bang in its production of "Wonderful Town" Wednesday night. (Last paragraph:) The show will be repeated at 8:15 p.m. through Sunday.
>
> His companion said Fennell dived from boat, swam away, went under and never came up. (Last paragraph:) Interment will be in Mt. Zion Cemetery.

(What was to be buried in lieu of the body that never came up?)

***Figure 4–2 At smaller newspapers, the copy desk also is the design desk.***

Photo by Philip Holman.

> The largest single cost of the trial was jury expenses, which total $3,807. (Later:) Another cost was $20,015 paid to extra guards and bailiffs.

A story concerned a robber. Part of it went like this: "The suspect apparently hid in the store when it closed at 9 p.m. About 11 p.m. he confronted a security guard, Paul H. Hogue, 57, of 5625 Lowell Blvd. as he was turning off the lights in the budget store of the basement." Last paragraph: "According to parole officials, Hogue's parole was suspended June 6 for failure to report and he was being sought as a parole violator." A correction sufficed in this case, but a correction does no credit to the reporter or to the editor.

The editor who edits to the final stop and who is ever alert to the kinds of problems described in this chapter will be a valued member of the newspaper's staff. By doing the job properly, that editor will be building credibility for the newspaper, not contributing to the demise of that credibility.

## BE COMPLETE

The greatest conflict editors face when dealing with the four C's is the one between *conciseness* and *completeness.* News is regarded one way in the newsroom and another way by readers or viewers. In the newsroom the reporter is constantly admonished to "keep it short."

"How much do you have on that hotel death?" the city editor asks.

"Enough for about 14 inches," answers the reporter.

"Hold it to 7," orders the superior. "We're short on space today."

So the reporter prunes the story to 7 inches. The story reaches the copy desk, and an editor goes to work to make it even tighter. All this, of course, is unknown to the reader who sits down to enjoy the newspaper or to the viewer watching the newscast.

A headline catches the eye: **80-FOOT FALL AT HOTEL ENDS ACTOR'S GRIM JOKE.** The reader begins the story: " 'Watch me do a trick,' said the 26-year-old actor to his companion, and he stepped out the eighth-floor window of their downtown hotel early Sunday."

Muses the reader, "I was downtown early Sunday morning. I wonder what hotel it was and what time?"

The story doesn't answer these questions. It does identify the victim and his companion. Near the end of the story is a brief description of the companion, Paul Lynde: "Lynde is a widely known actor, investigators said."

"Funny I never heard of him," the reader again muses. "Wonder what he appeared in?"

That was the story as sent by the AP. For the morning papers the story failed to identify the hotel, did not give the time of the fall and identified Lynde only as a widely known actor. Readers of afternoon papers got some of the missing details. The hotel was the Sir Francis Drake, the time was "the wee hours Sunday morning" and Lynde was identified as a comedian who appeared in the Broadway and film versions of "Bye Bye Birdie" with Dick Van Dyke and in the movie "Under the Yum Yum Tree" as Imogene Coca's henpecked husband.

The ability to give details is the newspaper's great advantage over its competitors in other media. Readers relate to the news. The more involved they become in the events, the more avid they become. In short, they demand the whole story down to the last detail. If the story says, "Hastings Banda, the leader of Nyasaland, received his education in Ohio," the reader wants to know, "Where in Ohio?" "At what university?" "When?" The reader relates to the news. "I wonder if that's the same Banda I knew when I was at Ohio State in the 1970s."

A wire story about a plane crash in New York City said the 79 passengers included "two young opera singers en route to a South Carolina concert, prominent Southern businessmen, and a former Virginia college beauty queen." The story as sent drew this protest from a client's managing editor: "Who the people are who die in these crashes is a point of interest equal to or more than the circumstances of the crash. We all identify with them—where they are going, where they are coming from. I want to know what opera those opera singers were going to sing in."

A story related that a man had gone to court to fight for a seat in the legislature, but it did not tell to which party he belonged. Another told of a woman mugged while waiting for a bus at Delaware and Woodlawn avenues, but did not give the time of the incident, which would be of interest to those who ride the bus. In a story about an airplane crash, the cost of the plane is an important piece of information and should be included.

One paper ran a story about a drunk chimpanzee that supposedly escaped and created havoc around the countryside by trying to break into homes. But the story failed to tell who owned the chimp, what he was doing in the county, how he got anything to drink and what finally happened after a game warden arrived on the scene. These were basic questions the reporter forgot to answer. The editor should have checked.

Another paper had a three-column picture and story about consecration ceremonies at the Cherry Hill Methodist Church's new "Harlan House." The story told that Harlan House was named for "Miss Mollie" Harlan, that the Rev. Dr. Darcy Littleton took part in the ceremonies, that "Miss Mollie" is now buried in the Cherry Hill Cemetery, that Littleton is now with Goodwill Industries in Wilmington, that G. Harlan Wells spoke and that the Rev. R. Jerris Cooke conducted the service. But when all this was said and the picture was examined, readers were still left to guess what Harlan House is or who Mollie Harlan was that the house should be named for her. Was she related to G. Harlan Wells?

A paper reported in detail the arrest of a minister on charges of operating a motor vehicle without a license, failure to carry a car registration card, disorderly conduct and disobeying a police officer. When three of the four charges were dismissed, the story failed to tell why. Answer: It is standard procedure to dismiss the license and registration charges when a driver has simply forgotten to carry the documents.

A Page One story told the fascinating details of a divorce decree upheld by the state Supreme Court but failed to mention the names of the parties in the case. Another story gave an account of the Senate's 78-8 approval of the president's trade bill but failed to tell who the eight opponents were and, even worse, how the senators from the paper's state voted on the measure. This was a revolutionary trade measure that had been in the news for months and was finally opposed by only eight senators. Wasn't anyone who handled the story curious about their names?

A story in March said, "The Bahais will celebrate New Year's Eve at the Bahai center. There will be readings and music." The story read, "New Year's Day tomorrow is known as Naw-Ruz." Couldn't the reporter have dropped the music and the readings and told instead who Bahais are and what the heck New Year's Day is doing in the middle of March?

A skindiver stayed under water for 31 hours and spent much of his time reading a paperback book. There was no word to explain what kept the pages from disintegrating. (The paper was a glossy stock.)

A woman won a fat prize in a magazine advertising contest. There was no hint what she did to win, a point made more important by the statement that the woman could neither read nor write.

Another story concerned a judge who reversed his own conviction of a union leader for breach of the peace. The reversal, said the story, was based on "new evidence" but failed to tell readers the nature of the new evidence.

The story said that 105 of the 120 blacks enrolled at Northwestern marched out of a building singing. It failed to mention the songs they sang, a detail that might have shed more light on their behavior.

If any part of the story is confusing, the editor should supply explanations. Obviously, the explanation should not be as hard to understand as the phrase itself. For instance:

> Congress in 1946 waived government immunity to suits in tort (a civil wrong in which a legal action may lie) and permitted suits on tort claims against the United States.

The parenthetical explanation hardly aids most readers. If an explanation is required, it should be one that really helps:

> As a rule, sovereign government may not be sued by its citizens unless the government consents. In 1946 Congress gave blanket permission to citizens to sue the U.S. government if they thought it was responsible for injuries to them or their property.

Then there was the story of a boy who died, apparently of suffocation after he choked on a hot dog in his home. Police said the boy left the table after dinner and was found choking in his bathroom. His mother slapped him on the back in a vain attempt to dislodge the obstruction. Firefighters took him to a hospital where he was pronounced dead. A few lines of first-aid instruction at the end of the story might have saved other lives.

A woman who was hospitalized twice in a short time asked to be transferred from one hospital to another to be near her husband. What ailed hubby? It wasn't explained.

The Royal Navy dropped the unit *fathom* and started measuring depths in meters. The story told all about it. All, that is, except how deep a fathom is.

A story contained the statement "where family income is below federal poverty levels" but neglected to tell the readers what the poverty level is by federal standards.

"A bell captain in a midtown hotel was arrested for scalping World Series tickets." Why the reluctance to name the hotel? A directive reminded editors, "In these days, when GIs, businessmen, students, school teachers et al. are traveling throughout the world, such identification is often of interest to many readers. The part of town where a news event occurs is sometimes pertinent too in stories from the big cities that are frequented by travelers."

Here is a complete story as one newspaper printed it:

> TALLAHASSEE–The simmering feud between Republican Gov. Claude Kirk of Florida and his Democratic cabinet erupted into a full-scale shouting battle today, and Kirk ordered an end to weekly cabinet meetings for the first time in state history.
>
> Cabinet members immediately declared they would go on meeting anyway.
>
> The stormy session began with the cabinet refusing to spend $35,000 on a federal liaison office that Kirk wants to open in Washington.

Some readers must have wondered how a Republican governor came to have a Democratic cabinet, what officials belong to the cabinet, what can be accomplished by cabinet meetings not attended by the governor and whether the weekly meeting is required by law.

The first rule in writing or editing a story is to ask yourself: Who will read the item, and what will they most want to know about the subject? Both the writer and the editor should pare the story for word economy. They should not pare it for fact economy.

# CHAPTER 5

# EDITING FOR PRECISION IN LANGUAGE

## ABUSING ENGLISH

The media, because of their mass appeal, have a significant impact on the development—and deterioration—of English usage in the United States. If the quality of English usage is deteriorating, as many have charged, then the media must share the blame.

The editor plays a major role in protecting the language against abuse. One who knows how to spell, makes certain a story is written in proper English, recognizes and clarifies fuzzy passages and protects the meaning of words is a valuable member of the staff. The editor who neglects those chores contributes to the erosion of the language and, ultimately, to the detriment of communication. Clear writing, correct spelling and proper grammar contribute to the ready communication of ideas.

Edwin Newman, television commentator and author of *Strictly Speaking* and *A Civil Tongue,* deplores the degradation of the language and asks, "Will America be the death of English?" He argues that it may be too much to hope for the stilted and pompous phrase, the slogan and the cliché to be banished, but he argues that they should not dominate the language.

The message was even more forcefully put by Wallace Carroll, former editor and publisher of the *Winston-Salem Journal & Sentinel:*

> The bastardization of our mother tongue is really a disaster for all of us in the news business. The English language is our bread and butter, but when ground glass is mixed with flour and grit with the butter, our customers are likely to lose their appetite for what we serve them.
>
> Our job is to interpret—to translate. Yet our translators—that is our reporters and editors—find it more and more difficult to do this basic job of translation. To

> begin with, they reach us from universities that are tending to become glorified jargon factories, and for four years or more they have been immured in a little cosmos where jargon is too often mistaken for knowledge or wisdom.

Carroll is correct in his assessment, but it is important to remember that language is always changing. As columnist James J. Kilpatrick writes:

> Ours is a constantly changing language. In speech, as in anything else, the acceptability of change is a matter of personal taste. [Some object] to "host" as a verb—"Johnny Carson hosted the show." I, too, would object. Neither have I yielded to "chair," as in "He chaired the committee." Yet my ear is no longer offended by "to service," "to intern," "to model," "to vacation."

Despite this, it is not the job of the editor to initiate change in the language. That role is better left to others. Editors serve their readers best by ensuring that copy conforms to accepted standards. With that in mind, let's review some of the common problems of grammar and usage.

## GRAMMAR

Grammatical problems help to destroy the credibility of a publication or broadcast station among well-educated readers, listeners or viewers. Grammatical errors can ruin otherwise clear writing or distort the meaning of a sentence. Just as the carpenter must learn to use a saw, hammer and nails, so, too, must journalists learn to use words. They are the tools of those respective trades.

### The Parts of Speech

To understand the English language and how it works, you must first know the eight parts of speech. They are *verbs, nouns, pronouns, adjectives* (including the articles), *adverbs, prepositions, conjunctions* and *interjections.* Let's review the uses of each.

#### VERBS

A verb is a word or group of words that expresses action or state of being. There are two kinds of verbs, finite and nonfinite (also called verbals). The finite verb works with the subject of the sentence to give a sense of completeness. A nonfinite verb works as a nominal, much like a noun, or as a modifier. Compare the two:

| FINITE | NONFINITE |
|---|---|
| The police *charged* him with robbery. | The police, *having charged* him with robbery… |

Let's discuss finite verbs now and take up the verbals later.

**Person and Number.** Finite verbs can be distinguished by *person* (first, second and third) and *number* (singular and plural). Most verbs have a different form only in the third person singular of the present tense:

| | SINGULAR | PLURAL |
|---|---|---|
| FIRST PERSON | I drink | we drink |
| SECOND PERSON | you drink | you drink |
| THIRD PERSON | he (she, it) *drinks* | they drink |

An exception is the verb *to be:*

| | SINGULAR | PLURAL |
|---|---|---|
| FIRST PERSON | I *am* | we are |
| SECOND PERSON | you are | you are |
| THIRD PERSON | he (she, it) *is* | they are |

Verbs also frequently have *auxiliary,* or helping, verbs. These can be:

- Modal auxiliaries. Examples of these auxiliaries, which are added to the main verb, are *will, would, can* and *must:* He *will go* to the store.
- Perfect auxiliaries. These consist of a form of the verb *have* followed by a verb with an *-en* or *-ed* ending: He *has walked* to the store (present). He *had walked* to the store (past).
- Progressive auxiliaries. These consist of a form of the verb *be* followed by a verb with an *-ing* ending: He *was going* to the store.
- Passive auxiliaries. These consist of a form of the verb *be* followed by a verb with an *-en* or *-ed* ending: He *was driven* to the store.

**Moods, Voices and Tenses.** Other properties of verbs, in addition to person and number, are moods, voices and tenses. The three moods, indicative, imperative and subjunctive, indicate differences in the intention of the writer or speaker. The indicative mood is used to make an assertion or to ask a question:

- The man *drove* to the store.
- Where *is* he *going?*

The imperative mood is used for commands, directions or requests:

- Command: *Drive* to the store.
- Direction: *Take* the next right turn.
- Request: Please *drive* me to the store.

The subjunctive mood expresses a condition contrary to fact: If I *were* wealthy, I would quit work. In this case, the speaker indicates an absence of wealth, so the expression is contrary to fact. The present tense forms of the subjunctive are *(if) I be, (if) you be, (if) he, she or it be, (if) we be, (if) you be* and *(if) they be.* The past tense forms are *(if) I were, (if) you were, (if) he, she or it were, (if) we were, (if) you were* and *(if) they were.* Unlike other verbs, which differ in the third person singular of the present tense, the subjunctive remains the same: *(if) he drive,* not *if he drives.*

Journalists prefer the *active voice* of verbs to the *passive voice* because in the active voice the subject of the sentence does the acting:

The *president fired* the secretary of state.

In the passive voice, the subject of the sentence is acted upon:

The *secretary of state was fired* by the president.

The passive voice therefore diverts attention from the person doing the acting to the person or thing being acted upon. In some cases, the writer or editor may find that preferable:

The *president was shot* today by a man who dislikes his foreign policy.

Clearly, in such a case the person acted upon is more important than the person doing the acting. But in most cases the active voice is better. It is more direct and less wordy.

The most common tenses in English are the *present tense,* the *past tense* and the *future tense.* These three tenses indicate the time *now,* the time *past* and the time *to come.* Tense can also indicate that an action has been completed at some definite time in the present, past or future. This is done by the three *perfect tenses,* which denote perfected, or finished, action or state of being. Finally there are the *progressive tenses,* which indicate action that is continuing or progressing in the present, past or future. Thus, various first-person forms of the verb *to see:*

- Present: *I see*
- Past: *I saw*
- Future: *I shall see*
- Present perfect: *I have seen*
- Past perfect: *I had seen*
- Future perfect: *I shall have seen*
- Present progressive: *I am seeing*
- Past progressive: *I was seeing*
- Future progressive: *I shall be seeing*
- Present perfect progressive: *I have been seeing*
- Past perfect progressive: *I had been seeing*
- Future perfect progressive: *I shall have been seeing*

## *Spelling Words to Know*

Some frequently misspelled words you should commit to memory:

accede
accommodate
accumulate
adherence
admissible
advertent
advertise
adviser*
all right
alleged
allotted, allotment
appall
asinine
ax*

balloon
battalion
bellwether
berserk
blond (adj., noun for male)
blonde (adj., noun for female)
boyfriend
buses (vehicles)
busses (kisses)

Canada geese
caress
cave-in
chaperon
cigarette
commitment
consensus
consistent
consul
controversy
council
counsel

demagogue
descendant
dietitian
disastrous
dissension
drunkenness

embarrass
emphysema
employee*
exhilarating

feud
firefighter
fraudulent
fulfill

gaiety
goodbye*
grammar
greyhound
grisly
guerrilla

hangar (aircraft shelter)
hanger (hanging device)
harass
hemorrhage
hitchhiker

impostor
incredible
indestructible

Because the progressives are wordy and awkward, they are seldom used. On occasion, however, they are helpful in indicating degrees of past or ongoing action. A verb that in its own meaning already expresses continuity does not need a progressive form. Thus, it is unnecessary to write, "I am living in Utah." In this case, "I live in Utah" is adequate. But compare these sentences:

> She *drove* to the store.
> She *was driving* to the store.

The second accurately stresses the continuity of the action.

*Irregular verbs* do not add *-ed* or *-d* to form the past tense and past participle.

indispensable
innocuous
inoculate
irresistible

judgment*

kidnapped*

lambastes
largess
liaison
likable
liquefy

mantel (shelf)
mantle (covering)
marijuana
marshal

nerve-racking

occasion
occurred

papier-mache
parallel
paraphernalia
pastime
penicillin
percent*
percentage
permissible
personnel
picnicking
playwright
politicking
pompon
preceding
principal (main)
principle (concept)
privilege
procedure
prostate
publicly

quandary
questionnaire
queue

rarefy
reconnaissance
restaurateur
rock 'n' roll (rock-and-roll)

sacrilegious
seize
separate
siege
sizable
skillful
stationary (not movable)
stationery (writing material)
strait-laced
strong-arm
subpoena
summonses
supersede

teen-age (adj.)
teen-ager (noun)
theater*
tumultuous

vacuum
vice versa
vilify

weird
whiskey* (whisky for Scotch)
wield
wondrous

X-ray (noun, verb and adj.)

*Preferred spelling. Words in parentheses also indicate common newspaper usage.

Some examples:

| INFINITIVE AND PRESENT TENSE | PAST TENSE | PAST PARTICIPLE |
|---|---|---|
| begin | began | begun |
| break | broke | broken |
| come | came | come |
| eat | ate | eaten |
| do | did | done |
| fly | flew | flown |

| INFINITIVE AND PRESENT TENSE | PAST TENSE | PAST PARTICIPLE |
|---|---|---|
| have (has) | had | had |
| leave | left | left |
| lie | lay | lain |
| sit | sat | sat |
| tear | tore | torn |
| win | won | won |

Verbs also are classified as *transitive* and *intransitive* to describe their use in a sentence. Transitive verbs require objects to complete their meaning:

The president *fired* the *secretary of state.*

Without the object, secretary of state, the sentence would be incomplete.

Intransitive verbs make assertions without requiring objects:

The moon *shines.*

The cowboy *is* on his horse.

Many verbs can be either intransitive or transitive. An example:

Intransitive: She *breathes.*

Transitive: She *breathes* foul air.

Used intransitively, the verb indicates that the woman is alive. Used transitively, it refers to an experience she is having.

**Verbals.** Verbals are verbs that have lost their ability to act as predicates in a sentence. Instead, they act as nouns, adjectives or adverbs. There are three kinds of verbals, *infinitives, present participles* and *past participles.*

Infinitives begin with the word *to.* Here are examples of infinitives in context:

To *forgive* is divine. (present, active voice)

I always seem *to be forgiving* you. (present progressive, active voice)

She was happy *to have been forgiven.* (perfect, passive voice)

An active verb can have two types of participles present and past. The present participle, such as *milking,* expresses action in progress. The past participle, *milked,* expresses finished action. *Having milked* is a modified form of the past participle known as the perfect participle. The passive forms are *being milked* for the present, *milked* for the past, and *having been milked* for the perfect participle.

When they are not part of a verb phrase, participles are used as adjectives and nouns. Here is an example of a participle used as an adjective:

Freedom *Enlightening* the Nation is the name of the statue on the campus. (*Enlightening* modifies Freedom.)

A present participle that functions as a noun is called a *gerund.* An example:

*Acting* like a gentleman is *acting* properly.
(The first *acting* is a gerund used as the subject of a sentence. The second is used as a predicate noun.)

## NOUNS

A noun names a person, place or thing. There are two kinds of nouns: *proper nouns* are titles for specific persons, places and sometimes things; *common nouns* are generic names such as *dog* or *wood.* Proper nouns are capitalized; common nouns are not. An example:

New York *City* is the nation's largest city.

The *city* was built beside the river.

In the first sentence, *city* is part of a title and therefore must be capitalized. In the second sentence, *city* is used as a common noun.

Nouns can be either singular or plural. Usually, the plural is formed by adding *-s* or *-es* to the end of the word: *jeep, jeeps; box, boxes.* But words ending in *y* and *f* sometimes change before a plural ending: *hoof, hooves; sky, skies.* Unfortunately, that's not always true: *roof, roofs.* Check a dictionary if you are in doubt. A complicating factor is that some nouns have irregular plural forms: *sheep, sheep; goose, geese.* And some nouns normally occur in the singular only: *dust,* not *dusts; courage,* not *courages.*

Nouns can also be possessive. Newspapers observe these rules for forming the possessive case:

- Plural nouns not ending in *s* take *'s: the alumni's contributions, women's rights.*
- Plural nouns ending in *s* take only an apostrophe: *the girls' pony, the ships' wake.*
- Nouns plural in form but singular in meaning take only an apostrophe: *measles' effects, mathematics' rules.*
- Singular nouns not ending in *s* take *'s: the church's needs, the girl's toys, Marx's theories.*
- Singular common nouns ending in *s* take *'s* unless the next word begins with *s: the hostess's invitation, the hostess' seat.*
- Singular proper nouns ending in *s* take only an apostrophe: *Ceres' rites, Socrates' life.*

For other rules that pertain to possessives, consult the wire service stylebooks.

Nouns have several possible uses in sentences. They can be used as:

- The subject, object or complement of a verb or verbal.

- The object of a preposition.
- A modifier following another noun, called an *appositive:* my brother *Wesley.*
- A modifier before another noun: a *windshield* wiper.
- The modifier of an adjective or verb:

  They were *battle* weary. (modifies the adjective weary)
  They left *Monday.* (modifies the verb left).

## PRONOUNS

A pronoun takes the place of a noun in a sentence. It often refers to a noun, a noun phrase or even a complete sentence mentioned earlier. The word to which the pronoun refers is known as the *antecedent* of the pronoun. Some examples:

> The man hit *his* head on the door. (*His* refers to *the man.*)
>
> I have suggested to my brother, *who* lives in Phoenix, that we consider a joint vacation. (*Who* refers to *my brother.*)
>
> *The Atlanta Braves have a good team this year. That* means the Braves could win the pennant.
> (*That* refers to the full sentence establishing that the Braves have a good team.)

There are several types of pronouns: *personal, relative, interrogative, demonstrative, indefinite* and *intensive.*

Personal pronouns have *case,* either *nominative, possessive* or *objective; person,* either *first, second* or *third;* and *number,* either *singular* or *plural:*

FIRST PERSON

| CASE | SINGULAR | PLURAL |
|---|---|---|
| Nominative | I | we |
| Possessive | my, mine | our, ours |
| Objective | me | us |

SECOND PERSON

| CASE | SINGULAR | PLURAL |
|---|---|---|
| Nominative | you | you |
| Possessive | your, yours | your, yours |
| Objective | you | you |

THIRD PERSON

| CASE | SINGULAR | | | PLURAL |
|---|---|---|---|---|
| | Masculine | Feminine | Neuter | |
| Nominative | he | she | it | they |
| Possessive | his | her, hers | its | their, theirs |
| Objective | him | her | it | them |

Relative pronouns are used to introduce adjectival or noun clauses:

The butter, *which* he bought yesterday, is already gone. (adjectival clause)
*Whatever* he bought yesterday is already gone. (noun clause)

The relative pronouns are *who, whom, whose, that, which, whoever, whomever, whatever* and *whichever.* Sometimes the relative pronoun can be omitted from an adjectival clause:

The butter he bought yesterday is already gone.

Interrogative pronouns are used to introduce questions. They are *who, whom, whose, which* and *what:*

*Who* will win the marathon?
*Whose* shirt is that in the washing machine?

The demonstrative pronouns, *this, these, that* and *those,* indicate nearness to or distance from the speaker:

*This* is my best chance to succeed.
*Those* are mine.

Indefinite pronouns have vague or unknown antecedents. They include words such as *somebody, everyone, whoever, all, each* and *either:*

*Somebody* stole my laundry basket.
*Each* of the girls in the class is smart.

Intensive pronouns, which end with *self* or *selves,* repeat and intensify the noun antecedent:

I shot *myself* in the leg.
Jim *himself* will pitch the ninth inning.

*Myself* should not be used in place of me:

Professor Barrett gave A's to Sandra and *me.* (not myself)

The *nominative, objective* and *possessive* cases mentioned earlier have various uses. The nominative pronouns (*I, we, you, he, she, it, they, who* and *whoever*) serve as the subjects of sentences or clauses—even when the verb is deleted—and act as subject complements following forms of the verb *to be:*

Marilyn and *I* are close friends. (subject of a sentence)
Marilyn and I, *who* are close friends, will go to the party together. (subject of a clause)
She is a better mathematician than *he.*
(subject of a clause with the implied verb *is*)
It is *I.* (subject complement following a form of the verb *be*)

The pronouns in the objective case are *me, us, her, him, them, who* and *whomever.* The objective case is used to express the object of a verb, verbal or preposition, and the subject of the infinitive:

She forced *me* to take the Fifth Amendment. (object of verb)

Many of *us* coal miners have black lung disease. (object of preposition)

She ranked him as high as *me.*
(object of *ranked* in understood clause, *she ranked me*)

The Tigers wanted *him* to be traded to the Red Sox. (subject of infinitive)

In the possessive case, also known as the *genitive,* pronouns show possession either in combination with specific nouns (as *regular possessives*) or alone (as *independent possessives*):

| REGULAR POSSESSIVES | INDEPENDENT POSSESSIVES |
|---|---|
| my | mine |
| our | ours |
| your | yours |
| her | hers |
| his | his |
| its | its |
| their | theirs |
| whose | whose |

Some examples of usage:

I think *my husband* will mow the grass today. (regular)

I like mine better. (independent)

Regular possessives can also show possession with gerunds, which act as nouns:

*Her* dropping the course was ridiculous, the professor said.

## ADJECTIVES

Adjectives add to the meaning of nouns or pronouns. In this sense an adjective describes, limits or modifies the noun or pronoun to make it clearer. The two major kinds of adjectives are *descriptive* and *limiting.* An adjective that answers the question What kind? is descriptive:

The *handsome* man stepped out of the coach.

*Handsome* in the sentence above is a *common adjective,* one of three kinds of descriptive adjectives. Others are *participial adjectives* (the *dried* fruit) and *proper adjectives* (the *Indian* princess). A descriptive adjective sometimes is used to complete the meaning of the verb *to be* or any of its forms. When used that way, it is known as a *predicate adjective.* Other verbs, including *seem, look, feel* and *taste,* also may take predicate adjectives:

The milkshake tastes *sweet.*

A descriptive adjective sometimes can act as the subject of a sentence when the article *the* is placed in front of it:

The *loneliest* call the hot line regularly.

A limiting adjective tells how many, how often, what number or what amount:

The *five* generals paid homage to George VI.

The articles—*a, an* and *the*—form a special group of limiting adjectives.

All adjectives except the four *demonstrative* ones, *this, that, these* and *those,* change form to show difference in amount or degree. There are three degrees of comparison, *positive, comparative* and *superlative.* Most adjectives of more than one syllable form the comparative by using *more* or *less* in front of the adjective. Similarly, the superlative of multisyllable words is formed by using *most* and *least:*

| POSITIVE | COMPARATIVE | SUPERLATIVE |
|---|---|---|
| thorough | more thorough | most thorough |
| willing | more willing | least willing |

The endings *-er* or *-est* are added to most adjectives of one syllable to form the comparative and superlative:

| POSITIVE | COMPARATIVE | SUPERLATIVE |
|---|---|---|
| clear | clearer | clearest |
| strong | stronger | strongest |

Some words, however, have irregular comparisons: *good, well; more, most; better, best.*

## ADVERBS

Adverbs change or add to the meaning of verbs, adjectives or other adverbs. They help the words they modify, just as adjectives help nouns. Adverbs are classified three ways by use. *Simple* adverbs are used solely to modify:

She looked *happily* at her elder son.

He was *perfectly* happy.

We were waiting *quite* contentedly.

*Conjunctive* adverbs are used to connect main clauses:

He took the bottle home; *however,* he did not drink.

The conjunctive adverbs include *therefore, consequently, moreover, nevertheless, accordingly, otherwise* and *wherefore.* And, finally, *interrogative* adverbs introduce a question:

*When* did he take the big step?

Adverbs also are classified by time, place, degree, manner and assertion. For details of those usages, consult a grammar book.

Like adjectives, some adverbs have comparative and superlative forms:

| POSITIVE | COMPARATIVE | SUPERLATIVE |
|---|---|---|
| fast | faster | fastest |
| soon | sooner | soonest |
| little | less | least |
| well | better | best |

## PREPOSITIONS

A preposition is a connecting word that shows a relationship between its object and some other word in the sentence. Common prepositions include *in, on, at, by, against, above, upon, from, of, without, under, over, through* and *during.* There are many others.

The object of the preposition must be in the objective case, except when the object is a double possessive:

The museum has several hats *of Washington's.*

The object may be a noun, an adjective used as a noun, an infinitive, a participle or a clause. All, however, must be used as nouns.

## CONJUNCTIONS

Another connecting word is the conjunction, which is used to connect words, phrases, clauses and sentences. *Coordinating conjunctions* connect two words, phrases, clauses or sentences that are grammatically equal. Two independent clauses connected by a coordinating conjunction form a *compound sentence:*

John will attend Harvard, *and* Mary plans to go to Yale.

Note that a coordinating conjunction must be preceded by a comma.

The coordinating conjunctions include *and, both, as well as, either, or, neither, nor, but, yet, so* and *so that.* Conjunctions that come in pairs, such as *either* and *or* and *neither* and *nor,* are known as *correlative conjunctions.*

A *subordinating conjunction* joins a subordinate clause to the principal clause of a sentence. Common subordinating conjunctions include *when, as, since, before, after, until, where, because, if, unless, although, though, even if, that, so that* and *in order that.*

Sentences with clauses joined by subordinating conjunctions are called *complex sentences.* In these constructions, the joined clauses are not grammatically equal:

> She can ride horses better *than* I can.

In this case, *than* connects the main clause, *She can ride horses better,* to the subordinate clause, *than I can* (*ride horses* is understood).

### INTERJECTIONS

An interjection is an exclamation that expresses surprise, pain or some other intense emotion. It has no grammatical relation to other words and can be used alone. Some examples: *Ouch!, Whew!, Darn!, Help!* and *Oh!*

## USAGE

Knowing the parts of speech is essential to understanding the English language, but that alone is not enough. The real difficulty lies in trying to make words and sentences work together to convey meaning. Often, what *sounds* correct may not *be* correct because of common misuse. Let's examine some of the common misuses of the language.

### Subject-Predicate Disagreement

The editors of *After the Fact,* internal publication of *The* (Louisville, Ky.) *Courier-Journal,* lamented, "Everybody knows that subjects come in matched pairs: singulars together and plurals together. But, sadly, we let too many unmarriageables slip into the paper." Some examples of incorrect agreement follow, with comments and corrections.

> There are two things that either Sloane or Hollenbach have ...

The thought is: "... things that Sloane has or that Hollenbach has." Remember, when nouns are connected by *either ... or, neither ... nor* or *or* alone, the noun closest to the verb should be considered in deciding whether a singular or plural verb is required. If the final noun in the subject is singular, use a singular verb, and vice versa.

> The continuing lag in industrial production and the rise in unemployment is a result of inflation. (Lag and rise *are* ...)
>
> An American Bar Association accreditation team, in addition to a citizens' panel, have both recommended this.
>
> (*Team* is the subject; the verb must be *has.* Parenthetical matter should be ignored in selecting the number of the verb.)

It's moved to a point where the anxiety and the concern is unrealistic.
(Anxiety and concern *are* . . . )

The monotony of the concrete walls painted in dull green and blue are broken only. . .
(Monotony *is* . . . )

A two-thirds vote by both houses of Congress and ratification by three-quarters of the states is necessary.
(Vote and ratification *are* . . . )

## Dangling Modifiers

The dangling modifier is one of the most common errors committed by beginning writers and by all who write in a hurry. The writer knows what is intended but doesn't say it, forcing the reader to rearrange the sentence to grasp its meaning. Some examples:

If convicted of the assault and battery charge, a judge may impose any sentence he sees fit on the defendants.
("If convicted" applies to the defendants, not to the judge.)

Besides being cut on the left cheek and bloodied in the nose, Zeck's purse was attached for $825.

Already hospitalized a month, doctors estimate it will be three or four months before he is out again.

An E-shaped building, the fire started in the southwest wing.

A "natural" fertilizer, he predicted that it would solve many problems

After blowing out the candles atop his birthday cake in three puffs, a movie camera flashed old fight films on the screen near the bar.

The fluoroscopic system makes moving pictures and tape recordings of the mouth and throat while speaking, chewing and swallowing.

Short and readable, I finished it off in about 45 minutes.

## Misuse of Relative Pronouns

No one can say what the most common grammatical errors in news writing are, but near the top must be the misuse of the relative pronoun.

Leon Stolz of the *Chicago Tribune* advised reporters and editors, "If you have trouble deciding whether the relative pronoun should be who or whom, you can usually find the right answer by remembering that *who* is nominative, like the personal pronouns *he, she* and *they. Whom* is objective, like *him, her* and *them.* Turn the clause into an independent sentence and substitute a personal pronoun for the relative pronoun."

Applying the Stolz formula:

After his decision to cancel the trip, he sent most of the officials who he had invited to attend. (He invited *they* to attend?)

The repeal gives property owners absolute freedom in deciding who they will rent or sell to. (They will rent to *they?*)

Miss Barbara Warren, who he met while they were medical students at Passavant Hospital ... (He met *she?*)

In his last eight games, covering $13\frac{2}{3}$ innings, the skinny Texan, who teammates call "Twiggy," has held opponents scoreless. (They call *he* Twiggy?)

The paper said two residents of the housing project were known to have been a young man whom they said looked like the description of the sniper.
(They said *him* looked like the description of the sniper?)

Reporters and editors frequently have trouble distinguishing between *that* and *which. That* is preferred when introducing restrictive (essential) clauses that refer to inanimate objects, places, ideas and animals; *which* introduces nonrestrictive (nonessential) clauses, which are nondefining and parenthetical. Two examples:

The river that flows northward through Egypt is the Nile.

The Missouri River, which flows into the Mississippi at St. Louis, is cleaner than it was 10 years ago.

Note that nonrestrictive clauses are set off with commas.

## Other Word Problems

### NON SEQUITURS

A non sequitur is an error in logic, the phrase means "it does not follow":

A guard at the Allied Kid Co., he died at 7:10 a.m., about five minutes after one of the youths implicated in the attack was taken into custody.
(This implies that guards die at 7:10 a.m. Can we infer that workers die at 8:10 and executives at 9:10?)

Worn on a chain with swivel and button, this model retails at $39.95.
(How much does it cost if I just carry it loose in my pocket?)

"Because breath is so vital to life," Burmeister explained, "the field of inhalation therapy and the development of breathing equipment have become increasingly important in medical science today."
(It may be true that these things are increasingly important, but not because breath is vital to life. Breath was just as important to life 3,000 years ago as it is today.)

Stored in an air-conditioned room in lower Manhattan, the tapes contain information on the reading habits of one million Americans.
(The nature of the information on these tapes is not in any way related to the place of their storage or the condition of the air there. An easy way to edit this sentence is to start with the subject: "The tapes, stored in an air-conditioned room in lower Manhattan, contain information...")

Planned by Jones, Blake and Droza, Detroit architects, the new school has 18 classrooms in addition to such standard facilities as cafeteria and library.
(This implies that it's a natural thing to expect a school planned by that particular firm to have 18 classrooms and the other features.)

Completed three years ago, the plant is 301 feet by 339 feet and is a one-story structure containing...
(A plant of exactly that size could have been completed 50 years ago or yesterday.)

This particular error crops up most frequently in obituaries:

Unmarried, Jones is survived by his mother, Mrs....
Born in Iowa, he worked on two newspapers in Illinois before coming to St. Louis.

## WORD CLUTTER

Clutter words dirty a newspaper, waste space and get in the reader's way. Every day, stories die because strings of clutter words didn't die at the computer or at the hand of the editor. Reporters should know better than to write:

She was on the operating table from 8 p.m. *Monday night* until 5 a.m. *Tuesday morning.*
(*Night* and *morning* are redundant because p.m. and a.m. mean the same thing.)

For mankind, *the biologists who study genetics* seem to offer...
(Try *geneticists.*)

He was a Democratic nominee for *U.S.* Congress from the 7th *Congressional* District but lost in the *final* election to *incumbent* Rep. Donald W. Riegle Jr.
(Not everyone can get four redundancies into one sentence.)

He said the USDA is *currently* spending...

Justice said Double Spring had been *in the process of* phasing out its operation...

Hayley has some 30 *different* (fish) tanks in his home.

Retired Adm. Jackson R. Tate slipped away to a secret retreat yesterday for his *first meeting* with the daughter *he has never met.*

Compression, a shortcut to ideas, is one of the cardinal virtues of good writing. "The field is 50 feet in length" should be "The field is 50 feet long." "He is said to be resentful" means simply, "He is said to resent."

Pacing is gained by substituting short words for long words and single words for phrases: *big* for *enormous, find* for *discover, approving* for *applying its stamp of approval, Smith's failure* for *the fact that Smith had not succeeded, field work* for *work in the field.*

Compression eliminates the superfluous: "pledged to *secrecy* not to disclose," "wrote a formal letter of resignation" (*resigned*), "read from a *prepared* statement," "go into details" (*elaborate*). A few words may say a lot; a lot of words may say little.

## WRONG WORDS

Good reporters are meticulous in presenting the facts for a story. Others are not so precise in their choice of words. By habit, they write *comprise* when they mean *compose, affect* when they want *effect, credible* for *creditable.* They use *include,* then list all the elements.

Each time a word is misused it loses some of its value as a precision tool. AP reported that "U.S. officials connived with ITT." There was no connivance, which means closing one's eyes to wrongdoing. The precise word would have been *conspired,* if, indeed, that is the charge.

Note these examples that got by the copy desk:

He has been an "intricate part of the general community."
(The writer meant *integral,* meaning essential.)

The two officers are charged with dispersing corporate funds.
(The word is *disburse*—to pay out, to expend. *Disperse* means to scatter in various directions; distribute widely.)

A story indicating that a man might not be qualified for his job says: "Miller refutes all that," and then he says why he is capable.
(*Denies* would have been much better.)

A football pass pattern was referred to as a "flair out."
(*Flair* means a natural talent or aptitude. The word is *flare* or expansion outward in shape or configuration.)

Mrs. Reece, a spritely woman ...
(She may be a sprite—an elf or pixie—but what the writer probably meant was *sprightly*—full of life.)

His testimony about the night preceding the crime was collaborated, in part, by his mother.
(*corroborated,* perhaps?)

The MSD could be the biggest benefactor in Kentucky under a ... reimbursement program.
(It should be *beneficiary,* one who receives; *benefactor* is the giver.)

Other terms frequently misused:

- Adopted, passed—resolutions are adopted or approved; bills are passed. In legislative jargon, *passed* also can mean passed by for the day or for that meeting.
- Aggravate, irritate—the first means to make worse. The second means to incite or provoke.
- Amateur, novice—an *amateur* is a nonprofessional. A *novice* is a beginner.
- Amount, number—*amount* indicates the general quantity. *Number* indicates an enumerable quantity.
- Avenge, revenge—*avenge* for another. *Revenge* for self.

## *Circumlocutions to Avoid*

Most experienced editors can add to this list of circumlocutions:

A bolt of lightning (lightning)
A great number of times (often, frequently)
A large number of (many)
A period of several weeks (several weeks)
A small number of (few)
A sufficient number of (enough)
Absolute guarantee (guarantee)
Accidentally stumbled (stumbled)
Advance planning (planning)
Advance reservations (reservations)
All of a sudden (suddenly)
As a general rule (usually)
Assessed a fine (fined)
At a later date (later)
At the conclusion of (after)
At the corner of 16th and Elm (at 16th and Elm)
At the hour of noon (at noon)
At the present time (now)
At 12 midnight (at midnight)
At 12 noon (at noon)

Bald-headed (bald)

Called attention to the fact (reminded)
Climb up (climb)
Commute back and forth (commute)
Completely decapitated (decapitated)
Completely destroyed (destroyed)
Completely surrounded (surrounded)
Consensus of opinion (consensus)
Cost the sum of $5 (cost $5)

Despite the fact that (although)
Disclosed for the first time (disclosed)
Draw to a close (end)
Due to the fact that (because)
During the winter months (during the winter)

End result (result)
Entered a bid of (bid)
Exact replica (replica)
Exchanged wedding vows (married)

Few in number (few)
Filled to capacity (filled)
First priority (priority)
First prototype (prototype)
For a period of 10 days (for 10 days)
Foreign imports (imports)
Free gift (gift)
Free pass (pass)

General public (public)
Grand total (total)

Heat up (heat)
Hostile antagonist (antagonist)

In addition to (and, besides, also)
In back of (behind)
In case of (if, concerning)
In order to balance (to balance)
In the absence of (without)
In the event that (if)
In the immediate vicinity of (near)
In the near future (soon)
In the not too distant future (eventually)
Incumbent governor (governor)
Introduced a new (introduced)
Is going to (will)
Is in the process of making application (is applying)

Is of the opinion that (believes)

Jewish rabbi (rabbi)

Kept an eye on (watched)

Large-sized man (large man)
Lift up (lift)

Made good his escape (escaped)
Major portion of (most of)
Married his wife (married)
Merged together (merged)
Midway between (between)

New bride (bride)
New construction (construction)
New innovation (innovation)
New record (record)

Off of (off)
Old adage (adage)
Old cliché (cliché)
On account of (because)
On two different occasions (twice)
Once in a great while (seldom, rarely)

Partially damaged (damaged)
Partially destroyed (damaged)
Past history (history)
Period of time (period)
Placed its seal of approval on (approved)
Possibly might (might)
Postponed until later (postponed)
Prior to (before)
Promoted to the rank of (promoted to)

Qualified expert (expert)

Receded back (receded)
Recur again (recur)
Reduce down (reduce)
Refer back (refer)
Remand back (remand)
Revise downward (lower)
Rise up (rise)
Rose to the defense of (defended)

Self-confessed (confessed)
Short space of time (short time)
Since the time when (since)
Sprung a surprise (surprised)
Started off with (started with)
Strangled to death (strangled)
Summer season (summer)
Sworn affidavits (affidavits)

Tendered his resignation (resigned)
There is no doubt that (doubtless)
Total operating costs (operating costs)
True facts (facts)

Underground subway (subway)
United in holy matrimony (married)
Upward adjustment (increase)

Voiced objections (objected)

Went up in flames (burned)
Whether or not (whether)
Widow of the late (widow)
With the exception of (except)

Young juveniles (juveniles)

- Bale, bail—a farmer's hay is *baled;* water is *bailed* out of a boat; a prisoner is released on *bail.* (*Bond* is cash or property given as a security for an appearance or for some performance.)
- Biannual, biennial—the first means twice a year. The second means every two years. The copy editor could help the reader by substituting "every six months" for *biannual* and "every other year" for *biennial.*
- Bills, legislation—"The president announced he will send Congress legislation aimed at liberalizing trade with Eastern Europe." *Legislation* is the law enacted by a legislative power. The president, of course, is not such a power. What he sends to Congress is proposed legislation or bills.
- Canvas, canvass—the first is a cloth. The second means to solicit.
- Celebrant, celebrator—a *celebrant* presides over a religious rite. A *celebrator* celebrates.
- Center around—something can be centered in, centered at or centered on, but it cannot be centered around.
- Collision—"Cars driven by Robert F. Clagett and Mrs. Lois Trant were damaged yesterday when they collided on Denison Avenue. Stonington police reported that Mrs. Trant stopped her car before making a turn into Isham Street and it was hit in the rear by the other vehicle." Two objects can *collide* only when both are in motion and going—usually but not always—in opposite directions. It is not a *collision* when one car is standing still.
- Compared with, compared to—the first uses specific similarities or differences: "He *compared* Johnson with Wilson." The second notes general or metaphorical resemblance: "You might *compare* him to a weasel."
- Comprise, compose—*comprise* is not synonymous with *compose,* but actually almost its opposite. "The secretaries of State, Defense, Interior and other departments compose the cabinet." That is, they constitute it. "The cabinet comprises the secretaries of State, Defense, Interior and other departments." That is, it includes, embraces, contains them.
- Concert, recital—two or more performers give a concert. One performer gives a *recital.*
- Continuous, continual—if it rains steadily every day for a week it rains *continuously.* If it rains only part of every day for a week it rains *continually* or intermittently.
- Ecology, environmental—*ecology* is concerned with the interrelationship of organisms and their environment. *Environmental* refers to conditions or forces that influence or modify surroundings.
- Farther, further—the distinction is between extension of space and expansion of thought.

- Flaunt, flout—the first means to wave or flutter showily. The second means to mock or treat with contempt. "The students *flouted* the authority of the school board."
- Flounder, founder—horses *flounder*—struggle, thrash about—in the mud. Ships *founder* or sink. Of course, horses can founder when they become disabled from overeating.
- Grant, subsidy—a *grant* is money given to public companies. A *subsidy* is help to a private enterprise.
- Grizzly, grisly—"Miss Karmel begins her work in a valley of shadows that deepen and darken as she heaps one grizzly happening upon the next." One *grizzly* heaped upon the next produces only two angry bears. The word the writer wants is *grisly.*
- Half-mast, half-staff—masts are on ships. Flagstaffs are mounted on buildings or in the ground.
- Hardy, hearty—a story of four visiting policemen from Africa said they expressed appreciation for their hardy welcome. If that's what they said, they meant *hearty.*
- Hopefully—incorrect for "it is hoped" or "I hope." It really means, "in a hopeful manner."
- Impassable, impassible—the first is that which cannot be passed. The second is that which can't suffer or be made to show signs of emotion.
- Imply, infer—the speaker does the *implying,* and the listener the *inferring.*
- Mean, median—*mean* is the average. If the high is 80 and the low is 50, the mean is 65. *Median* means that half are above a certain point, half below.
- Oral, verbal—all language is *verbal*—"of words." But only *oral* language is spoken.
- People, persons—*person* is the human being. *People* are the body of persons—"American people." There is no rule saying a large number can't be referred to as people—"61 million people."
- Sewage, sewerage—*sewage* is human waste, sometimes called municipal or sanitary waste. *Sewerage* is the system to carry away sewage. They are sewerage (not sewage) plants. Industrial waste is the waste from factories. Some cities have storm sewers to carry away rain water and sanitary sewers for sewage.
- Sustenance, subsistence—"The two survived despite little besides melted snow for subsistence." No wonder they almost starved. The word is *sustenance.*
- Tall, high—properly, a building, tree or man is *tall.* A plane, bird or cloud is *high.*

## WRONG WORD ORDER

An insufficient water supply problem for firefighting at Fitch Senior School will be discussed next Thursday.

(Try this: "The problem of insufficient water supply for firefighting at Fitch . . .")

White segregationists waving Confederate flags and black integrationists marched past each other Tuesday.

(Or did white segregationists waving flags march past black integrationists Tuesday?)

Joseph H. Hughes Jr. of Los Angeles wrote to many of his late son's, Coast Guard Ensign Joseph H. Hughes III, friends.

(Translation: "wrote to many friends of his late son.")

## CONFUSING ANTECEDENTS

Miss Adele Hudlin agreed to give the dog a home, even though she already had two of her own. (Does she have two homes or two dogs?)

They (the Smiths) have been married 24 years and have two children. Both are 53.

## MISUSED ADJECTIVES

Informative, rather than merely descriptive, adjectives strengthen nouns: "7-foot Steve Stipanovich" is more effective than "towering Steve Stipanovich."

Many adjectives are redundant, editorial, incorrect or misplaced:

- Redundant—armed gunmen, chilly 30 degrees below zero, exact replica, foreign imports.
- Editorial—blistering reply, cocky labor leader, so-called liberal, strong words.
- Incorrect—"Whirring or grinding television cameras": Television cameras are electronic devices and do not whir or grind. "A Pole with an unpronounceable name": Every name is pronounceable by somebody. "An unnamed man": Every man has a name; the adjective should be unidentified.
- Misplaced—"Unfair labor practices strike": The practices, not the strike, are unfair. "The treacherous 26-mile Arkansas down-river race": The river, not the race, is treacherous.

Searching for the right word takes time. Omit an adjective rather than rely on a shoddy one.

## MISUSED IDIOMS

Careless use of the idiom (the grammatical structure peculiar to our language) occurs frequently in the news report. Usually the fault lies in the prepositions or conjunctions.

Three times as many Americans were killed than [*as*] in any similar period.

It remains uncertain as to when the deadline for the first payment will be. (Omit *as to.*)

She had always been able to get through the performance on [*of*] this taxing role.

The economist accused him with [*of*] failing to make a decision.
(You charge somebody with blundering, but you accuse him of it.)

He said the guns are against the law except under [*in*] certain specified situations.
(But *under* conditions or circumstances.)

Dressen is no different than [*from*] other experts.
(*Different* may be followed by *than* when introducing a clause: "The patient is no different than he was yesterday.")

Five men were pelted by [*with*] stones.

The reason for the new name is because the college's mission has been changed.
(Is *that* the college's mission has been changed.)

He said he would not call on [*for*] assistance from police except as a last resort.
(Call the police or call on the police for assistance.)

At other times the fault lies in entire phrases which must be used verbatim to be idiomatic:

These men and women could care less about Bush's legislative magic.
(The correct phrase is "couldn't care less.")

Gerunds, but not present participles acting as adjectives, require the possessive:

It was the first instance of a city [*city's*] losing its funds.

He said he didn't know anything about Hollenbach [*Hollenbach's*] interceding in his behalf.

An example of a present participle used in a similar context is

I do not like the man standing on the corner.

The possessive is not required because *standing on the corner* is used to identify man.

## UNNECESSARY JARGON

A university press release announcing a significant engineering meeting on the campus reported that one of the major papers would be on "the aerodynamic heating of blunt, axisymmetric, re-entry bodies with laminar boundary layer at zero and at large angles of yaw in supersonic and hypersonic air streams." To the consumer of news, that title is *gobbledygook.* Translated, the topic is "how hot a spaceship gets when it swings back into the air around the earth."

Doctors, lawyers, educators, engineers, government officials, scientists, sociologists, economists and others have their professional jargon or shoptalk peculiar to

## *Clichés to Avoid*

A good writer uses a fresh and appropriate figure of speech to enhance the story. The editor should distinguish between the fresh and the stale. This isn't always easy because some words and phrases are used repeatedly in the news report.

The Associated Press ran almost 400,000 words of its copy through a computer to determine which of the tired words and phrases were used most frequently. The result: *hailed, backlash, in the wake of, informed, violence flared, kickoff, death and destruction, riot-torn, tinder dry, racially troubled, voters marched to the polls, jam-packed, grinding crash, confrontation, oil-rich nation, no immediate comment, cautious (or guarded) optimism, limped into port.*

Editors can add to the list of tired expressions:

Acid test
Area girl
Average (reader, voter, etc.)

Banquet (never a dinner)
Belt tightening
Bitter (dispute)
Blistering (accusation)
Bloody riot
Bombshell (announcement, etc.)
Boost
Briefing
Brutal (murder, slaying)

Cardinal sin
Caught the eye of
Controversial issue
Coveted trophy
Crack (troops, etc.)
Cutback

Daring (holdup, etc.)
Deficit-ridden
Devastating (flood, fire)
Devout (Catholic, etc.)
Do your own thing
Dumped

-ees (trainees, escapees)
Experts
Eye (to see)
Eyeball to eyeball

Fiery holocaust
Fire broke out, swept
Fire of undetermined origin
Foot the bill
Freak accident

Gap (generation, credibility, etc.)

Hammer out
Hard-core, hard-nosed
Hike (for *raise*)

the profession. Sometimes this jargon is used to impress the uninitiated; sometimes it is a cover-up.

A judge's ruling on a case involving an actress contained this sentence: "Such vanity doubtless is due to the adulation which the public showers on the denizens of the entertainment world in a profusion wholly disproportionate to the intrinsic contribution which they make to the scheme of things." That's pretentious verbosity. So, too, is this from an educator:

Historical document
Hobbled by injury
Hosted
Hurled

Identity crisis
In terms of
Initial (for *first*)
-ize (finalize, formalize)

Junket

Keeled over

Led to safety
Luxurious (apartment, love nest, etc.)

Made off with
Miraculous (cure, escape, etc.)
Momentous occasion

Name of the game

Opt for
Overwhelming majority

Passing motorist
Plush (hotel, apartment, etc.)
Police were summoned
Pressure (as a verb)
Probe

Relocate (for *move*)
Reportedly, reputedly

Seasoned (observers, etc.)
Senseless murder
Shot in the arm
Staged a riot (or protest)
Standing ovation
Stems from
Stinging rebuke
Sweeping changes
Swing into high gear

Task force
Tense (or uneasy) calm
Terminate (for *end*)
Thorough (or all-out) investigation
Timely hit
Top-level meeting
Tragic accident
Turn thumbs down

Uneasy truce
Unveiled
Upcoming

Vast expanse
Verbalize
Violence erupted
Violent explosion

Whirlwind (tour, junket)

Young boys

The educator will hold a practicum for disadvantaged children who are underachieving in reading.
(Try "slow learners who can't read.")

Translation is needed when a story on education contains "professional terms" such as *paraprofessionals, academically talented, disadvantaged* (culturally deprived, impoverished students) and *ungraded* and *nongraded classrooms.*

In a special study of state wire reporting, the AP found that unintelligible jargon appeared in legislature stories *(resolves, engrossment, tucked in committee)*; in alphabet-soup references to agencies and organizations (SGA, the UCA, CRS, LTA and MMA); in Weather Service forecasts; and in market reports. AP then noted, "Neither weather reports nor markets are sacrosanct to editorial pencils."

The editor can help the reader by substituting simple words for technical terms and by killing on sight words like *implement* and the *-ize* words. Here are translations of some technical terms that frequently appear in the news report:

| TERM | TRANSLATION |
|---|---|
| Motivated or motivationed | Moved |
| Object | Aim |
| Mentality | Mind |
| Percentage | Part |
| Assignment | Task or job |
| Astronomical | Big |

Some common euphemisms:

| TERM | TRANSLATION |
|---|---|
| Container | Can |
| Continental breakfast | Juice, roll and coffee |
| Dialogue, conversation | Talk, discussion |
| Planned parenthood | Birth control |
| Revised upward | Raised |
| Withdrawal | Retreat |

## EXCESSIVE SLANG

Many editors will agree with this advice from the AP: "Use of slang should be a rarity in the news report." Some editors might even dream that use of slang can be reduced in sports stories, in the signed columns, in comic strips and in ads.

Slang in direct quotations helps reveal the speaker's personality. Readers expect the gangster to use terms of the underworld. They do not expect the reporter to resort to slang, such as "The Brinton household is a go-go preparing . . . for guests."

Some slang words should be avoided because they are offensive *(cops* for police officers, *gobs* for sailors, *wops* for Italians); others are avoided because they reveal a writer's carelessness *(got clobbered* for *was defeated)*.

A few examples from a wire service show how an editor can overcome the slang:

The Supreme Court ruled today that a lower court goofed.
(What's wrong with the proper word *erred?*)

A Los Angeles story spoke of a couple's getting "a few belts in one of the local bars."
(What's wrong with *drinks* if that's what they got?)

A Washington reporter wrote that "well-heeled admirers of the senator have shelled out $7,000."
(Translation: *Well-heeled* means wealthy and *shelled out* means contributed.)

## LACK OF PARALLELISM

Similar ideas or elements in a sentence should be phrased in a similar structural or grammatical form. You would say, "I like gardening, fishing and hunting," not "I like gardening, fishing and to hunt." In the following, the word *requiring* makes a nonparallel construction: "Instead of requiring expensive cobalt drill bits, disposable brass pins are used."

Comparisons should compare similar things. Here is a sentence that compares an apple (the increase) with a pumpkin (the sales): "Consolidated sales of Cottontex Corp. for the first six months of this year were $490 million, an increase of $27 million compared with the first half of last year." Use "an increase of $27 million over the previous year's first half." "The soldier was ragged, unshaven, yet walked with a proud step." Make it read, "The soldier was ragged and unshaven, yet walked with a proud step."

## MIXED METAPHORS

Mixed metaphors add confusion to news stories:

Legislative Hall here was swarming with lobbyists as the second session of the 121st General Assembly got under way Monday.

With lawmakers treading water while awaiting Gov. Elbert N. Carvel's state and budget messages, due Wednesday, lobbyists had a field day.

(In two paragraphs the story pictured Legislative Hall as a beehive, a swimming pool and an athletic field.)

Breaking domestic ties with gold would make the nation's gold stock a real barometer of international fever for gold.

(Do you shove that barometer under your tongue or what?)

They hope to unravel a sticky turn of events that was further complicated recently.

(Did you ever try to unravel glue, molasses, maple syrup or other similar strings or yarns?)

## READER-STOPPERS

An ear for language is as important as an eye for grammar. "This doesn't sound right," the editor protests after spotting fuzzy passages. Careful reading of copy and careful editing will enable the editor to ferret out unclear or nonsensical expressions.

As explained by one engineer to Reed, one of the reasons for the high cost of repairing the streets is that the space between the concrete and the ground presents problems of pumping liquid concrete between. (We can only hope that Reed understands.)

Gangs of white rowdies roamed the area last night attacking cars bearing blacks with baseball bats, bricks and stones. (Who had the bats?)

Three counties, Meigs, Pike and Vinton, get more than 85 percent from the state. Morgan gets 90.3 percent. (How many counties?)

Many of the 800 executives and clerical people will be transferred, and some probably will be eliminated. (That's rough on people.)

Victims in the other cars were not hurt. (Then why were they victims?)

Monday will be the first day of a new way of delivering an expanded hot-meal program to senior citizens in Genesse County.
(Would the "program" or the "meals" be delivered?)

The delegation ... was welcomed by 20,000 mostly black supporters.
(What color was the other part of each supporter?)

Many of the players hovered around 5 feet 2.
(They must have looked funny up in the air like that.)

Indignant at being arrested after waiving extradition, Graham lashed out.
(Was he indignant because of the facts leading to his extradition, or to his arrest?)

A misplaced time element leads to awkward construction.

Parents protesting the closing of Briensburg School Monday tried to ...
(Did they protest Monday only? Was the school closed Monday? Or did they "try to" Monday?)

## MOCK RURALISMS

Here is some advice from Leon Stolz of the *Chicago Tribune:* "If you hold your quota to one mock ruralism a century, your readers will not feel deprived." Expressions such as "seeing as how" or "allowed as how" are supposed to give a folksy touch. They don't. They merely make the writer sound stupid.

## FOREIGN WORDS

A foreign expression has its place in the news report if it supplies a real need or flavor or has no precise native substitute (*blasé, chic, simpatico*). When editors come across foreign expressions they should be sure of the spelling, the use and the translation. Unless it is a commonly known expression, the editor provides the translation if the reporter has not done so.

The number in Latin words can cause trouble. For instance, *data* is plural and *datum* is singular. But *datum* is rarely used and *data* can be either singular (as a synonym for information) or plural (as a synonym for facts). *Trivia* is always plural; *bona fides* is always singular. *Media, criteria, insignia* and *phenomena* are plural.

Editors frequently are confronted with problems of translation, not as a rule directly from a foreign language but from a foreign correspondent's translation. Translations made abroad are often hurried; many are the work of men and women more at home in another language than in English. The translations may be accurate but not idiomatic.

An AP dispatch telling of a factory explosion in Germany read, "Most of the victims were buried when the roof of a large factory hall came down following the explosion.... The blast... damaged five other halls." What is a "factory hall"? The editor would have saved readers a puzzled moment if the dispatch had been changed to read, "Most of the victims were buried when the factory roof fell on them. The blast... damaged five other sections of the plant."

# CHAPTER 6

# EDITING FOR STYLE

## WHY STYLE MATTERS

More than one newspaper reporter has been known to say, "Let the copy desk people worry about style. It's not important to me." More than likely, that same reporter also can't spell and frequently makes grammatical errors. Style *is* important, and it is just as important in broadcasting as in print.

The writing in newspapers and magazines, and the writing for radio and television, must be clear, concise and free of annoying inconsistencies. Adherence to style rules provides the print media with a sense of consistency that would be absent if *goodbye* were used in one story and *goodby* in the next. Style rules provide the broadcast media with similar consistency and dictate conventions for reading the story aloud. What newspapers or magazines would write as *$50 million,* for example, becomes *FIFTY-MILLION DOLLARS* in a broadcast script.

Chapter 15 highlights the differences in print and broadcast style. But the basis of both are the rules in the stylebooks of The Associated Press and United Press International. At one time, the two wire services collaborated on their stylebooks. While that no longer is true, most of their rules remain identical.

Media outlets embrace wire service style as their own, although many have exceptions to cover the vagaries of local usage. A few large metropolitan newspapers, and a few large magazines, write their own stylebooks that differ substantially from the common rules. But in most cases, wire service rules prevail. For that reason, it's a good idea to commit as many of them as possible to memory.

Unfortunately, that's not an easy chore. The AP and UPI stylebooks are voluminous, and few editors can memorize their content. But it helps to learn the main rules of style—those that editors encounter almost daily in their work. That helps the editor minimize the amount of time spent looking things up in the stylebook or on the computer, where digitized versions of the stylebook now reside.

The best way to learn the main rules of style is to commit to memory as much as possible of Appendix I. There, with the permission of the wire services, we have organized the most important rules of style into categories that simplify the learning process—capitalization; abbreviations and acronyms; punctuation and hyphenation; numerals; and grammar, spelling and word usage. The purpose of this chapter is not to dwell on the rules listed there but to highlight some style-related problems.

## IDENTIFICATIONS

Shakespeare knew the value of a name. In *Othello* he had Iago say, "But he that filches from me my good name / Robs me of that which not enriches him / And makes me poor indeed." A name misspelled is a person misidentified. Of all the errors a newspaper is capable of making, one of the most serious is a misspelled or a misused name. In radio and television it is the mispronounced name.

One of the important roles of the editor is to make sure that all names in the copy are double-checked. The proper form is the form the person uses. It may be Alex rather than Alexander, Jim rather than James, Carin rather than Karen. The individual may or may not have a middle initial, with or without a period. Former President Harry S. Truman had no middle name. Some newspapers once used his middle initial without a period, but most now include it. Truman expressed indifference to the issue. Men's first names are seldom used alone, except in sports copy. The same should be true for first names of women.

Anyone resents an attempt at cleverness when a name is involved. Such "cuteness" should be felled on sight:

> Orange County will have a lemon as district attorney. Jack Lemon was elected to the job Tuesday.
>
> Of the patrolmen on the staff, two are crooks. (The last name was Crook.)

A short title generally precedes a name, but a long title follows. It is Susan F. Taylor, principal of Philip C. Showell School, rather than Philip C. Showell School Principal Susan F. Taylor.

Nor should the story make the reader guess at the identification. Here is an example:

> Albert A. Ballew took issue with Mayor Locher today for announcing in advance that the post of administrative assistant in the Safety Department will be filled by a black.
>
> The president of the Collinwood Improvement Council commended the mayor for creating the post, but added…

Now then, who is the president of the Collinwood Improvement Council? Will readers assume it is Ballew? The solution is simple: "Ballew, president of the Collinwood Improvement Council, commended the mayor."

In recent years editors have grappled with the difficult problem of whether to abandon use of the courtesy titles traditionally given to women. Most newspapers use the last name only to refer to men on second reference but use the courtesy titles *Miss, Mrs.* or *Ms.* on second reference to women. Feminists and others argue that men and women should be treated equally in news columns and insist that newspapers abandon this practice. For editors caught in the middle of this debate, it is a no-win situation. If they continue to use courtesy titles, feminists and others protest. If they drop courtesy titles, older people, in particular, are offended. Many elderly women prefer to be called *Miss* or *Mrs.,* followed by a family name. They consider it demeaning to be called by the last name only. The number of people who object to this practice is not large, however, so many editors have decided to drop courtesy titles altogether. A few newspapers, including *The Kansas City Star,* have done the opposite by using courtesy titles for men as well as women.

It is unlikely that courtesy titles will disappear altogether. In a story about a married couple, it is useful to distinguish between the two on second and subsequent references by using, for example, *Mr. Rodriguez* and *Mrs. Rodriguez.* Using first or given names would be cumbersome and in some cases may give the reader the impression that the reporter is "talking down" to the couple. Some newspapers also use courtesy titles in obituaries and editorials.

Newspaper style may dictate that titles belonging to certain positions be nonsexist *(chairperson* for *chairman),* but this rule has sometimes led to the absurd *(personkind* for *mankind).* The wire service stylebooks disallow forms like *chairman* unless they are formal titles. Editors can be fair to both sexes by eliminating purely sexist adjectives and by using plural pronouns when possible.

Newspaper style calls for females to be called girls until they are 18 and women thereafter. Males are boys until they are 18 and men thereafter. Appropriateness should determine whether males and females 18 and older or even those under 18 should be referred to by the familiar given name.

A married woman's original name can cause trouble: "He married the former Constance Coleman in 1931." This is incorrect; Constance Coleman was Constance Coleman when he married her. He married Constance Coleman. His wife is the former Constance Coleman.

*Woman* is used as a general descriptive possessive—woman's rights. *Women's* is used as a specific—women's club (but Woman's Christian Temperance Union). It is women fliers, Young Women's Christian Association, women workers, but woman suffrage. It is never *the Smith woman.*

Foreign names are tricky. In Spanish-speaking countries, individuals usually have two last names, the father's and the mother's—Adolfo Lopez Mateos. On second reference Lopez should be used. Newspapers have adopted the official Chinese spelling known as Pinyin. Thus, *Foochow* becomes *Fuzhou.* Familiar names of places and people, including Peking and Mao Tse-tung, still take the familiar American spelling.

In Arabic names, *al* generally is hyphenated—al-Sabah, al-Azhar. Some Arabs drop the article—Mamoun Kuzbari, not al-Kuzbari, but it should be used if the individual prefers. Compound names should be left intact—Abdullah, Abdel, Abdur. Pasha and Bey titles have been abolished. Royal titles are used with first names—Emir Faisel, Sheik Abdullah. *Haj* is used with the first name in both first and subsequent references—Haj Amin al-Hussein, Haj Amin.

The *U* in Burmese names means uncle, our equivalent of *Mr.,* or master. *Daw* means *Mrs.* or *Miss.* Many Burmese have only one name—U Thant. If a Burmese has two names, both should be used—U Tin Maung, Tin Maung.

Some Koreans put the family name first—Park Chung Hee. The second reference should be Park, not Chung Hee, the given name.

Many Indonesians have only one name—Sukarno, not Achmed Sukarno.

Swedish surnames usually end in *-son,* and Danish names usually end in *-sen.*

## TRADE NAMES

Few editors have escaped letters that begin something like this: "Dear Editor: The attached clipping from your paper of July 14th contains a mention of our product and we very much appreciate this unsolicited publicity. However, the name of our product was used with a lowercase 'c.' "

Makers of trade-name products want to protect their rights under the Lanham Trademark Act of 1947 and insist that in any reference to the name of the product the manufacturer's spelling and capitalization be used. This is to protect the trade name from becoming generic, as happened to aspirin, cellophane, escalator, milk of magnesia, zipper, linoleum and shredded wheat.

Much of the confusion and protest can be eliminated simply by using a generic term rather than the specific trade name—*petroleum jelly* for *Vaseline, freezer* for *Deepfreeze, fiberglass* for *Fiberglas, artificial grass* for *AstroTurf.* When the product is trade-named and there is no substitute, the trade name should be used, especially if it is pertinent to the story. Withholding such information on the ground of free publicity is a disservice to readers.

Institutions should also be labeled correctly—Lloyd's, not Lloyd's of London; J.C. Penney Co. (Penneys in ads, Penney's in other usages); American Geographical Society; National Geographic Society.

## RELIGION

Jewish congregations should be identified in news stories as Orthodox, Conservative or Reform. To help readers, the editor should insert "branch of Judaism" or whatever other phrase might be necessary to convey the proper meaning. The terminology of the congregation concerned should be followed in naming the place of worship as a temple or a synagogue. Most Orthodox congregations use *synagogue.* Reform groups use *temple* and Conservative congregations use one word or the other, but *synagogue* is preferred. The generic term is *Jewish houses of worship.* It is never *church,* which applies to Christian bodies.

*Sect* has a derogatory connotation. Generally it means a church group espousing Christianity without the traditional liturgical forms. *Religion* is an all-inclusive word for Judaism, Islam, Christianity and others. *Faith* generally is associated with Protestants. *Denomination* should be used only when referring to the church bodies within the Protestant community.

Religious labels can be misleading. *Jews* and *Judaism* are general terms. Israelis are nationals of the state of Israel, and Jews are those who profess Judaism. The state of Israel is not the center of or the voice of Judaism. Some Jews are Zionists; some are not.

Not all denominations use *Church* in the organization's title. It is the First Baptist Church but the American Baptist Convention. It is the Church of Jesus Christ of Latter-day Saints (but Mormon Church is acceptable); its units are missions, stakes and wards. It is the Episcopal Church, not the Episcopalian Church. Its members are Episcopalians, but the adjective is *Episcopal:* Episcopal clergymen.

Mass may be *celebrated, said* or *sung.* The rosary is *recited* or *said.* The editor can avoid confusion by making the statement read something like this: "The mass (or rosary) will be at 7 p.m." The Benediction of the Blessed Sacrament is neither *held* nor *given;* services close with it.

The order of the Ten Commandments varies depending on the version of the Bible used. Confusion can be spared if the commandment number is omitted. Also to be deleted are references to the burning of a church mortgage unless there actually is a burning ceremony. It is an elegant but ridiculous way of saying the mortgage has been paid off.

The usual style in identifying ministers is *the Rev.,* followed by the individual's full name on first reference and only the surname on second reference. If the minister has a doctorate, the style is *the Rev. Dr.,* or simply *Dr.* on subsequent references. *Reverend* should not be used alone, nor should plural forms be used, such as *the Revs. John Jones and Richard Smith.* Churches of Christ do not use the term *reverend* in reference to ministers. They are called *brothers.*

Rabbis take *Rabbi* for a title. Priests who are rectors, heads of religious houses or presidents of institutions and provinces of religious orders take *the Very Rev.* and are addressed as *Father.* Priests who have doctorates in divinity or philosophy are identified as *the Rev. Dr.* and are addressed either as *Dr.* or *Father.* For further guidance, consult the stylebook, which contains rules pertinent to usages in all religions.

The words *Catholic* and *parochial* are not synonymous. There are parochial schools that are not Catholic. The writer should not assume that a person is a Roman Catholic simply because he is a priest or a bishop. Other religions also have priests and bishops.

Not all old churches merit the designation of *shrine.* Some are just old churches. *Shrine* denotes some special distinction, historic or ecclesiastical. Usually, shrines are structures or places that are hallowed by their association with events or persons of religious or historic significance, such as Lourdes or Mt. Vernon.

Use *nun* when appropriate for women in religious orders. The title *sister* is confusing except with the person's name *(Sister Mary Edward).*

## DEATH STORIES

People die of heart *illness,* not *failure;* after a *long* illness, not an *extended* illness; *unexpectedly,* not *suddenly; outright,* not *instantly; following* or *after* an operation, not *as a result of* an operation; *apparently of a heart attack,* not *of an apparent heart attack.* A person dies *of* a disease, not *from* a disease.

The age of the person who died is important to the reader. The editor should check the age given against the year of birth. Generally, the person's profession or occupation, the extent of the illness and the cause of death are recorded, but without details. The length of the story is dictated by the fame of the person. Winston Churchill's obit ran 18 pages in *The New York Times.*

A person *leaves* an estate; that person is *survived* by a family. A man is survived by his *wife,* not his *widow.* A woman is survived by her *husband,* not her *widower.* A man or woman is survived by *children* if they are children and by *sons* and *daughters* if they are adults.

If the family requests that the story include the statement that memorial donations may be made to a particular organization, the statement should be used. Whether such a statement should contain the phrase "in lieu of flowers" is a matter of policy. A few papers, in deference to florists, do not carry the phrase, but that practice is widely discredited as currying favor with advertisers.

A straightforward account of a death is better than a euphemistic one. The plain terms are *died,* not *passed away* or *succumbed; body,* not *remains* or *corpse; coffin,* not *casket; funeral* or *services,* not *obsequies; burial,* not *interment,* unless interred in a tomb above the ground. Flowery expressions such as "two of whom reside in St. Louis" and "became associated with the company shortly after college" show no more respect for the dead than do the plain expressions "live in St. Louis" or "went to work for the company."

Few stories in a newspaper are more likely to be written by formula than the obituary. There isn't much the desk can do about the conventional style except to contrast it with those that take a fresh approach. Here is a lead from an Associated Press story:

> NEW YORK (AP)—If you are a movie fan, you will remember Mary Boland as the fluttery matron, the foolishly fond mother, the ladylike scatterbrain.
>
> The character actress who died Thursday at 80 was none of these in real life.

The editor should be on guard for the correct spelling of all names used in the death story and for slips such as *cemetary* for *cemetery* and *creamation* for *cremation.* Errors are inexcusable:

> A postmorten failed to disclose the cause of death because the girl's body was too badly decomposed. (The correct spelling is *post-mortem,* or, preferably, use *autopsy.*)
>
> Thousands followed the cortege.
> (The thousands must have been *in,* not *following,* the cortege[the funeral procession]).

Even after death, a medal won by a serviceman is awarded to him. It may be *presented* to his widow, but it is not *awarded* to his widow.

If the service is at a mortuary, the name of the mortuary should be included for the convenience of mourners. A funeral is *at* a place, not *from* it. A funeral mass usually is *offered;* a service is *held.* Even so, *held* often is redundant: "The service will be at 2 p.m. Thursday." The passage should leave no doubt for whom the service was held. This one did: "Services for 7-year-old Michael L—, son of a Genoa Intermediate School District official who was struck and killed by a car Monday in Bay City, will be held. . . ."

People are *people,* not *assaults* or *traffic deaths* or *fatals* or *dead on arrivals:*

> A youth stabbed at a downtown intersection and a woman pedestrian run down by a car were among assaults on six persons reported to police during the night.
>
> **HUGO MAN AMONG NINE TRAFFIC DEATHS**
>
> Dead on arrival at Hurley after the crash was Oscar W. . .

The events in a person's life that should be included in the obituary pose a problem for the desk. One story told of a former school administrator who died at 87. It said he had been the first principal of a high school and had served in that capacity for 17 years. Then the story noted that he resigned two months before he was found guilty of taking $150 from the school yearbook fund and was fined $500. Should an account of a minor crime committed a quarter century ago be included in the obituary? To those who knew the former principal intimately, the old theft was not news. Those who didn't know him personally probably would not care about the single flaw in an otherwise distinguished career. Sometimes, however, it is necessary to include unsavory details in obituaries. Good editors do not hesitate to do so.

## MEDICAL NEWS

Reporters and editors have no business playing doctor. If a child is injured in an accident, the seriousness of the injury should be determined by medical authorities. To say that a person who was not even admitted to the hospital was "seriously injured" is editorializing.

Hospitals may report that a patient is in a "guarded condition," but the term has no meaning for the reader and should be deleted. The same goes for "he is resting comfortably."

A story described a murder suspect as "a diabetic of the worst type who must have 15 units of insulin daily." The quotes were attributed to the FBI. An editor commented, "In my book, that is a mild diabetic, unless the story means that the suspect is a diabetic who requires 15 units of regular insulin before each meal. A wire service should not rely on the FBI for diagnosis of diabetes and the severity of the case."

*Doctor* and *scientist* are vague words to many readers. *Doctor* may be a medical doctor, a dentist, a veterinarian, an osteopath, a minister or a professor. The story would be clearer if it named the doctor's specialty or the scientist's specific activity, whether biology, physics, electronics or astronautics. Here are more examples of preferred usage in medical reports:

- One may not sustain a "fractured leg," which seldom causes death, but one may sustain a heart attack. A person may suffer a *fracture of the leg* or a *leg fracture* or, better still, simply a *broken leg.* Injuries are *suffered* or *sustained,* not *received.*
- Medical doctors diagnose the *illness,* not the patient. The proper term to use in determining the remedy or in forecasting the probable course and termination of a disease is *prognosis.*
- Mothers are *delivered;* babies are *born.*
- Everyone has a temperature. *Fever* describes above-normal temperature.
- Everyone has a heart condition. It is news only if someone's heart is in bad condition.
- "The wife of the governor underwent major surgery, and physicians reported she apparently had been cured of a malignant tumor." It is unlikely that any doctor said she was *cured* of a malignant tumor. They avoid that word with malignant growths.
- "A team of five surgeons performed a hysterectomy, appendectomy and complete abdominal exploration." Why the unnecessary details? It would have been enough to say, "Five surgeons performed the abdominal operation."
- Unless they are essential to the story, trade names of narcotics or poisons should be avoided. If a person dies of an overdose of sleeping pills, the story should not specify the number of pills taken.
- Use *Caesarean section* or *Caesarean operation.*
- Usually, no sane person has a leg broken. Use the passive voice: "Her leg was broken."

- Use the expression *physicians and dentists,* not *doctors and dentists.* The second suggests that dentists are not doctors.
- A doctor who specializes in anesthesia is an anesthesiologist, not an anesthetist.
- A person may wear a sling on his or her right arm. That person doesn't wear his or her right arm in a sling.
- "He suffered a severed tendon in his right Achilles' heel last winter." It was the Achilles' tendon in his right heel or his right Achilles' tendon.
- "A jaundice epidemic also was spreading in Gaya, Indiana health officials said. The disease claimed 30 lives." Jaundice is not a disease but a sign of the existence of one or another of a great many diseases.
- Technical terms should be translated:

| TERM | TRANSLATION | TERM | TRANSLATION |
|---|---|---|---|
| Abrasion | Scrape | Suturing | Sewing |
| Contusion | Bruise | Hemorrhaging | Bleeding |
| Lacerations | Cuts | Obese | Overweight |
| Fracture | Break | Respire | Breathe |

## WEATHER

An editor said, "Ever since the National Weather Service started naming hurricanes after females, reporters can't resist the temptation to be cute." He then cited as an example the lead, "Hilda—never a lady and now no longer a hurricane—spent the weekend in Louisiana, leaving behind death, destruction and misery." That, the editor said, is giddy treatment for a disaster causing 35 deaths and millions of dollars in property damage. The potential for passages in poor taste is not lessened by the fact that men's names are now used for hurricanes, too.

Another editor noted that a story referred to "the turbulent eye of the giant storm." Later in the story the reporter wrote that the eye of the hurricane is the dead-calm center.

A story predicted that a hurricane was headed for Farmington and was expected to cause millions of dollars in damage. So the Farmington merchants boarded their windows, the tourists canceled their reservations and the hurricane went around Farmington. This is the trouble when an editor lets a reporter expand a prediction into a warning.

The headline **FREEZE TONIGHT EXPECTED TO MAKE DRIVING HAZARDOUS** was based on this lead: "Freezing temperatures forecast for tonight may lead to a continuation of hazardous driving conditions as a result of Monday night's snow and freezing rain." The story did not mention that there would still be dampness on the

## *Using Percentages and Numbers*

Two types of errors appear frequently in stories dealing with percentages. One is the failure to distinguish between percentage and percentage points; the other lies in comparing the change with a new figure rather than the original one. For example, when a tax rate is increased from $5 per $100 of assessed valuation to $5.50, the increase is 10 percent. The new figure less the old figure is divided by the old figure:

$$\$5.50 - \$5 = .50; \frac{.50}{\$5} = .10 \text{ or } 10 \text{ percent}$$

"Jones pointed out that the retail markup for most other brands is approximately 33 percent, whereas the markup on Brand J is 50 percent, or 17 percent higher." No. It is 17 percentage points higher but 51.5 percent higher. Divide 17 by 33.

"Dover's metropolitan population jumped from 16,000 just 10 years ago to more than 23,000 last year, an increase of better than 70 percent." Wrong again. It's a little less than 44 percent.

Is the figure misleading or inaccurate? The story said, "A total of $6,274 was raised at each of the four downtown stations." This adds up to $25,096. The writer intended the sentence to mean, "A total of $6,274 was raised at four stations." A not-so-sharp editor let this one get by: "Almost 500,000 slaves were shipped in this interstate trade. When one considers the average price of $800, the trade accounted for almost $20 million."

Are terms representing figures vague? In inheritance stories, it is better to name the amount and let the reader decide whether the amount is a "fortune." One editor noted, "Fifteen thousand might be a fortune to a bootblack, but $200,000 would not be a fortune to a Rockefeller."

For some reason, many stories contain gambling odds, chances and probabilities. When a princess gave birth to a son, reporters quickly latched on to the odds on his name. Anthony was 1-2, George was even money and Albert was 3-1. One headline played up the third in the betting. The name chosen was William.

All this shows the foolishness of editors who play into the hands of gamblers. If odds must be included in the story, they should be accurate. The story said that because weather records showed that in the last 86 years it had rained only 19 times on May 27, the odds were 8-1 against rain for the big relay event. The odds mean nothing to readers except to those who like to point out that in the story just mentioned the odds actually were 3 ½-1.

"Dr. Frank Rubovits said the children came from a single egg. He said the chance of this occurring 'probably is about 3 million to 1.' " He meant the odds against this occurring. The chance of this occurring is 1 in 3 million.

"The Tarapur plant will be the world's second largest atomic generator of electricity. The largest will be the 500-ton megawatt plant at Hinkley Point in Britain." A 500-ton megawatt plant makes no sense. What the writer meant was a 500-megawatt plant.

Some readers may rely on the idiom and insist that "five times as much as" means the same as "five times higher than" or "five times more than." If so, five times as much as $50 is $250 and five times higher than $50 is still $250. Others contend that the second should be $300. If earnings this year are 3 ½ times as large as last year's, they are actually 2 ½ times larger than last year's.

Insist on this style: 40,000 to 50,000 miles, not 40 to 50,000 miles; $3 million to $5 million, not $3 to $5 million.

"The committee recommended that a bid of $26,386.60 be accepted. After recommending the higher bid, the committee also had to recommend that an additional $326.60 be appropriated for the fire truck, because only $26,000 was included in the budget." The sum is still $60 short of the bid.

Equivalents should be included in stories that contain large sums. Most readers cannot visualize $20 billion, but they can understand it if they are

*(Continued)*

## *Using Percentages and Numbers (continued)*

told how much the amount would mean to each individual.

Here is one editor's advice to the staff:

> We can do a service for those important people out there if we use terms they are most acquainted with. For example, to most of our readers a ton of corn is more easily visualized if it is reported as bushels, about 36 in this case. We normally report yields and prices in bushels and that is the measurement most readers know. The same goes for petroleum; barrels is probably more recognizable than tons. When the opportunity presents itself, translate the figures into the best-known measurement.

Nothing is duller or more unreadable than a numbers story. If figures are the important part of the story, they should be related to something—or at least presented as comparisons.

Two of the most common mathematical errors in news copy are the use of millions for billions, and vice versa, and a construction such as "Five were injured..." when only four names are listed.

ground when freezing temperatures arrived. There wasn't, and driving was unimpeded.

A lead said, "One word, 'miserable,' was the National Weather Service's description today of the first day of spring." The head was **SNOW PREDICTED; IT'S SPRING, MISERABLE.** It was the weatherman's prediction, not his description. The sun stayed out all day, the clouds stayed away, and readers of the paper must have wondered where this forecaster was located.

In flood stories, the copy should tell where the flood water came from and where it will run off. The expression *flash flood* is either a special term for a rush of water let down a weir to permit passage of a boat or a sudden destructive rush of water down a narrow gully or over a sloping surface in desert regions, caused by heavy rains in the mountains or foothills. It is often used loosely for any sudden gush of water.

Weather stories, more than most others, have an affinity for the cliché, the fuzzy image, overwriting, mixed metaphors, contrived similes and other absurdities. For example:

> A Houdini snow did some tricks Thursday that left most of the state shivering from a spine-tingling storm.

Houdini gained fame as an escape artist, not as an ordinary magician. Did the snow escape, or was it just a tricky snow? After the lead, the 22-inch story never mentioned the angle again. *Spine-tingling* means full of suspense or uncertainty or even terror. Sports writers are fond of using it to describe a close game, called a *heart stopper* by more ecstatic writers, often in conjunction with a *gutsy performance.* If a cliché must be used, a very cold storm is *spine-chilling,* not *spine-tingling.*

> Old Man Winter Wednesday stretched his icy fingers and dumped a blanket of snow on the state.

How would reporters ever write about the weather without Old Man Winter, Jack Frost, Icy Fingers and Old Sol? Why do rain and snow never *fall?* They are always *dumped.*

> At least two persons were killed in Thursday's snowstorm, marked at times by blizzardlike gales of wind.

By Weather Service standards, this is an exaggeration and a contradiction. By any standard, it is a redundancy. A blizzard is one thing. Gales are something else. Gales of wind? What else, unless maybe it was gales of laughter from discerning readers.

An editor's moral: Good colorful writing is to be encouraged. But a simply written story with no gimmicks is better than circus writing that goes awry. To quote a champion image maker, Shakespeare, in Sonnet 94, "Lilies that fester smell far worse than weeds."

Blizzards are hard to define because wind and temperatures may vary. The safe way is to avoid calling a snowstorm a *blizzard* unless the Weather Service describes it as such. Generally, a blizzard occurs when there are winds of 35 mph or more that whip falling snow or snow already on the ground and when temperatures are 20 degrees above zero Fahrenheit, or lower.

A severe blizzard has winds that are 45 mph or more, temperatures 10 degrees above zero or lower and great density of snow either falling or whipped from the ground.

The Weather Service insists that ice storms are not sleet. Sleet is frozen raindrops. The service uses the terms *ice storm, freezing rain* and *freezing drizzle* to warn the public when a coating of ice is expected on the ground. The following tips will help editors use the correct terms:

- Temperatures can become *higher* or *lower,* not *cooler* or *warmer.*
- A *cyclone* is a storm or system of winds rotating about a moving center of low atmospheric pressure. It is often accompanied by heavy rain and winds.
- A hurricane has winds above 74 mph.
- A *typhoon* is a violent cyclonic storm or hurricane occurring in the China Seas and adjacent regions, chiefly from July to October.
- The word *chinook* should not be used unless so designated by the Weather Service.

Here is a handy table for referring to wind conditions:

| | |
|---|---|
| Light | up to 7 mph |
| Gentle | 8 to 12 mph |
| Moderate | 13 to 18 mph |
| Fresh | 19 to 24 mph |
| Strong | 25 to 38 mph |
| Gale | 39 to 54 mph |
| Whole gale | 55 to 75 mph |

Temperatures are measured by various scales. Zero degrees Celsius is freezing, and 100 degrees Celsius is boiling. Celsius is preferred over the older term *centigrade.* On the Fahrenheit scale, 32 degrees is freezing, and 212 degrees (at sea level) is boiling. On the Kelvin scale, 273 degrees is freezing, and 373 degrees is boiling. To convert degrees Celsius to Fahrenheit, multiply the Celsius measurement by nine-fifths and add 32. To convert degrees Fahrenheit to Celsius, subtract 32 from the Fahrenheit measurement and multiply by five-ninths. Thus, 10 degrees Celsius is 50 degrees Fahrenheit. To convert degrees Kelvin to Celsius degrees, subtract 273 from the Kelvin reading. (Kelvin is used for scientific purposes, not for weather reporting.)

Avoid these weather clichés:

| | |
|---|---|
| Fog rolled (crept or crawled) in | Fog-shrouded city |
| Winds aloft | Mercury dropped (dipped, zoomed, plummeted) |
| Rain failed to dampen | Biting (bitter) cold |
| Hail-splattered | Hurricane howled |
| Storm-tossed | |

## DISASTER

Conjecturing about possible damage to settlements from forest fires is as needless as conjectures on weather damage. The story should concentrate on the definite loss. Stories of forest fires should define the specific area burned, the area threatened and the type of timber. Be wary of death estimates because officials tend to overstate during the initial shock of the event.

Most stories of earthquakes attempt to describe the magnitude of the tremor. One measurement is the Richter scale, which shows relative magnitude. It starts with magnitude 1 and progresses in units with each unit 10 times stronger than the previous one. Thus, magnitude 3 is 10 times stronger than magnitude 2, which in turn is 10 times stronger than magnitude 1. On this scale the strongest earthquakes recorded were the South American earthquake of 1906 and the Japanese earthquake of 1933, both at a magnitude of 8.9. Intensity generally refers to the duration or to the damage caused by the shock.

In train and plane crashes the story should include the train or flight number, the place of departure, the destination, and times of departure and expected arrival. Airplanes may collide on the ground or in the air (not *midair*). Let investigators *search* the wreckage, not *comb* or *sift* it.

In fire stories, the truth is that in nine of 10 cases when people are "led to safety," they're not. Except for an occasional child or infirm adult, they simply have the common sense to leave the building without waiting for a firefighter to "lead them to safety." Eliminate terms such as *three-alarm fire* and *second-degree burns* unless they are explained.

In both fire and flood stories the residents of the area are rarely taken from their homes or asked to leave. Instead, they're always told to "evacuate" or they are "evacuated." What's wrong with *vacate?*

"An estimated $40,000 worth of damage was done Jan. 29." Damage isn't worth anything. Quite the contrary.

"The full tragedy of Hurricane Betsy unfolded today as the death toll rose past 50, and damages soared into many millions." *Damage* was the correct word here. You collect *damages* in court.

Avoid these disaster clichés:

| | |
|---|---|
| Rampaging rivers | Weary firefighters |
| Fiery holocaust | Flames licked (leaped, swept) |
| Searing heat | Tinder-dry forest |
| Raging brush fire | Traffic fatals or triple fatals (police station jargon) |

## LABOR DISPUTES

Stories of labor controversies should give the reasons for the dispute, the length of time the strike has been in progress and the claims by both the union and the company. Editors should be on guard against wrong or loaded terms. Examples: In a *closed shop* the employer may hire only those who are members of the union. In a *union shop* the employer may select employees, but the workers are required to join the union within a specified time after starting work. A *conciliator* or *mediator* in a labor dispute merely recommends terms of a settlement. The decision of an *arbitrator* usually is binding. There is a tendency in labor stories to refer to management proposals as *offers* and to labor proposals as *demands.* The correct word should be used for the correct connotation.

*Strikebreaker* and *scab,* which are loaded terms, have no place in the news if used to describe men or women who individually accept positions vacated by strikers. The expression "honored the picket line" frequently appears in the news even though a more accurate expression is "refused (or declined) to cross a picket line." *Union leader* is usually preferred to *labor leader.* A *stevedore* usually is considered an employee.

On estimates of wages or production lost, the story should have authoritative sources, not street-corner guesses. An individual, however voluble, does not speak for the majority unless authorized to do so. Statements by workers or minor officials should be downplayed until they have been documented.

If a worker gets a 10-cent-an-hour increase effective immediately, an additional 10 cents a year hence and another 10 cents the third year, that worker does not receive a 30-cent-an-hour increase. The increase at the time of settlement is still 10 cents an hour. It also is common to read, "The company has been on strike for the last 25 days." No. The employees are on strike. The company has been struck.

Criminal court terms should not be applied to labor findings unless the dispute has been taken to a criminal court. The National Labor Relations Board is not a court, and its findings or recommendations should not be expressed in criminal court terminology. In most settlements, neither side is *found guilty* or *fined.* A finding or a determination may be made, or a penalty may be assessed.

## FINANCIAL NEWS

A news release from a bank included the following: "The book value of each share outstanding will approximate $21.87 on Dec. 31, and if the current yield of 4.27 percent continues to bear the same relationship to the market price, it should rise to $32 or $33, according to..." The editor changed the ambiguous "it" to "the book value." Actually, the release intended "it" to refer to the market price, which shows what can happen when editors change copy without knowing what they are doing.

Another story quoted an oil company official as saying that "the refinery would produce $7 million in additional real estate taxes." It should have been obvious that this was a wholly unrealistic figure, but for good measure there was an ad in the same paper that placed the total tax figure at about $200,000 and read, "The initial installation will add about $7 million a year to the economy of the state, not including taxes."

A story and headline said the interest on the state debt accounted for 21 percent of the state government's spending. An accompanying graph showed, however, that the figure was for debt service, which includes both interest and amortization, or payments on principal.

All who edit copy for financial pages should have at least elementary knowledge of business terms. If they can't distinguish between a balance sheet and a profit-and-loss statement, between earnings and gross operating income, and between a net profit and net cash income, they have some homework to do.

This reality was emphasized by a syndicated financial columnist who cautioned business news desks against using misleading headlines such as **STOCKS PLUMMET–DOW JONES AVERAGE OFF 12 POINTS.** It may be a loss, the columnist noted, but hardly a calamity. The Dow Jones industrial average may indicate that the market is up, but in reality it may be sinking. Freak gains by a few of the 30 stocks used in compiling the Dow average may have pushed up that particular indicator. Nor does a slight market drop call for a headline such as **INVESTORS LOSE MILLIONS IN MARKET VALUE OF STOCKS.** They lost nothing of the sort. On that day, countless investors the nation over had substantial paper profits on their stocks. If they sold, they were gainers on the buying price in real terms; if they held, they had neither gains nor losses.

The Dow Jones average is one of several indexes used to gauge the stock market. Each uses its own statistical technique to show market changes. The Dow Jones bases its index on 30 stocks. It is an index-number change, not a percentage change.

Reporters of dividends should use the designation given by the company (regular, special, extra, increases, interim) and show what was paid previously if there is no specified designation such as regular or quarterly. The story should say if there is a special, or extra, dividend paid with the regular dividend and include the amount of previous added payments. When the usual dividend is not paid, or reduced, some companies issue an explanatory statement, the gist of which should be included in the story.

Wire service stylebooks recommend that news of corporate activities and business and financial news be stripped of technical terms. There should be some explanation of the company's business (plastics, rubber, electronics) if there is no indication of the nature of the business in the company's name. The location of the company's home office should be included.

Savings and loan companies object to being called banks. Some commercial banks likewise object when savings and loan companies are called banks. There need be no confusion if the institution is identified by its proper name. In subsequent references the words *company* or *institution* are used. Some newspapers permit **S&L FIRM** in tight headlines. Actually a *firm* is a partnership or unincorporated group. It should not be used for an incorporated company. *Concern* is a better word for the latter.

Jargon has no place in the business story. "Near-term question marks in the national economy—either of which could put a damper on the business expansion—are residential housing and foreign trade, the Northern Trust Co. said in its

December issue of 'Business Comments.'" Are near-term question marks economy question marks? If so, can they put a damper on anything? Isn't all housing residential?

This jargon-filled story should have been heavily edited:

> Major producers scrambled today to adjust steel prices to newly emerging industrywide patterns. . . . The welter of price changes was in marked contrast with the old-time industry practice of posting across-the-board hikes.
>
> This approach apparently breathed its last in April when it ran into an administration buzz saw, and a general price boost initiated then by United States Steel Corp., the industry giant, collapsed under White House fire.

Readers of financial pages read for information. False color is not needed to retain these readers. This story should have been edited heavily on the desk:

> Stock of the Communications Satellite Corp. went into an assigned orbit Friday on three major stock exchanges, rocketing to an apogee of $46 a share and a perigee of $42 and closing at $42.37, unchanged.
>
> It was the first day of listed trading on the exchanges. The stock previously was traded over-the-counter.
>
> The countdown on the first transaction on the New York Stock Exchange was delayed 12 minutes by an initial jam of buy and sell orders.

## SHIPS AND BOATS

Do not use nautical terms unless they're used properly. "Capt. Albert S. Kelly, 75, the pilot who manned the Delta Queen's tiller Monday . . ." What he manned was the helm or wheel. Few vessels except sailboats are guided with a tiller.

A story referred to a 27-foot ship. Nothing as small as 27 feet is a *ship. Ship* refers to big seagoing vessels such as tankers, freighters and ocean liners. Sailors insist that if it can be hoisted onto another craft it is a boat, and if it is too large for that it is a ship. Specific terms such as *cabin cruiser, sloop, schooner, barge* and *dredge* are appropriate.

"A rescue fleet ranging from primitive bayou pirogues to helicopters prowled the night." That should send the editor to a dictionary so he or she can explain to readers that a *pirogue* is a canoe or a dugout. Better yet, simply use the more common term.

"The youths got to the pier just before the gangplank was lowered." When a ship sails, the gangplank is *raised.*

Commercial ships are measured by volume, the measurement of all enclosed space on the ship expressed in units of 100 cubic feet to the ton. Fuller description gives passenger capacity, length and age. The size of vessels is expressed in tonnage, the weight in long tons of a ship and all its contents (called displacement).

A long ton is 2,240 pounds. All this is confusing to many readers. Editors should translate into terms recognized by readers, who can visualize length, age and firing power more readily than tonnage: "The 615-foot Bradley, longer than two football fields, . . ."

A *knot* is a measure of speed, not distance (nautical miles an hour). A nautical mile is about $1\frac{1}{7}$ land miles. *Knots per hour* is redundant.

## SPORTS

Some of the best writing in American newspapers appears in the sports pages. So does some of the worst. Sports pages should be, and are, the liveliest in the paper. They have action photos, a melange of spectator and participant sport and an array of personalities. Sports writers have more latitude than do other reporters. The good ones are among the best in the business; the undisciplined ones are among the worst.

Attractive pages and free expression mean little if the sports section is unintelligible to half the paper's readers. Too often, the editing reflects the attitude that if readers don't understand the lingo they should look elsewhere in the paper for information and entertainment.

The potential for attracting readers of the sports section is greater than ever because of the growing number of participants in golf, bowling, fishing, boating, soccer and tennis. The spectator sports, especially automobile racing, football, golf, basketball, baseball and hockey, attract big audiences, thanks to the vast number of television viewers. Thus the sports pages, if edited intelligently, can become the most appealing section in the paper. But first, writers and editors must improve their manners.

A report of a contest or struggle should appeal to readers if it is composed in straightforward, clear English. The style can be vigorous without being forced. Sports fans do not need the sensational to keep their interest whetted. Those who are only mildly interested won't become sports page regulars if the stories are confusing.

### Know the Game

One of the elementary rules in sports writing is to tell the reader the name of the game. Yet many stories talk about the Cubs and Pirates but never say specifically that the contest is a baseball game. Some writers assume that if the story refers to the contest as a "dribble derby," all sports page readers must understand that the story concerns a basketball game. It could be soccer, to which the term also applies.

The story may contain references to parts of the game yet never mention the specific game. Here is an example:

---

**THREE TEAMS TIED IN SLICEROO**

Three teams tied for low at 59 in the sweepstakes division as the 11th annual Sliceroo got under way Thursday at Lakewood Country Club.

Deadlocked at 59 were the teams of . . .

In the driving contest, it was . . .

In the putting and chipping contest . . .

A best ball is set for Friday and a low net for Saturday, final day of the Sliceroo. A $5,000 hole-in-one competition on the 124-yard 11th hole is set for both final days.

---

Golfers will understand the story. But nongolfers, even many who enjoy watching golf matches on television, should be told outright that the story concerns a golf tournament. The added information would not offend the golfers. It might encourage a nongolfer to read on.

Some stories fail to state categorically who played whom. The writer assumes that if the opponents' managers are named, all hard-core sports readers will recognize the contestants. Perhaps so, but the casual reader might like to know, too. The caption under a two-column picture read, "They can't believe their eyes. Coach Andy S—, left, and Manager May S—, right, showed disbelief and disgruntlement as the Braves belt Pitcher Don C— for five runs in the eighth inning of their exhibition baseball game Wednesday at Clearwater, Fla. The Braves won, 10-2." Now, whom did the Braves play?

Not all readers understand the technical terms used to describe a sports contest. It might be necessary to explain that a seeded team gets a favored placement in the first round, and that if Smith beats Jones 2-1 in match play it means that golfer Smith is two holes ahead of golfer Jones with only one hole left to play and is therefore the winner. The name of the sport should be used in reference to the various cups. The Davis Cup is an international trophy for men tennis players. The Heisman Trophy is an award presented annually to the outstanding college football player in the nation. America's Cup refers to yachting, and America Cup to golfing.

## Unanswered Questions

Answering more questions is one way to win more readers for the sports department. The key questions frequently overlooked are how and why. Why did the coach decide to punt on fourth down instead of trying to make one foot for a first down? How do tournament organizers get the funds to award $200,000 in prizes?

"The shadow of tragedy drew a black edge around a golden day at Sportsman's Park Thursday, bringing home the danger of horse racing with an impact that cut through the $68,950 Illinois Derby like a spotlight in darkness." So, what happened?

The best training for editors on the sports desk is a stint on the news copy desk. But before they go on the sports desk they should become familiar with the intricacies of all sports so they can catch the technical errors in sports copy. Here are examples:

"Center fielder Tony Cafar, whose fine relay after chasing the ball 'a country mile' held Ripley to a triple . . ." Unless Tony also made a throw of "a country mile," another player, a shortstop or second baseman, made the relay throw after taking a good throw from Tony.

When a writer covering a basketball game refers to a "foul shot," the reference should be deleted on the sports desk rim. The fouled player gets a free throw from the free-throw line, not the foul line or the charity line.

The editor also has to be alert for some of the wild flights of imagination used by sports writers. "The Tar Heels hurdled their last major obstacle on the way to an unbeaten season but still had a long row to hoe." Is this a track meet or a county fair?

The following passage is a sure way to discourage sports page readers:

> His trouble in Sunday's 27-6 victory over the Dallas Cowboys before 72,062, largest crowd ever to see the Chiefs, was the reason Moore was in the trenches to receive a shattering kick with 6:24 left in the game.

This example suggests another tendency in sports copy—turning the story into a numbers game. Box scores, league standings and records have a place in the sports story, but generally they should have a subordinate rather than a dominant role.

## Abbreviated Sports

The addiction to abbreviation is strong in both sports copy and headlines. **BRONCOS GET FIRST AFC WIN OVER NFC,** announces a headline. It means that Denver's professional football team scored the first victory of the season by an American Football Conference team over a National Football Conference team.

## "Color" in Sports

An editor told his colleagues, "There is nothing more exciting than a good contest. There is nothing duller than reading about it the next day." Yet many spectators who watch a Saturday contest can't wait to read about it the next morning. What were the coaches' and players' reactions to the game? What was the turning point? How long was the pass that won the game? What's the reporter's comment on the crowd's behavior?

If there is an audience for the report of a contest, the story needs no special flourishes. Loaded terms such as *wily mentor, genial bossman, vaunted running game, dazzling run* and *astute field general* add little or nothing to the story. Adjectives lend false color. The Associated Press reported a "vise-tight race," "the red-hot Cardinals" and the "torrid 13-4 pace" as if these modifiers were needed to lure readers.

If all editors were permitted to aim their pencils at copy submitted by the prima donnas of the sports world, there would be no sentences like the following ones:

> Jones, who has been bothered by a sore rib this spring, played eight innings Tuesday, collected two singles and drove in a run. (He didn't collect them; he hit them.)
>
> Ortiz threw his first bomb in the second round when he nailed Laguna with a left and right to the jaw. (How can you nail something with a bomb?)
>
> Left-hander Glavine, who started on the mound for Atlanta, recovered from a shaky start and pitched six-hit ball for eight innings before a walk and a botched-up double play caused the manager to protect a 4-1 lead, as a result of a three-run homer by Dave Justice in the seventh.
> (The sentence is hard to understand because it is overstuffed and because the facts are not told in chronological order. Revised: "Left-hander Glavine started on the mound for Atlanta. He recovered from a shaky start and pitched six-hit ball for eight innings. Then a walk and a missed double play caused the manager to bring in a new pitcher. Atlanta was leading, 4-1, as a result of a three-run homer by Dave Justice in the seventh.")

Here is how to tell a story upside down:

> Waldrop's 17-yard explosion for his eighth touchdown of the season punctuated a 65-yard march from Army's reception of the kickoff by a courageous Falcon team, which had gone ahead for the second time in the game, 10-7, in the ninth minute of the final period.

## Synonym Sickness

Not even the best editing can convert some sports writers into Homers and Hemingways. At least, though, editors can try to help writers improve their ways of telling a story.

Editors can excise clichés such as paydirt, turned the tables, hammered (or slammed) a homerun, Big Ten hardwoods, circuit clout, gridder, hoopster, thinclads, tanksters, sweet revenge, rocky road, freeloads (free throws), droughts (losing streaks), standing-room-only crowds, put a cap on the basket, as the seconds ticked off the clock, unblemished records, paced the team, outclassed but game, roared from behind, sea of mud, vaunted defense, coveted trophy and last-ditch effort.

They can insist on the correct word. Boxers may have *altercations* (oral) with their managers; they have *fights* with other boxers.

They can tone down exaggerated expressions like "mighty atom of the ring," "destiny's distance man" or "Northwestern comes off tremendous effort Monday"

(Northwestern tried hard). They can cut out redundancy in phrases such as "with 30,000 spectators looking on."

They can resist the temptation to use synonyms for verbs *wins, beats* and *defeats:* annihilates, atomizes, batters, belts, bests, blanks, blasts, boots home, clips, clobbers, cops, crushes, downs, drops, dumps, edges, ekes out, gallops over, gangs up on, gouges, gets past, H-bombs, halts, humiliates, impales, laces, lashes, lassoes, licks, murders, outslugs, outscraps, orbits, overcomes, paces, pastes, pins, racks up, rallies, rolls over, romps over, routs, scores, sets back, shades, shaves, sinks, slows, snares, spanks, squeaks by, squeezes by, stampedes, stomps, stops, subdues, surges, sweeps, tops, topples, triggers, trips, trounces, tumbles, turns back, vanquishes, wallops, whips, whomps and wrecks.

They will let the ball be *hit,* not always banged, bashed, belted, blooped, bombed, boomed, bumped, chopped, clunked, clouted, conked, cracked, dribbled, drilled, dropped, driven, hacked, knifed, lashed, lined, plastered, plunked, poked, pooped, pumped, punched, pummeled, pushed, rapped, ripped, rocked, slapped, sliced, slugged, smashed, spilled, spanked, stubbed, swatted, tagged, tapped, tipped, topped, trickled, whipped, whistled, whomped and whooped.

They will let a ball be *thrown* and only occasionally tossed, twirled, fired and hurled; they will let a ball be *kicked,* occasionally punted and never toed or booted.

They will resist the shopworn puns: Birds (Eagles, Orioles, Cardinals) soar or claw; Lions (Tigers, Bears, Cubs) roar, claw or lick; Mustangs (Colts, Broncos) buck, gallop, throw or kick.

They will insist on neutrality in all sports copy and avoid slanting the story in favor of the home team.

## LIFESTYLE PAGES

One of the brighter changes in newspapers in recent years has been the transformation of the society pages, with their emphasis on club and cupid items, to the family or lifestyle section, which has a broader-based appeal. Such sections still carry engagement announcements and wedding stories, but their added fare is foods, fashions, finance, health, education, books and other cultural affairs. They are edited for active men and women in all ranks, not solely for those in the top rank of society.

The better editors regard readers of lifestyle sections as alert individuals who are concerned with social problems such as prostitution, racism, civil disorders, prisons, alcoholism and educational reforms. Such editors strive to make their pages informative as well as entertaining.

Some papers now handle engagements, weddings and births as record items. And a few charge for engagement and wedding stories and pictures unless the event is obviously news, such as the wedding of a prince. Most, however, recognize the value of these items and continue to publish them as news.

To add some spice to its pages devoted to wedding accounts, the Gannett Rochester Newspapers started a "Wedding Scrapbook" page in its Saturday section, "Brides Book for Greater Rochester." Included were features on how the couple met, amusing incidents of the participants on the way to the ceremony and pictures of the couple in faraway places.

Wedding and engagement stories are with us to stay, but some customs, such as not giving the bridegroom much recognition, may change. This point was delightfully argued by Paul Brookshire in his column in the *South Dade* (Homestead, Fla.) *New Leader.* Here are some excerpts:

---

In these days when the world is quaking in its boots and news of great significance is daily swept into newspaper trash cans for lack of space, it is sickening to read paragraph after paragraph about some [woman] changing HER name to HIS.

The groom? He apparently wasn't dressed at all ... if he was even there. But Mother and Mother-in-Law? Yes. They were fashion plates in beige ensembles and matching accessories or something.

I ask you. Is it a wedding or a fashion show?

If it is a fashion show, why isn't it held in a hotel ballroom and why isn't the groom given a tiny bit of credit for showing up with his clothes on?

The blackout of the bridegroom in wedding accounts is an unpardonable sin. If the groom is mentioned at all he is afforded as much space as an atheist gets on the church page.

And pictures. Did you ever see a photograph of a bridegroom? Maybe in the post office but not in the newspaper.

I'm going on record right now in favor of wedding announcements being run as legal notices—payable in advance by the father of the bride.

Better still, if the bride insists on giving a minute, detailed description of every inch of clothing she happens to have on her person, I suggest she take out a paid display advertisement.

In this manner, trade names may be used and shops that sold the girl all her glorious gear could get equal space on the same page.

Newspapers would reap untold profits from this arrangement, and readers might be able to get some world news for a change instead of bouffant skirts highlighted with tiers of lace and aqua frocks with aqua-tipped orchids and maize silk linen ensembles with ... whatever you wear with maize silk linen ensembles.

---

Even though many newspapers have changed the society section to a lifestyle section, some retain a static style, especially on engagement and wedding stories. Wedding story leads too often read like these:

First Methodist Church in Littleton was the setting for the double-ring wedding rites of ...

Miss Cynthia Jones has become the bride of Joe Smith, it was announced by her parents ...

After a wedding trip to Las Vegas, Nev., Mr. and Mrs. Jones will live in ...

All Saints Roman Catholic Church was the setting for the single-ring rites ...

Newspapers continue to use the following:

- Stock words and phrases—holy matrimony, high noon, exchanged nuptial vows.
- Descriptive adjectives—attractive, pretty, beautiful, charming, lovely.
- Non sequiturs—"Given in marriage by her parents, the bride wore a white silk organza gown with a sabrina neckline and short sleeves." "Wearing a gown of white lace cotton over taffeta with empire waistline, square neckline and short sleeves, the bride was given in marriage by her father."
- Details—gown and flower descriptions and social affiliations that reflect status.
- Roundabout headlines—**BETROTHAL TOLD, HOLY VOWS EXCHANGED, WEDDING CEREMONIES SOLEMNIZED.**

Excerpts from an error-filled account of a wedding reveal the unbelievable triteness of such stories:

---

A petite white United Methodist Church nestled below the towering Rocky Mountains in Lyons, Colo., was the setting for the marriage of Shirley Ann M– and Wesley William B– on Aug. 4. A happy sun brought warmth and color through the old-fashioned peaked stained glass windows at precisely 3 p.m. when Gloria L– played The Wedding March, and guests stood from hand-carved oak seats, curved into an intimate half-circle to witness the double-ring ceremony....

The Rev. D. L. N– officiated at the afternoon ceremony amid arrangements of mint-green carnations, white gladiolas, and white daisies with giant white satin bows which adorned the altar. Pews were delightfully enhanced with waterfall baskets of fresh greenery and wild mountain flowers plucked early that morning from beside the St. Vrain River and tied with lime and powder blue satin ribbons....

The bride tossed her bouquet from the stairway and "went away" all dressed in white and yellow.

---

Talented editors will not allow writers to single out women's achievements with sexist terms such as housewife (rather than homemaker) and woman doctor (simply use doctor). They will delete phrases such as *is affiliated with, refreshments will be served, featured speaker, special guests, noon luncheon, dinner meeting.* They will refuse to let a person *host (or hostess) a party, gavel a meeting* or *chair a committee.*

They will not let reporters go out of their way to use *female, feminine* and *ladies* in all manner of sentences where the word *women* would be proper and more appropriate, if, indeed, even that word is necessary.

They will catch slips such as *Mrs. Richard Roe, nee Jane Doe. Nee* means *born,* and people are born only with their surnames. The first (not Christian) name is given later.

They will remain on guard for awkward sentences:

> Do you keep track of your weight and lose the first 5 or 10 pounds too much?
>
> Seniors realize the importance of proper dress more than younger students, but after a while they catch on.

All copy for lifestyle sections, as well as for other sections of the paper, should be edited for its news value. This should apply to the syndicated features as well as to the locally produced copy. Similarly, the headlines should reflect as much care and thought as do those on Page One.

# CHAPTER 7

# EDITING FOR LIBEL, TASTE AND FAIRNESS

## LAW AND THE MEDIA

The editor who lives in constant fear of a damage suit, the copy editor who sniffs libel in every story and tries to make the safe safer, and the broadcast reporter who thinks it is cute to refer to an inept council member as a simian have no place in the media. The first procrastinates and vacillates, the second makes the story vapid and the third lands the owner in court.

The reporter and copy editor need not be lawyers, but both should know enough about the legal aspects of journalism to know when to consult one. This chapter discusses some of the trouble spots.

The U.S. media use their immense freedom vigorously. The print media need no license to establish a press and start publishing. The broadcast media are licensed, but the government's primary purpose in doing so is to assign frequencies. The media can criticize the government and its officials severely and have no fear that the doors to the publication or broadcast station will be padlocked. In our system, no government—federal, state, county or municipal—can be libeled. The media are not public utilities. They can accept or reject any story, advertisement, picture or letter.

Courts generally cannot exercise prior restraint to prevent publication of information, although one lower court did in the Pentagon Papers case, and the Supreme Court declined to review a judge's prior restraint on Cable News Network's decision to broadcast tapes of former Panamanian dictator Manuel Noriega's conversations with his lawyer. Punishment, if any, comes after publication. The notion that the greater the truth, the greater the libel was rejected long ago.

The media can report, portray or comment on anyone who becomes newsworthy. Not even the president is immune from press coverage.

References to a half-dozen U.S. Supreme Court decisions will indicate the scope of the freedom the press enjoys. Some early cases helped to establish the principles that truth is a defense in libel and that the jury may determine both the law and the fact. The court has prohibited a discriminatory tax on the press. The court has told judges that neither inherent nor reasonable tendency is sufficient to justify restriction of free expression and that contempt of court is to be used only when there is a clear and present danger of interfering with the orderly administration of justice. The court has held that comment on or about public officials is privileged, even if false, provided there is no "actual malice." The court defined actual malice as publication with knowledge that the information is false or with reckless disregard of whether it is false. This privilege now covers public figures—those in the public limelight—as well as public officials.

This brief review is intended to remind editors of the unusual liberties they enjoy. It should not deter editors and publishers from maintaining a constant vigil to preserve and extend these freedoms. The problems of news management at all levels of government still exist. There are still some judges and attorneys who would dry up most news of crime until after the trial. In some jurisdictions cameras still cannot be used in the courtroom. There are still those who would like to censor what we read, hear or view. The problem of what constitutes obscenity and who is to decide what is obscene still vexes our society. Worst of all, we have many in our society who care little about press freedom. If these people could have their way, they would return to 16th-century England and the Court of Star Chamber where any criticism of the realm was promptly punished. What some people don't realize is that the freedom to read, to listen and to view is their right, not the special privilege of the media.

## EDITING FOR LIBEL

Publishers and broadcasters face risks far greater than do most other professional or business executives. More than a century ago a London editor, John T. Delane of the *Times*, said, "The Press lives by disclosures." All disclosures are hazardous. If errors occur, they are public and may subject the error maker to liability.

The day is rare when any publisher or broadcaster doesn't commit errors—wrong facts, wrong names and identifications, wrong addresses, wrong dates, wrong spelling or pronunciation, wrong grammar or wrong headlines. Fortunately, only a handful of such errors are serious enough to prompt lawsuits. Few libels are deliberate. Almost all result from erroneous reporting, misunderstanding of the law or careless editing.

## Misunderstanding the Law

In an attempt to foster understanding of the law, here is a review of some common misconceptions:

- There is a common assumption that if a statement originated from an outside source, it is safe. That is wrong because a newspaper or broadcast station is responsible for whatever it publishes or broadcasts from whatever source—advertisements, letters, feature stories.
- There is a feeling that if a person is not named, he or she may not sue. That is not true because a plaintiff sometimes can be identified by means other than name.
- There is a feeling that if the harmful statement concerns a group, individual members cannot sue. That is wrong because some groups are small enough (juries, team members, council members) that each may have a case.
- Another common error is misjudging the extent of privilege in an arrest. Statements by the police as to the guilt of the prisoner or that the prisoner "has a record a mile long" are not privileged, although some states do grant a privilege to publish fair and accurate stories based on police reports. All persons are presumed innocent until they are proved guilty.

## Carelessness in Reporting and Editing

These situations may prove to be dangerous:

- **Mistaken identity.** Similarity of names doesn't necessarily mean similarity of identity. People in trouble often give fictitious names. Identification should be qualified by phrases explaining the situation such as "who gave his name as," "listed by police as" or "identified by a card in her purse as." In listing addresses in crime and court stories, some papers use the block instead of a specific number. Several families might live at the same address.
- **Clothing the damaging statement with *alleged* or *allegedly*.** Qualification with these words carries little protection.
- **"Needled" headlines.** Qualifications are difficult in a headline because of the limited character count. It is wrong to assume that as long as the story is safe the head can take liberties. Many readers read only the headline. A picture caption also may be libelous. Within the story itself, statements usually cannot be taken out of context to create a libel. The story usually must be considered in its entirety.
- **The assumption that a person with an unsavory reputation can't be libeled.** A man may be a notorious drunk, but that doesn't necessarily make him a thief.

- **Confession stories used before the confession has been admitted as evidence in court.** In pretrial stages, this poses dangers. It is better to say merely that the prisoner has made a statement.
- **The assumption that any statement one person makes about another is protected if the reporter can prove that the first person actually made the statement about the second.** That is not true in most states. If A reports that B is a liar, the reporter must be prepared to prove not that A made the statement, but that B is, in fact, a liar.

## Libelous Statements

The legal definition of libel is damage to a person's reputation caused by bringing that person into hatred, contempt or ridicule in the eyes of a substantial and respectable group. Anything printed or broadcast is libelous if it damages a living person's reputation or has an adverse effect on that person's means of earning a living. The same applies to businesses and to institutions.

A story is defamatory if it accuses a living person of a crime or immorality or imputes a crime or immorality; if it states or insinuates that a person is insane or has a loathsome or contagious disease; if it tends in any way to subject the victim to public hatred, contempt or ridicule or causes others to shun that person or refuse to do business with that person; or if it asserts a want of capacity to conduct one's business, occupation or profession.

Wrong assumptions sometimes can make a statement defamatory. A person who sets fire to a dwelling is not necessarily an arsonist. A person who kills another is not necessarily a murderer. Some items in a newspaper are false but not necessarily defamatory. A false report that a man has died usually is not libelous.

A statement may cause someone pain and anguish, but mere vituperation does not make a libel; it must be substantial. It is not enough that the statement may disturb that person. It must damage that person in the estimation of those in the community or of those with whom the person does business.

Libel can be avoided if the staff exercises responsibility in accuracy, exactness and judgment. But even when libel does occur, it need not terrify the staff. Some cases are not serious enough to entice a lawyer to take the case to court.

Only the person libeled has cause for action. Relatives, even though they may have suffered because of the false and defamatory statements, have no recourse in libel. The offended person must bring suit within the statutory period (ranging from one to six years depending on the jurisdiction). If the person should die before or during the trial, there may be no continuation of the case by survivors. In the United States, only a living person can be libeled.

A person who has been libeled may ask the publication to print a correction. This could satisfy the individual because it tends to set the record straight. In states having retraction laws, the plaintiff can collect only actual damages—and no punitive damages—if the retraction is made on request and within a certain time limit.

Sometimes newspapers, magazines and broadcast stations offer to run a correction, possibly offer a nominal payment, and then obtain a release from further liability. This procedure saves the time and cost of a trial and may eliminate the possibility of major judgments against the newspaper.

Suppose the plaintiff insists on taking the rascal editor into court. The plaintiff must hire a lawyer and pay the filing fee. The person should be advised of the defenses available to the newspaper—constitutional defense, truth, privilege and fair comment. Because libel concerns reputation, the plaintiff's reputation, good name and esteem can be put at issue. A person with a skeleton in the closet may hesitate to have the past revealed in court. A public figure or public official will have the burden of proving the material was published with actual malice.

Suppose the plaintiff wins in lower court. That person may get damages of hundreds of thousands of dollars—or only a few cents. If the defending publisher or station owner loses, an appeal is likely, even to the U.S. Supreme Court if the question involves a constitutional issue. Is the plaintiff able to pay appeals court costs if the case is lost? As a final protection, most publishers buy libel insurance.

Most of the larger dailies and broadcast stations have their own lawyers to advise them on sensitive stories. Some lawyers urge, "When in doubt leave it out." But the publisher's attitude is, "This is something that should be published. How can it be published safely?" On extra-sensitive stories in which the precise wording has been dictated by an attorney, the desk should make no changes. The headline must be as carefully phrased.

Libel may involve businesses as well as individuals. A corporation, partnership or trust, or other business may be damaged if untrue statements tend to prejudice the entity in the conduct of its trade or business or deter others from dealing with it. Nonprofit organizations likewise may collect damages resulting from a publication that tends to prejudice them in the public estimation and thereby interferes with the conduct of their activities.

## Libel Defenses

### CONSTITUTIONAL

In a landmark decision in 1964, the U.S. Supreme Court ruled that the constitutional provisions of the First and Fourteenth amendments could be used as a defense against libel if the defamatory words were used to describe the public acts of public officials and were published without actual malice (New York Times Co. v. Sullivan, 376 U.S. 254, 1964). The court argued that "debate on public issues should be uninhibited, robust and wide-open, and that it may well include vehement, caustic and sometimes unpleasantly sharp attacks on government and public officials."

In the *Times* decision the court defined actual malice as knowledge that a statement is false or reckless disregard of whether it is false. The burden of proving actual malice was placed on the plaintiff. The ruling was later extended to include public figures, those in the public eye but not in public office, who thrust themselves into the vortex of public debate (AP v. Walker, 383 U.S. 130, 1967) or pervasive public figures (Curtis Publishing Co. v. Butts, 388 U.S. 130, 1967). It was used to permit robust discussion in criminal libel cases (Garrison v. Louisiana, 379 U.S. 64, 1964) and in privacy cases where the issue is in the public interest (Hill v. Time, 385 U.S. 374, 1967). Finally, the *Times* rule was extended to include private persons involved in public interest issues (Rosenbloom v. Metromedia, 403 U.S. 29, 1971).

In 1974 the court overturned the Rosenbloom plurality decision and ruled that the Constitution does not require private persons involved in public issues to prove actual malice in suits seeking actual damages for defamation (Gertz v. Welch, 418 U.S. 323, 1974). One effect of the Gertz decision was to permit the states to make their own interpretation of libel defense standards for private individuals seeking actual damages. Since the ruling, 28 states and the District of Columbia have given private individuals more protection against damaging statements than accorded to public officials or public figures by adopting a negligence test. Those jurisdictions are Arizona, Arkansas, Delaware, the District of Columbia, Hawaii, Illinois, Kansas, Kentucky, Maryland, Massachusetts, Michigan, Minnesota, Mississippi, New Hampshire, New Mexico, North Carolina, Ohio, Oklahoma, Oregon, Rhode Island, South Carolina, Tennessee, Texas, Utah, Vermont, Virginia, Washington, West Virginia and Wisconsin. The negligence test requires reporters to use the same care in reporting and writing as any reasonable reporter would use under the same or similar circumstances.

Four states—Alaska, Colorado, Indiana and New Jersey—use the same standards for private citizens as for public figures and public officials. New York requires private citizens to prove that the reporter exercised "grossly irresponsible conduct." Standards have not yet been set in other states.

## TRUTH

Truth is an absolute defense to libel. The truth must be as broad and as complete as the publication upon which the charge was made. Truth offered in evidence need not mean the literal accuracy of the published charge but rather the substance or gist of the charge.

A defending publisher or station owner who relies on a document as evidence to show truth must be sure the document can be produced at the trial and be admitted in evidence. A publisher or station owner who relies on a witness to give testimony as to truth must be assured the witness is qualified to testify. To take an extreme example, the media could not rely on the testimony of a doctor when that testimony would violate the doctor-patient relationship.

## PRIVILEGE

Reports of judicial, legislative and executive proceedings—federal, state or municipal—may be published and successfully defended on the grounds of qualified privilege. The qualifications are that the report be fair, substantially accurate, complete or a fair abridgment.

For example, a food inspector may make an official report to a board of health describing conditions found at a certain establishment. The information, even though false, may be reported safely as long as the media's accounts are full, fair and accurate. If truth were required in the accounts, then privilege would be worthless as a defense. What a food inspector may say about Sunday school teachers at a meeting of a service club is not privileged. Only statements made by the official acting in an official capacity can be defended as privileged.

Statements of attorneys, except those made in a courtroom, or by civic organization officials usually are not privileged, nor are press releases from government bureaus.

In many states the mere filing of a complaint, petition, affidavit or other document is not privileged. Anyone can go to the court clerk and file a complaint containing false, scandalous and damaging statements about another merely upon payment of a filing fee. Proof that libelous statements are contained in the document is not a basis for privilege in many states.

## FAIR COMMENT AND CRITICISM

The media are free to discuss public affairs and to comment on the conduct of all public officials, even a low-ranking park board member. This defense has three qualifications:

- The comment is founded on facts or what the publisher or station owner had reasonable grounds to believe are facts.
- The comment is not made with intent to damage. Here, the burden of proof is on the plaintiff.
- The comment does not involve the private life or moral character of a person except when it has a direct bearing on qualifications or work.

Those who put themselves or their work before the public are subject to public assessment of their performance, however strong the terms of censure may be. Decisions of the U.S. Supreme Court suggest that, short of malice and reckless disregard of the truth, the *Times* rule still holds true: All debates on public issues should be uninhibited, robust and wide-open, and that such debates may well include vehement, caustic and sometimes unpleasantly sharp attacks on government and public officials. The same freedom could well apply to comments on anyone in the public eye. The only limitation on the press is that the comments cannot be based on misstatements.

This should not be construed as license. Character and public reputation are priceless possessions and are not good hunting grounds simply because a person holds public office, aspires to public office or in any manner offers talents to the public. There is a difference between assessing the fitness of a candidate or commenting on the products of a public performer and a reckless attack on character and reputation.

### CORRECTIONS

The publication or broadcast of a correction technically admits the libel and therefore negates truth as a defense. But when the defense of truth is not clearly evident, the publisher or station owner should make the decision to correct. When made, the correction should be full and frank and be displayed or broadcast as conspicuously as the story complained of.

A reporter obtained her story over the phone from the judge's secretary. She took her notes in shorthand. When she transcribed her notes, she mistook DWS (driving while under suspension) for DWI (driving while intoxicated) and thus wrote falsely that a certain person had pleaded guilty to driving while intoxicated. Even if this story had been edited by another, it is unlikely the editor would have caught the error. The paper should have printed a correction to indicate lack of malice (ill will or intent to damage) and to escape punitive damages should the injured person sue for libel.

But the follow-up story was not a clear correction, and this should have been caught by an editor. The headline read **Ex-sheriff's patrolman admits count.** The lead: "A man who five months ago was suspended from the Franklin County Sheriff's Patrol was arraigned in Municipal Court here yesterday, pleading guilty to driving for the past nine years on a suspended license." Later, the story said that the patrolman had been dropped from the force for "misuse of authority" and then qualified that statement with another to the effect the patrolman had to resign on order of the Office of Strategic Information, Fort Ethan Allen, Vt.

The patrolman sued for libel. The defense tried to argue that the crime of driving while the license was suspended was as serious as the crime of driving while intoxicated and therefore the newspaper should not be held accountable for a minor error. The jury disagreed and returned a judgment of $3,500 for the former patrolman.

## Right of Privacy

A libel action is brought to protect a person's reputation against defamation. A privacy action is brought to protect a person's right to be left alone. The distinction between the two is not as clear as it once was because privacy is now being used as an alternative to libel or in conjunction with libel.

The right of privacy is an expanding legal doctrine that is so vague and has such ill-defined limits that the press has difficulty in knowing where it stands legally. Violation of the right of privacy encompasses four torts:

- Intrusion on the plaintiff's seclusion or solitude, or into private affairs.
- Public disclosure of embarrassing private facts about the plaintiff.
- Publicity that places the plaintiff in a false light in the public's eyes.
- Appropriation, for the defendant's advantage, of the plaintiff's name or likeness.

Media accounts generally concern newsworthy subjects who have forfeited, voluntarily or involuntarily, their rights of privacy. The right of privacy does not protect a person or that person's actions if they are a matter of legitimate public interest. Newsworthiness is based on three basic components: public interest, public figures and public records.

Generally, a person who voluntarily participates in a public event abandons the right of privacy. Thus, a publication or broadcast station may legitimately display the types of persons who join the Easter parade. But a newspaper in Alabama had to pay damages to a woman photographed while her skirts were blown above her head by an air jet in a fun house at a country fair. She was recognizable because her children were with her.

Some risk is involved in printing photographs taken of people without their consent in their homes or in hospital beds. Pictures showing ways to beat the summer heat may be humorous to readers but not to the woman fanning herself under a tree in her own backyard.

Truth usually is not accepted as a defense in a privacy invasion case, but truth, combined with publishing information released to the public in official court records, is a defense. In other words, the defense is adequate if the information is truthful, comes from an open public trial or comes from court documents that are open for public inspection. Another defense against "private facts" invasion of privacy is that the information is in the public domain because it is widely enough known.

The definition of newsworthiness is so broad and its use as a defense against "private facts" invasion of privacy is so powerful that some scholars wonder if "private facts" invasion of privacy still exists. Newsworthiness, however, is not a defense to intrusion into seclusion.

Appropriation of name or likeness is an expanding threat. Plaintiffs have won appropriation of voice suits, and more jurisdictions are permitting surviving relatives to bring privacy actions against the exploitation of the names or likenesses of deceased relatives.

## Juveniles

It is not illegal for a newspaper to publish the name of a juvenile. But some states do not allow officials to release news or pictures of juveniles, defined in most states as anyone under the age of 18. Many officials insist that names of juvenile offenders be withheld on the theory that there is greater opportunity for rehabilitation if the youth is not stigmatized by publicity that may affect him or her for years.

In some states the children's code gives exclusive jurisdiction to the juvenile court over offenders under 14, regardless of the acts committed, and gives concurrent jurisdiction to the district court over youngsters between 16 and 18, unless the crime involved is punishable by either death or life imprisonment if committed by an adult. In murder cases involving youths 14 and over, the district court has original jurisdiction.

Coverage of juvenile matters may be allowed under the following circumstances:

- **Public hearings.** A U.S. Supreme Court decision extends to juveniles the same guarantees to due process of law provided adults in criminal proceedings. The juvenile has the same rights against self-incrimination, to representation and even to a jury trial. Even in juvenile court, the youngster has a right to a public trial if requested. But after the finding of the jury at a public trial, the juvenile could still be placed before a juvenile court for disposition, out of sight of the press and the public.
- **Permission of the court.** If the code permits it, the judge may, at his or her discretion, allow coverage concerning the hearing of a juvenile. Frequently, the judge of a juvenile court allows reporters to attend juvenile court sessions but does not permit identification of the youthful offender. There may be publication of news of such cases, provided that any reference to any child involved is so disguised as to prevent identification.
- **Traffic cases.** Names of persons of any age may be used in traffic cases.

Editors should be alert to the distinction between a juvenile and a minor. In most states a minor is defined as anyone under the age of 21, although some states have reduced the age of majority to 18.

## Plagiarism and Copyright Infringement

Plagiarism and copyright infringement are still other areas of concern to the editor. Only news stories, not the news itself, can be copyrighted. Even if a newspaper does not protect itself by copyrighting the entire paper or individual stories, it still has a property right in its news and can prevent others from "lifting" the material. It is assumed that editors will be so thoroughly familiar with the contents of opposition papers that they will be able to spot material that copies or paraphrases too closely the work of others.

If a wire service sends out a story based on the story of another member or client, editors should not delete the wire service credit to the originator of the story. Nor should they delete any credit on stories or pictures. They may, if directed, compile stories from various sources into one comprehensive story, adding the sources from which the story was compiled.

If their own paper publishes a story to be copyrighted, editors should ensure that the notice is complete—the notice of copyright, the copyright date and who holds the copyright.

In editing book review copy, editors should have some notion of the limits of fair use of the author's quotations. The problem is relatively minor because few copyright owners would object to the publishing of extracts in a review, especially if the review were favorable. If the review has to be trimmed, the trimming probably should come in the quoted extracts.

## Lotteries

A lottery is any scheme containing three elements—consideration paid, a prize or award and determination of the winner by chance. This includes all drawings for prizes and raffles, and games such as bingo and keno. It is immaterial who sponsors the scheme. Pictures and advertising matter referring to lotteries and similar gift enterprises are barred from the mail.

Newspapers may not be permitted to announce them or to announce results. Federal law now exempts newspapers and magazines in states with legal lotteries as long as they confine themselves to reporting state lotteries. Stories also may be used in cases in which something of news value happened as a result of the lottery. An example would be a story concerning a laborer who became wealthy overnight by having a winning ticket on the Irish Sweepstakes. This would probably be considered a legitimate human interest story rather than a promotion for horse races and lotteries.

## Crime and Courts

No longer do American editors play crime by the standards of past generations. They print and broadcast crime news, but they do not rely on crime stories, even a sex-triangle murder, to boost street sales or ratings. Topics such as space and ocean exploration compete with crime for the attention and interest of today's more sophisticated readers and viewers.

That may not have been evident during the celebrated O.J. Simpson trial of 1995, in which the famous former football player and television commentator was

***Figure 7–1*** *O. J. Simpson reacts to his acquittal in the murder of his wife, Nicole Brown Simpson. Defense attorneys F. Lee Bailey, left, and Johnnie Cochran Jr. were key members of Simpson's all-star defense team. The trial was the most heavily covered of the century.*
Photo by Associated Press.

accused of brutally murdering his wife and another man (see Figure 7–1). That trial received daily coverage in the national media, and live television reported every sordid detail of the murders.

The blanket coverage of the Simpson trial, particularly on television, refueled the long-running debate about conflicts between the First and Fourteenth amendments to the U.S. Constitution—the right to a free press vs. the right to a fair trial. Many questioned whether someone undergoing such massive public scrutiny in the media could possibly get a fair trial. But despite concerns that publicity would convince the jury of Simpson's guilt, he was acquitted.

Only major crimes get such attention. Unless they have unusual angles, minor crimes generally are merely listed or even ignored. When presented in detail, the crime story should be written with the same thoroughness and sensitivity that experts give other subjects. Some observers argue that the media should offer more news of criminal court activities—but with the constructive purpose of showing the community the origins and anatomy of crime and how well or how poorly the justice system is functioning.

Editors who handle crime stories should make sure that their reports contain no prejudicial statements that could deprive the defendants of fair trials. One editor admonished his staff, "We should be sensitive about assumption of guilt, not only to avoid libel but to avoid criticism and a bad impression on readers." The caution was occasioned by this lead: "With the dealer who sold a .32-caliber pistol to Mrs. Marian C— apparently located, Shaker Heights police today were using handwriting expert Joseph Tholl to link the accused slayer of Cremer Y—, 8, to the weapon purchase." The whole tone of this lead is an assumption of guilt and the effort of police to pin the crime on somebody. It should have said the police were trying to determine whether Mrs. C— was linked to the gun purchase—not trying to link her.

Here is a conviction lead:

> FAYETTEVILLE, N.C. (AP)–Two Marines are being held without bond after terrorizing a family, stealing a car and trading shots with officers.
>
> They were identified as…
>
> They told officers after their Saturday capture they were members of the National Abolitionist Forces, which they described as a militant black group.

In reference to this story, the general news editor said in part, "We do not have a formal set of guidelines for handling crime news, but this story certainly does not conform to regular AP practice. It makes us authority for that statement that the two men held had terrorized a family, stolen a car and traded shots with officers. All we should have said was that they were charged with doing all those things, and who had made the charge."

## Correct Terminology

Newsworthy crimes and trials are covered in detail so that essential information may be conveyed to the public. Editors have some knowledge of legal terms and the legal process if they are to make the story and headline technically correct yet meaningful to the public.

*Arrested* is a simple verb understood by all readers. It is better than *apprehended* or *taken into custody.* It is equal to *captured.* A person who is cited, summoned or given a ticket is not arrested.

An *arraignment* is a formal proceeding at which a defendant steps forward to give the court a plea of guilty or not guilty. It should not be used interchangeably with *preliminary hearing,* which is held in a magistrate's court and is a device to show probable cause that a crime has been committed and that there is a likely suspect.

*Bail* is the security given for the release of a prisoner. The reporter reveals ignorance when writing, "The woman is now in jail under $5,000 bail." She can't be in jail under bail. She can be free on bail, or she can be held in lieu of bail.

A *parole* is a conditional release of a prisoner with an indeterminate or unexpired sentence. *Probation* allows a person convicted of some offense to go free, under suspension of sentence during good behavior and generally under the supervision of a probation officer.

The word *alleged* is a trap. Used in reference to a specific person (Jones, the alleged gambler), it offers little or no immunity from libel. Jones may be charged with gambling or indicted for gambling. In both instances, *alleged* is redundant. The charge is an allegation or an assertion without proof but carries an indication of an ability to produce proof.

A jail sentence does not mean, necessarily, that a person has been jailed. The individual may be free on bail, free pending an appeal or on probation during a suspended imposition of sentence.

Listing the wrong name in a crime story is the surest route to libel action. Thorough verification of first, middle and last names, of addresses and of relationships is a necessity in editing the crime story.

Names of women or children in rape cases or attempted rape cases generally should not be used. Nor should the story give any clue to their addresses in a way by which they can be identified. Increasingly, though, feminists are arguing that rape victims' names should *not* be withheld. They argue that rape victims have no reason to feel shame.

Prison sentences may be *consecutive* or *concurrent.* If a man is sentenced to two consecutive three-year terms, he faces six years of imprisonment. If his sentences are concurrent, he faces three years. But why use these terms? The total sentence is what counts with the readers and the prisoner. Note also that if a person has been sentenced to five years but the sentence is suspended, he or she is given a *suspended five-year sentence,* not a five-year suspended sentence.

Juries are of two kinds—*investigative* (grand) and *trial* (petit). If a grand jury finds evidence sufficient to warrant trial, it issues a *true bill* or indictment. If sufficient evidence is lacking, the return is a *no-bill* or *not-true bill.* "Jones was indicted" means as much as "the grand jury indicted Jones." To say "the grand jury failed to indict Jones" implies it shirked its duty.

A *verdict* is the finding of a jury. A judge renders decisions, judgments, rulings and opinions, but seldom verdicts, unless the right to a jury trial is waived. Although verdicts are returned in both criminal and civil actions by juries, a guilty verdict is found only in criminal actions. Judges *declare,* but do not order, mistrials. Attorneys general or similar officials give *opinions,* not rulings.

*Corpus delicti* refers to the evidence necessary to establish that a crime has been committed. It is not restricted to the body of a murder victim; it can apply as well to the charred remains of a burned house.

*Nolo contendere* is a legalistic way of saying that a defendant, although not admitting guilt, will not fight a criminal prosecution. *Nolle prosequi* means the prosecutor or plaintiff will proceed no further in the action or suit. Most readers will understand the translation more readily than the Latin expression.

The Fifth Amendment guarantees the due process of law protection for all citizens. The report should not suggest that the use of this protection is a cover-up for guilt. Phrases such as "hiding behind the Fifth" should be eliminated.

The story should distinguish between an act itself and an action. *Replevin,* for example, is an action to recover property wrongfully taken or detained. Trouble will arise if the editor lets the reporter translate the action too freely: "Mrs. March filed the replevin action to recover furniture stolen from her home by her estranged husband." So, too, with the tort of *conversion.* "Wrongful conversion" may imply theft, but neither the copy nor the head should convey such implication.

Keeping track of the plaintiff and the defendant should pose no problem except in appellate proceedings in which the original defendant may become the appellant. The confusion is not lessened by substituting *appellee* and *appellant.* The best way is to repeat the names of the principals.

In some civil suits the main news peg is the enormous sum sought by the plaintiff. Whether the same angle should be included in the headline is questionable. In some damage claims the relief sought is far greater than the plaintiff expects to collect. The judgment actually awarded is the news and the headline.

### MISUSED TERMS

Editors can "tidy up" the crime and court report by watching for the following:

- All narcotics are drugs, but many drugs are not narcotics.
- A defendant may plead guilty or not guilty to a charge or a crime. There is no such plea as innocent, but "not guilty by reason of insanity" is a plea. A defendant may be judged not guilty by reason of insanity. He or she is not innocent by reason of insanity. An acquittal means the defendant has been found not guilty. The danger of dropping the *not* has caused some editors to insist on using *innocent* rather than *not guilty,* although some lawyers argue that *innocent* is inappropriate for verdicts.
- All lawsuits are tried in courts. *Court litigation* is therefore redundant.
- Statements are either written or oral (not verbal).
- "Would-be robber" has no more validity than "would-be ballplayer."
- The word *lawman* has no place in the report. It can mean too many things—a village constable, a sheriff's deputy or the sheriff, a prosecutor, a bailiff, a judge, an FBI agent, a revenue agent and so on. It is also sexist. A more precise word is almost always more suitable in a newspaper.
- Use *sheriff's deputies* rather than *deputy sheriffs.*

- Divorces are granted or obtained. Medals are won or awarded.
- "Hit-and-run," "ax-murder," "torture-murder" and the like are newspaper and television clichés. They should be changed to "hit by an automobile that failed to stop," "killed with an ax," "tortured and murdered."
- There's no such thing as an *attempted holdup.* A holdup is a robbery even if the bad guy got nothing from the victim.

## LEGAL JARGON

Legal jargon also should be avoided: "The case was continued for disposition because the attorney requested no probation report be made on the boy before adjudication."

W.J. Brier and J.B. Rollins of Montana State University studied some Missoula, Mont., adults and their understanding of legal terms. The following terms were correctly defined by more than half the respondents. The correct explanation is given here:

- *Accessories before the fact*—those charged with helping another who committed the felony.
- *Extradition*—surrender of the prisoner to officials of another state.
- *Arraigned*—brought to court to answer to a criminal charge.
- *Indicted*—accused or charged by a grand jury.
- *Civil action*—pertaining to private rights of individuals and to legal proceedings against these individuals.
- *Extortion*—oppressive or illegal obtaining of money or other things of value.
- *Remanded*—sent the case back to a lower court for review.
- *Felony*—a crime of a graver nature than a misdemeanor, usually an offense punishable by imprisonment or death. Misdemeanors, however, can carry sentences of up to one year.
- *Continuance*—adjournment of the case.
- *Writ of habeas corpus*—an order to bring the prisoner to court so the court may determine if the prisoner has been denied rights.
- *Administratrix*—female administrator.
- *Stay order*—stop the action or suspend the legal proceeding.
- *An information*—an accusation or a charge filed by a prosecutor.
- *Venire*—those summoned to serve as jurors.
- *Demurrer*—a pleading admitting the facts in a complaint or answer but contending they are legally insufficient.

Other misunderstood terms include:

- *Released on her personal recognizance*—released on her word of honor to do a particular act.
- *Make a determination on the voluntariness of a confession*—decide whether a confession is voluntary.
- *The plaintiff is . . .* —the suit was filed by. . .

Terms that should be translated include: *bequest* (gift), *debenture* (obligation), *domicile* (home), *in camera* (in the judge's office), *liquidate* (settle a debt), *litigant* (participant in lawsuit), *paralegals* (legal assistants), *plat* (map), *res judicata* (matter already decided).

Lawyers are fond of word-doubling: *last will and testament, null and void, on or about, written instrument.* Another is *and/or.* "The maximum sentence is a $20,000 fine and/or 15 years' imprisonment." The maximum would be the fine and 15 years.

Few readers understand the meaning of the word *writ* (a judge's order or a court order). "In a petition for a writ of mandamus, the new bank's incorporators asked the court" should be changed to "The new bank's incorporators asked the court to." Or if the term *mandamus* was essential, it should have been explained (a court order telling a public official to do something).

Euphemisms include *attorney* for *lawyer, sexually assaulted* or *sexually attacked* for *raped.* Not all jurists, who profess to be or are versed in the law, are judges, and certainly not all judges are jurists.

## BE EXACT

A *robber* steals by force. A *thief* steals without resorting to force. Theft suggests stealth. A *burglar* makes an unauthorized entry into a building. If a burglar is caught in the act, pulls a gun on the homeowner and makes off with the family silverware, that person is a robber.

*Theft* and *larceny* both mean taking what belongs to another. *Larceny* is the more specific term and can be proved only when the thief has the stolen property. Pickpockets and shoplifters are thieves.

Words such as *looted, robbed* and *swindled* should be used properly. It is incorrect to write: "Two men were fined and given suspended sentences yesterday in Municipal Court for stealing newsracks and looting money from them" or "Thieves broke into 26 automobiles parked near the plant and looted some small items." Money or other property is not looted. That from which it is taken is looted. Nor is money robbed. A bank is robbed; the money is stolen. "A man in uniform swindled $1,759 from a woman." No, the person is swindled, not the money.

Here are some final points on terminology related to legal reporting:

- *Statutory grounds for divorce* is redundant. All grounds for divorce are statutory in the state where the divorce is granted.
- *Charge* has many shades of meaning and is often misused. "The psychologist charged last night that black high school students generally do not think of the university as a friendly place." The statement was more an observation than a charge.
- Members of the Supreme Court are justices, but not judges or supreme judges. The title of the U.S. Supreme Court's chief justice is Chief Justice of the United States.
- Some papers object to saying that fines and sentences are *given,* on the ground that they are not gifts.

## EDITING FOR TASTE

A story can be free of libel but still be in bad taste. An editor of a morning newspaper said his newspaper likes to protect those who read during breakfast against the incursions of unpalatable news. How then, he asked, did this sentence get to the breakfast table? "Plans to take still other samples were canceled when Hutchinson became ill and threw up." Actually, he *vomited.* Had the story said he became nauseated, anyone who is familiar with 10 ½ beers—Hutchinson's load in less than 2 ½ hours—would have gotten the point.

The *Los Angeles Times* and other metropolitan newspapers have adopted a screen code to control and avoid lewd advertising in entertainment copy. One of the advertising executives of the *Times* said, "It is not our intention to be either picayunish or prudish in our evaluation, but we are convinced that moral and social values have not decayed as frequently as portrayed, and we trust that together we can find a better standard of values in the area of good taste." Among subjects banned are bust measurements, compromising positions, double meanings, nude figures or silhouettes, nymphomania, perversion, and suggestive use of narcotics or alcohol. Words avoided include *cuties, girlie, lust, nymph, party girls, scanty panties, sexpot, strippers* and *third sex.*

The caution should apply equally to amusement promotion copy and to all other copy. Both wire services direct their editors to downplay anatomy. Editors should apply heavy pencils to stories about the "10 best undressed women" and about an actress hired because of her uncommonly ample bosom. Better yet, those stories should not be used. There is no need to run everything turned in as news by the staff, the wire services or the syndicates. There is an obligation to print the news. There also is an obligation to edit it.

Some vulgarisms get into the report, usually when they are said by a public figure at a public gathering and in a justifiable news context. Most member papers used the following lead from London even though the AP headed the dispatch with a cautionary note: " 'Gentlemen,' said Prince Philip, 'I think it is time we pulled our fingers out.' "

But the editor of a Dayton, Ohio, newspaper was forced to resign when management panicked because 50 callers protested after the editor had approved the inclusion of dialogue, including words connoting sexual intercourse, in a Page One murder story. The fact that the dialogue was from a direct transcript of testimony during the hearing had no effect on management's decision. Nor did the fact that 110,000 or so other readers didn't complain.

Is *s.o.b.* milder than the full expression? If the president of the United States refers to a syndicated columnist as an "s.o.b.," that's news. The columnist in question passed off the slur by saying the president obviously meant "sons of brotherhood." Another president used the phrase "sons of business." When Jack Ruby shot Lee Harvey Oswald, accused of assassinating President John F. Kennedy, Ruby is purported to have exclaimed, "I hope I killed the son of a bitch." The quote appeared in the news dispatches from Dallas. There was a day when editors would have substituted dashes or asterisks for the words. Some bannered the quote, but with initials: **JACK RUBY—"I HOPE I KILLED THE S.O.B."**

Frankness used in good taste is preferable to yesterday's euphemisms, such as *social disease* for *syphilis, intimate relationship* for *sexual intercourse, assault* for *rape.* Why refer to washrooms and toilets in public buildings, such as schools, as *bathrooms?* Ever try to take a bath in one?

This story was published in a daily under a two-column headline:

---

**COED REPORTS RAPE**

An 18-year-old university student claimed she was raped early Sunday morning by a man she met in a local nightclub, sheriff's officers said.

The woman said she was drinking and dancing with the man at the Sweet Lass lounge before accepting a ride to the man's apartment. The man invited her to his apartment for some drugs, she told officers.

When the couple arrived at the apartment, the man invited the woman into the bedroom and the woman accepted. The man then partially undressed the woman (she completed the undressing) and attempted to have sexual relations with her, she said.

The woman told police she said no to the man but did not resist his advances.

After a brief period, the woman said the man "gave up, rolled over and went to sleep."

While the man slept, the student said she got dressed and went to her dormitory. Because of her intoxicated condition, the woman said she was unsure whether the sexual act was completed.

---

The reporter should never have submitted this nonstory. The story should never have passed the city desk and certainly should have been challenged on the copy desk. No one was arrested or charged, so there was no news value. It is, at best, but idle chatter and has no place in a newspaper.

It is important to remember, though, that no two cities are alike. What may be considered tasteless in Hays, Kan., may be quite acceptable in Miami. So, unless the passage is clearly tasteless, ask a superior for a ruling. Junior editors are wise not to make such decisions. If that's passing the buck, so be it. But most top editors would rather make those calls themselves than be surprised when reading the paper or watching the newscast.

## EDITING FOR FAIRNESS

What may not be libelous or tasteless may still be unfair. One Midwest editor tells this story of a bad decision he made:

> Our reporter had been working for weeks on a story about our public administrator, the elected official who administers the estates of people who die without wills or of people judged incompetent to manage their own affairs. We got word that soon after his election, the new public administrator had moved all the bank accounts he was to manage into a bank in which he was a shareholder.
>
> We confirmed that this indeed had occurred, and we called the administrator to get his explanation. He refused comment, so the next morning we printed the story on Page One. It was a good story, we thought. A public scandal of minor proportion but still quite a significant one in a midsized community.
>
> The morning of publication, I got a call from the administrator. "I have no problem with the facts of your story," he said, much to my relief. Then he administered the zinger: "By the way, I calculated that I would profit from that transaction, due to my bank stock holdings, by about $1.13 this year. I moved the accounts for my convenience, not to profit from the income."

The editor was devastated. It would have been easy to excuse the gaffe by blaming it on the administrator, who had refused to speak before publication. Nevertheless, the editor felt responsible for what was really an unfair story. The newspaper, he reasoned, should have made that calculation itself. A correction followed.

Increasingly, the media are concerned about fairness. It's often a topic of discussion at editors' meetings. Stories like the one just described merely fuel the public's perception that the media tilt the news report to make things appear as bad as possible. When unfair stories are published or broadcast, media credibility suffers. Good editors increasingly are willing to admit such mistakes. They run corrections when a story has been determined to be unfair just as quickly as when a libel has been committed.

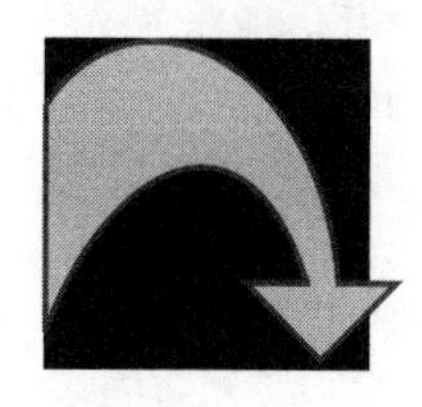

# PART 3

# EDITING FOR THE PRINT MEDIA

# CHAPTER 8

# WRITING THE NEWSPAPER HEADLINE

## HEADLINE FUNCTIONS

A copy editor's first task is to correct and refine copy. This means, as outlined in earlier chapters, checking copy for accuracy, clarity, conciseness, tone and consistency of style. A second task is to write a headline that:

- Attracts the reader's attention.
- Summarizes the story.
- Helps the reader index the contents of the page.
- Depicts the mood of the story.
- Helps set the tone of the newspaper.
- Provides adequate typographic relief.

Those are the major functions of headlines. Not every headline can accomplish each of those tasks, but editors who write headlines with these goals in mind will write better headlines than those who ignore them.

Good headlines attract the reader's attention to stories that otherwise may be ignored. The day's best story may have little or no impact if the headline fails to *sell* it, or attract the reader's attention. Headlines sell stories in many ways, but often they do so by focusing on how the reader's life will be affected. For example, if the city council has approved a city budget of $30 million for the coming year, one approach is to headline the story:

**COUNCIL APPROVES $30 MILLION BUDGET**

Another approach, which does a better job of selling the story, might be this:

**CITY TAX RATE TO REMAIN UNCHANGED**

That approach answers the question the reader is most likely to ask about the council's action: How will it affect me?

Some headlines attract attention because of the magnitude of the event they address:

**EARTHQUAKE IN ALGERIA KILLS 20,000**

Others attract attention because the headline is clever or unusual:

**HUNGER PANGS**
**Thief finds sandwich goodies,**
**wine provide appetizing loot**

Each story requires a different approach, and the headline writer who is able to find the correct one to attract the reader's attention is a valued member of the newspaper's staff.

Most headlines that appear over news stories are designed to inform, not entertain, so the headline that simply summarizes the story as concisely and accurately as possible is the bread and butter of the headline writer:

**U.S., CHINA TO SIGN MAJOR GRAIN DEAL**

Such headlines seldom win prizes for originality or prompt readers to write letters of praise. But a newspaper full of headlines that get right to the point is a newspaper that is easy to read. The reader knows what the story is about and can make an intelligent decision about whether to read more. The headlines summarize the news, much as five-minute radio newscasts do.

If the headlines on a page do a good job of summarizing the stories, the editors have created for their readers a form of index to the page. This also helps the reader determine what to read and what to bypass. In one sense, good headlines help readers determine what *not* to read. And, while that may seem counterproductive to the newspaper's objectives, it is realistic to recognize that a reader will partake of only a small percentage of the newspaper's offerings. Newspapers help make those choices easier by providing a choice of fare, much as supermarket managers offer their customers various brands of green beans. That may not be an appealing comparison to those who view newspapers as entities above that sort of thing, but it *is* realistic. To ignore that reality is a mistake.

The headline also sets the mood for the story. The straightforward news headline indicates that the story it accompanies is a serious one. Similarly, a headline above a how-to-do-it story should reflect the story's content:

**IT'S EASY TO SAVE BY CHANGING YOUR CAR'S OIL REGULARLY**

Setting the mood is even more important when writing headlines for humorous stories. One newspaper hurt readership of a bright story during the streaking craze by using a straight headline:

**JUDGE LECTURES STREAKER**

The story was a humorous account of the court appearance of a group of college students who had run across a softball diamond in the nude. In the second edition, the headline writer did a much better job:

**STREAKER GETS THE PITCH**
**It's a whole nude ball game**

The mood was set for the reader to enjoy the story.

Headlines probably reveal as much about the tone, or character, of a newspaper as anything it contains. If the top story on the front page is headlined **COPS SEEK LOVER IN AX MURDER** and the second story carries the headline **LIZ FLIPS OVER NEW BEAU,** there can be little doubt about the nature of the publication. Serious tones, as well as sensational ones, can be set with headlines.

Finally, headlines provide typographic relief (see Figure 8–1). They separate stories on the page and relieve the tedium that would exist with masses of text-sized type. This function will be discussed in detail in Chapter 12.

## THE HEADLINE WRITING PROCESS

Readers read the headline first, then the story. Copy editors work in reverse; they first read the story, then write the headline. This often leads to confusing heads because copy editors mistakenly assume that if readers will only read the story they will understand what the headline is trying to convey. But except in rare cases deliberately designed to tease the reader, the headline must be instantly clear. In most cases, the reader will not read a story simply to find out what the headline means.

Headline writing, then, involves two critical steps:

- Selecting which details to use.
- Phrasing them properly within the space available.

The copy editor exercises editorial judgment in completing the first step in the process. Most use the *key-word method* in which the copy editor asks: Which words must be included in the headline to convey to the reader the meaning of the story? In its simplest form, this involves answering the question: Who does what? Thus, most good headlines, like all good sentences, have a subject and predicate, and usually a direct object:

**TORNADO STRIKES JONESBORO**

That done, the copy editor tries to make the headline fit. Synonyms may be necessary to shorten the phrase, and more concise verbs may help:

**TWISTER RIPS JONESBORO**

**TICKS**

**The best defense? A good, strong dose of prevention.**

**Mayor orders investigation of park police**

*5 Days of Testimony End*

**Jury Gets Zimmer Case**

**Police Expand Task Force**

**Missing Persons Bureau Placed Under Redding's Command**

***Figure 8–1*** *Headlines come in all sizes and typefaces. Sizes are measured in points, a printer's unit of measurement equal to 1/72 of an inch. Because there are 72 points in an inch, 24-point type would be about one-third inch in height. Typefaces are given distinctive names, such as Bodoni and Helvetica. For more on the characteristics of type, see Chapter 12.*

That, in simplified form, is the essence of headline writing. But it is seldom that easy. All newspapers have rules to define the limits of what is acceptable, and consideration must be given to such factors as the width of the column and the width of the characters in the typeface to be used. In the sections that follow, the complexity of headline writing will become apparent. Through it all, however, it may be useful to keep in mind the two critical steps previously outlined.

## THE HEADLINE ORDER

Typically, the copy editor receives a headline order when assigned to edit a story. The editor responsible for layout will have determined the headline size and style based on the length and significance of the story and its placement on the page. (In some cases, the copy may be marked "HTK," or "hed to kum," indicating that a layout decision has not been made and therefore no headline size has been determined. When this occurs, the copy editor merely edits the story; the headline will be written later.)

Each newspaper has its own schedule showing headline designations and line count, the maximum number of units that will fit on a line in the specified typeface

## The Evolution of Headlines

Styles of headlines, like fashions, change constantly, even though their functions remain the same. Because style is an important factor in determining what can and cannot be included in a headline, it may be useful to review the historical development of headline styles.

Newspapers' first news display lines were short and slender, usually a single crossline giving little more than a topical label: **LATEST FROM EUROPE.** By adding more lines or by varying the length of the lines, designers created the hanging indention, the inverted pyramid and the pyramid:

XXXXXXXXXXXX
XXXXXXXXX
XXXXXXXXX

XXXXXXXXXXXX
XXXXXXXXXX
XXXXXX

XXXXXX
XXXXXXXX
XXXXXXXXXX

Later, by centering the second line and making the third flush with the right-hand margin, they developed the stepline. It became one of the most popular styles of headlines and is still in use at a few newspapers:

**HEAVY RAIN**
**SHUTS DOWN**
**ALL BEACHES**

The next move was to combine these elements—a stepline, an inverted pyramid, a crossline, then another inverted pyramid. The units under their introductory head became known as *banks* or *decks*. An article in "The Quill" cited one found in a western newspaper describing a reporter's interview with Gen. Phil Sheridan in 1883:

**FRISKY PHIL**

**GAZETTE REPORTER HOLDS**
**INTERESTING INTERVIEW**
**WITH HERO OF WINCHESTER**

**THE GREAT WARRIOR RECEIVES THE**
**NEWSPAPERMAN WITH OPEN ARMS;**
**HE IS MORE OR LESS BROKEN UP ON**
**THE CRAFT ANYWAY**

**HE TRAVELS IN A SPECIAL**
**MILITARY COACH AND LIVES**
**ON THE FAT OF THE LAND**

**SHERIDAN IS MANY MILES AWAY,**
**BUT THE CHAMPAGNE WE DRANK**
**WITH HIM LINGERS WITH US STILL**

**WE FEEL A LITTLE PUFFED**
**UP OVER OUR SUCCESS**
**ATTENDING OUR RECEPTION BY**
**LITTLE PHIL, BUT MAN IS**
**MORTAL**

**MAY HE WHO WATCHES OVER THE**
**SPARROWS OF THE FIELD NEVER**
**REMOVE HIS FIELD GLASSES FROM**
**THE DIMINUTIVE FORM AND GREAT**
**SOUL OF PHIL SHERIDAN**

Throughout most of America's history, newspaper headlines have tended to depict the mood of the times as well as the tone of the paper. **JERKED TO JESUS,** shouted the *Chicago Times* on Nov. 27, 1875, in headlining the account of a hanging. Another classic:

**AWFUL EVENT**

---

**President Lincoln**
**Shot by an Assassin**

---

The Deed Done at Ford's
Theatre Last Night

---

**The Act of a Desperate Rebel**

---

The President Still Alive at
Last Accounts

---

No Hopes Entertained of His
Recovery

---

*Attempted Assassination of*
*Secretary Seward*

---

**Details of the Dreadful Tragedy**

*The New York Times*

Big type and clamoring messages still weren't enough for some newspapers in the late 1800s. According to Gene Fowler, an executive told the owners of the *Denver Post,* "You've got to make this paper look different. Get some bigger headline type. Put red ink on Page One. You've got to turn Denver's eyes to the *Post* every day, and away from the other papers." So the *Post* ran its headlines in red to catch readers' attention. The message had to be gripping. According to Fowler's version, Harry Tammen, co-owner of the *Post,* was so incensed over a lifeless banner that he grabbed a piece of copy paper and composed one of his own: **JEALOUS GUN-GAL PLUGS HER LOVER LOW.** When the copy desk protested the headline wouldn't fit, Tammen snapped, "Then use any old type you can find. Tear up somebody's ad if necessary." Still, the desk wasn't satisfied. "It isn't good grammar," the desk chief argued. But Tammen wouldn't budge. "That's the trouble with this paper," he is quoted as saying. "Too damned much grammar. Let's can the grammar and get out a live sheet."

The battle for circulation was hot. So were the headlines. Many were also colorful:

**DEMON OF THE BELFRY**
**SENT THROUGH THE TRAP**

**DONS PLANNED TO SKEDADDLE IN THE NIGHT**

**DOES IT HURT TO BE BORN?**

**CONDUCTORS ROBBING LITTLE GIRLS**
**OF THEIR HALF-FARE TICKETS**

**DO YOU BELIEVE IN GOD?**

During and after the Spanish-American War, some newspapers used as many as 16 decks, or headline units, to describe the story. Frequently the head was longer than the story.

With improved presses and a greater variety of type available, designers were able to expand the headline. Eventually the main Page One head stretched across the page and became known as the *banner, streamer* or *ribbon.* On some papers it was called, simply, the *line.* This headline sometimes called for the largest type available in the shop. When metal type wasn't adequate for the occasion, printers fashioned letters from wood (called *furniture*). A 12-liner meant that the line was 12 picas, or 144 points (two inches).

During this period, the names of headline forms were derived from their use or position on the page. A story placed above the nameplate and banner headline is called a *skyline,* and the accompanying headline is known as a *skyline head.* Sometimes the skyline head stands alone but carries a notation about where the story can be found.

A headline may have several parts—the main headline and auxiliary headlines known as *decks, dropouts* or *banks.* These are not to be confused with *subheads* or lines of type (usually in boldface) sometimes placed between paragraphs in the story.

A *kicker* headline is a short line of display type, usually no larger than half the point size of the main headline and placed over the main part of the headline. On some papers the kicker is termed the *eyebrow* or *tagline.*

A *stet head* is a standing headline such as **TODAY IN HISTORY.** A *reverse plate* headline reverses the color values so that the letters are in white on a black background. A *reverse kicker,* in which one line in larger type is above the deck, is called a *hammer* or a *barker.*

As the tone of the newspaper was moderated after the turn of the century, so were the headlines. Banner headlines still shout the news, occasionally in red ink, but gloom and doom headlines have virtually disappeared. Understating is more likely to be found in headlines today than overstating. Extra editions have been out of fashion for a long time. And no longer do circulation managers hurry into the city room to demand a banner headline that will increase the newspaper's street sales.

Between World Wars I and II the cult of simplification, known as streamlining, brought changes in the newspaper headline. Designers put more air or white space into the head by having each line flush left, with a zigzagged, or ragged, right margin:

XXXXXXXX
XXXXXXX
XXXXXX

XXXXXX
XXXXXXX
XXXXXXXX

XXXXXXXX
XXXXXXX
XXXXXXXX

*(Continued)*

## *The Evolution of Headlines* *(continued)*

Urged by this spirit of simplification, they abolished the decorative *gingerbread,* such as fancy boxes, and reduced the number of banks or eliminated them altogether except for the deck reading out from a major head—called a *readout* or *dropout.* They argued that the *flush left* head was easier to read than the traditional head and that it was easier to write because the count was less demanding.

Another part of the streamlining process was the introduction of modern sans serif typefaces to challenge the traditional roman typefaces such as Century, Caslon, Goudy and Garamond (see Chapter 12). Advocates of the new design contended that sans serif faces such as Helvetica and Univers were less ornate than the roman ones, gave more display in the smaller sizes, contained more thin letters (thus extending the count) and allowed a greater mixture of faces because of their relative uniformity.

Headlines in all-capital letters gradually gave way to capital and lowercase letters, which are easier to read. In modern headline design, only the first word of the headline and proper names are capitalized. This form of headline capitalization is known as *downstyle.*

The wider columns in contemporary newspaper design give headline writers a better chance to make meaningful statements because of a better count in one-column heads. The trend away from vertical makeup and toward horizontal makeup provides more multicolumn headlines on the page. Such *spread heads* can be written effectively in one line.

Traditionally, the headline has headed the column and hence its name. But the headline need not necessarily go at the top of the news column.

Increased emphasis on news display in recent years has prompted designers to discard established rules in favor of headline styles that complement the story. More and more, they are borrowing design concepts from magazines. Thus, newspapers now contain flush-right headlines, hammer heads are proliferating, and decks, or dropouts, are returning. Through it all, the flush-left headline remains dominant.

and size (see Figure 8–2). The practice of designating headlines varies. Here are four methods:

1. **Number designation.** Headlines of one or two columns are assigned a number. A 2 head, for example, might call for a one-column, two-lined head in 24-point Bodoni, capitals and lowercase. A 22 head would be the same as a 2 head but in two columns. Sometimes the number corresponds to the type size and style. A 24 head would be in 24 points, and a 25 would be in 24 points italic.
2. **Letter designation.** Here letters are used to show the type family and size. If C calls for 30-point Vogue extrabold, the head might be indicated as follows: 2C =. This means two columns, two lines of 30-point Vogue extrabold. The letter may also indicate the headline style. If D, for instance, means a one-column head in three lines of type, ½D would be one column in two lines. Or if D is a headline with a deck, the ½D would be the same head but without the deck.

# Headline Schedule

This headline schedule lists the unit count (maximum count) per line of headlines of the indicated column width and type size. It does not list each headline by number of lines since, for example, the count per line for a 1-24-2 would be the same as that for a 1-24-3.

**BODONI BOLD AND BODONI BOLD ITALIC**

| Headline | Maximum | Headline | Maximum | Headline | Maximum |
|---|---|---|---|---|---|
| 1-14 | 21 | 4-36 | 39 | 4-60 | 23.5 |
| 1-18 | 18 | 2-48 | 14 | 5-60 | 30 |
| 1-24 | 13 | 3-48 | 21 | 6-60 | 36 |
| 2-24 | 26.5 | 4-48 | 29 | 2-72 | 7 |
| 1-30 | 10.5 | 5-48 | 37 | 3-72 | 14.5 |
| 2-30 | 23 | 6-48 | 44 | 4-72 | 20 |
| 1-36 | 9 | 2-60 | 11.5 | 5-72 | 24.5 |
| 2-36 | 19 | 3-60 | 17.5 | 6-72 | 30 |
| 3-36 | 29 | | | | |

Avoid using headlines not listed above. For example, 14BB would not be used across two columns.

**BODONI LIGHT AND BODONI LIGHT ITALIC**

| Headline | Maximum | Headline | Maximum | Headline | Maximum |
|---|---|---|---|---|---|
| 1-14 | 22 | 4-36 | 42 | 4-60 | 25.5 |
| 1-18 | 19 | 1-48 | 7 | 5-60 | 32 |
| 1-24 | 14 | 2-48 | 15 | 6-60 | 38.5 |
| 2-24 | 29 | 3-48 | 23.5 | 2-72 | 7.5 |
| 1-30 | 11 | 4-48 | 31.5 | 3-72 | 15.5 |
| 2-30 | 24 | 5-48 | 40 | 4-72 | 21.5 |
| 1-36 | 9.5 | 6-48 | 48 | 5-72 | 26.5 |
| 2-36 | 20.5 | 2-60 | 12.5 | 6-72 | 32 |
| 3-36 | 31 | 3-60 | 19 | | |

**POSTER BODONI**

| Headline | Maximum | Headline | Maximum | Headline | Maximum |
|---|---|---|---|---|---|
| 1-14 | 16 | 1-30 | 8 | 1-60 | 4 |
| 1-18 | 13 | 1-36 | 6.5 | 1-72 | 3 |
| 1-24 | 10.5 | 1-48 | 5 | | |

**Note:** To figure the maximum count for Poster Bodoni headlines of more than one column, multiply the maximum for one column by the number of columns.

Italic and Roman versions of the same headline type size should count the same.

In all cases, count spaces between words as one-half (0.5). Count numerals (except 1), the dollar sign ($), the percentage symbol (%) and devices of similar width as one and one-half (1.5).

*Figure 8–2 A sample headline schedule for a newspaper. Printed schedules such as this one are disappearing. Most newspapers now count headlines on computers, rendering unnecessary the old method of estimating fit by counting characters. The "maximum" referred to here is the maximum average character count, or unit count, that will fit on each line of a headline of the specified column width and point size. Computers make an exact calculation instead.*

3. **Designation by numbers and type family.** A designation such as 1-24-3 Bod. means one column, 24 points, three lines of Bodoni; 3-36-1 (or 3361) TBCI means three columns of 36 point, one line of Tempo bold condensed italic. At some papers, the first number designates the column width, the second number the number of lines and the third number the size of type (2/3/36).
4. **Computer code designations.** The code F (for font) 406, could well mean the 400 series (Gothic) and 6 could indicate the size of type (48 point). The code numbers thus identify the size and family of type, line width, number of lines and choice of roman or italic.

Here are some of the common abbreviations used in headline designations:

- *X* for extra (VXBI means Vogue extrabold italics).
- *x* for italics (2-30-2) SP *x* means two columns, two lines of 30-point Spartan italics).
- *It., I.* or *ital.* for italics.
- *K* for kicker line or eyebrow (2-30-2K).
- *H* for hammer and inverted kicker (3-36-1H); also called barker head.
- *J* for jump or runover head.
- *RO* for readout, or *DO* for dropout, under a main multicolumn headline.
- *Sh* for subhead.
- *R* for roman (VXBR is Vogue extrabold roman).

## COUNTING THE HEADLINE

Most newspapers now have computer systems that allow the copy editor to press a button and determine almost instantly whether a headline will fit. This is possible when computers are programmed with the width value of each character available in the typeface. At split-second speed the computer can add the width values of the characters the copy editor has assembled on the computer screen and determine whether that total exceeds the maximum width value of the line in the specified typeface, size and column width. The availability of this feature simplifies the copy editor's work. If it is not available, the copy editor turns to time-tested manual methods of calculation.

The easiest way to count a headline manually is with the typewriter system—one letter for all letters, figures, punctuation and space between words. If a line has a maximum of 18 units as specified by the newspaper's headline schedule (see Figure 8–2) and the head has 15 counts, it will fit, unless it contains several fat letters (such as M and W). In that case, the headline writer recounts the line using the popular Standard Method:

| STANDARD HEAD-COUNT METHOD | |
|---|---|
| Lowercase letters | 1 unit |
| Uppercase letters | $1\frac{1}{2}$ units |
| **EXCEPTIONS** | |
| Lowercase f, l, i, t and j | $\frac{1}{2}$ unit |
| Lowercase m and w | $1\frac{1}{2}$ units |
| Uppercase M and W | 2 units |
| Spaces | $\frac{1}{2}$ unit |
| Punctuation | $\frac{1}{2}$ unit |
| Uppercase I or L and numeral 1 | $\frac{1}{2}$ unit |

In the Standard Method system of headline counting, each character and space in each line of the headline is assigned a unit value based on the preceding chart. The unit values are then added to determine the count of the line. If the count is below the maximum (see Figure 8–2), the headline should fit. If it exceeds the maximum, the headline is too long and must be rewritten. In this example, the unit count is $23\frac{1}{2}$, assuming that only the first letter of the first word is capitalized.

**MAYOR ORDERS INVESTIGATION**

Some desk editors insist that each line of the head, even in a flush-left head, take almost the full count. Others argue there is merit in a ragged-right edge. Unless a special effect is desired, each line should fill at least two-thirds of the maximum type-line width, and some editors insist that each line come within two units of the maximum.

The copy editor should not write several heads for the same story and invite the editor to choose. The editor should submit only the best head possible within the available count.

Manual counting of heads is obviously a slow, inefficient system largely outdated by the arrival of computers. However, a few publications still rely on it, and that's why we mention it here. Only a few small magazines and newspapers still use it.

## FORMING THE HEADLINE

Occasionally the desk editor or some other executive will suggest an angle that should go into the head or call attention to an angle that should be avoided. Usually, however, headline writers are on their own. They know that within a matter of minutes they must edit the copy and create a headline that will epitomize the story and make a statement in an easy-to-digest capsule.

By the time copy editors have edited the copy they should have an idea brewing for the head. They begin by noting key words or phrases. These are their *building blocks.* With these blocks copy editors try to make an accurate and coherent statement. First they try to phrase the statement in the active voice. If that fails, they use the passive. If possible, they try to get key words and a verb in the top line.

Take a routine accident story:

> Three Atlantans were killed Friday when their car collided with a garbage truck on Highway 85 at Thames Road in Clayton County.
>
> The victims were:...
>
> Patrolman C.F. Thornton said the truck was driven north on Highway 85 by...
>
> The auto pulled into the highway from Thames Road and was hit by the truck, Thornton said.

The headline order calls for a one-column, three-line head with a maximum of nine units. In this story the lead almost writes the head. The obvious statement is "Three Atlantans killed as car and truck collide." **3 ATLANTANS** won't fit, so the writer settles for **3 KILLED.** For the second line the writer tries **IN TRUCK,** and for the third line **CAR COLLISION.** The headline writer discards **CAR COLLISION** because it is too long; **CAR CRASH** will fit. By changing the second line from **IN TRUCK** to **AS TRUCK** the writer gains another verb in the third line. The head now reads **3 KILLED AS TRUCK, CAR CRASH.**

Key headline words are like signposts. They attract the reader's attention and provide information. Such words, meaningfully phrased, produce effective headlines.

Note how quickly the key words *(Dutch, prince, born)* emerge in the lead of a wire story: "Crown Princess Beatrix gave birth to a son last night and Dutchmen went wild with joy at the arrival of a king-to-be in a realm where queens have reigned since 1890." One copy editor used the key words this way:

**DUTCH TREAT**
**A prince is born,**
**first in century**

Many dull heads can be improved if the writer will make the extra effort. Sometimes the first idea for the head is the best; often it is not. By rejecting the lifeless heads and insisting on better ones, an editor inspires everyone on the desk to try harder. Furthermore, a good performance on news heads is likely to generate better headlines in all departments of the paper.

The job of the copy editor is to create effective headlines for all copy (see Figure 8–3). The big story is often easier to handle than the routine one because the banner story has more action and thus more headline building blocks.

The usual death and wedding stories offer little opportunity for original headlines. There are only so many ways to announce a wedding or a death in a head, and the writer dare not try to be clever in handling these topics. The standing gag on the copy desk is, "Let's put some life in these obit heads." If Jonathan Doe dies,

*Figure 8–3 A copy editor tries to craft a headline for a news story.*
Photo by Philip Holman.

that is all the headline can say, except to include his age: **JONATHAN DOE DIES AT 65** or **JONATHAN DOE DEAD AT 65.** If he were a former mayor, that fact would be used: **JONATHAN DOE, EX-MAYOR, DIES.**

## THE TITLE LESSON

Beginning headline writers might start with what Carl Riblet Jr., an expert on the subject, calls the title lesson. Learners start by listing the titles of all the books they have read. This is to demonstrate that readers can recall titles even if they have forgotten the contents. It also demonstrates the effectiveness of an apt title. A good title helps sell the product, as illustrated by those revised by alert publishers: "Old Time Legends Together with Sketches, Experimental and Ideal" *(The Scarlet Letter),* "Pencil Sketches of English Society" (*Vanity Fair),* "The Life and Adventures of a Smalltown Doctor" *(Main Street),* "Alice's Adventures Underground" *(Alice in Wonderland).*

The beginner limited to a one-line head with a maximum of 15 or 20 units is forced to pack action into a few words, to merge as many elements as possible from the story, to indicate the tone or mood of the story and, finally, to compose an interesting statement. Riblet gives this example:

> A young man said he and a middle-aged business agent had been drinking in a saloon. They went to the older man's apartment. There they quarreled over a gambling debt. The younger man told police he was threatened with a .30-caliber rifle. In self-defense, he picked up a bow and arrow and shot the older man in the stomach. The younger man tried to pull out the arrow but it broke and the wounded man died.

The copy editor got the story with instructions to write a one-line, 20-count head. Knowing the paper's rule that every headline must contain a verb, the editor wrote: **BUSINESS AGENT SLAIN.** In this instance it would seem wiser to bend the rule and get the key words in the headline: **BOW AND ARROW KILLING.** This lesson has practical value in writing the one-line head, such as the banner **MAYOR CENSURED** or **STEELERS NO. 1,** the kicker **SMELLY PROBLEM** or **CIRCUS MAGIC,** the column heading **THE GOLDKEEPERS,** the filler head **POSTAL AUCTION** or the magazine feature **ACROBAT ON SKIS.**

## HEADLINE RULES

Practices vary, but on some desks no headline writer is permitted to hang conjunctions, parts of verbs, prepositions or modifiers on the end of any line of a headline. Like many newspaper rules, this one can be waived, but only if an exceptionally better headline can be created.

Rigid rules can bring grief to copy desks. One newspaper group has two ironclad rules: No head may contain a contraction. Every head must contain a verb. This leads to some dull and tortured heads, especially on offbeat feature stories.

The *Los Angeles Times* evidently has no policy against splitting ideas in heads. It may be argued that readers see the one-column headline as a unit and do not have to read each line separately as they would in a multicolumn head. Eventually experiments may be designed to test whether readers can comprehend a split head as easily as they can one that breaks on sense. Until the results of such tests are available, headline writers should phrase headlines by sense. The practice can't hurt them if they join a desk that tolerates splits. A talented copy editor will take no longer on a phrased headline than on a split headline.

Good phrasing in a headline helps the reader grasp its meaning quickly. Each line should be a unit in itself. If one line depends on another to convey an idea, the headline loses its rhythm. It may cause the reader to grope for the meaning. Note the differences in the original and the revised heads:

| ORIGINAL | REVISED |
|---|---|
| **MEN NEED TO<br>SLEEP LONGER,<br>REPORT SHOWS** | **MEN REQUIRE<br>MORE SLEEP,<br>STUDY REVEALS** |

| | |
|---|---|
| **STAY CALM, BROWNS, YOU'LL<br>SOON KNOW YOUR TITLE FOE** | **JUST STAY CALM BROWNS;<br>YOU'LL KNOW FOE SOON** |
| **JONES LOSES NEW<br>BID FOR FREEDOM** | **JONES AGAIN LOSES<br>BID FOR FREEDOM** |
| **FOUR RUSSIAN<br>WOMEN TOUR<br>DENVER AREA** | **FOUR WOMEN<br>FROM RUSSIA<br>TOUR DENVER** |
| **ORDNANCE STATION TO<br>WELCOME COMMANDER** | **NAVAL ORDNANCE STATION<br>TO WELCOME COMMANDER** |
| **U.S. EYEING NEW MEXICO<br>SITE FOR NUCLEAR WASTES** | **U.S. FAVORS NEW MEXICO<br>FOR NUCLEAR WASTE BURIAL** |
| **POLICE WORK DRAWS<br>IN JAPANESE WOMEN** | **POLICE WORK LURES<br>JAPANESE WOMEN** |

What follows is a list of headline rules one editor distributed to his staff. He did so with the warning: "Rule No. 1 is that any other rule can be broken if you have a valid reason for doing so." That's good advice, even though many of these rules are common in newsrooms around the country. His list:

- Draw your headline from information near the top of the story. If the story has a punch ending, don't give it away in the headline.
- Build your headline around key words—those that must be included.
- Build on words in the story, paraphrasing if necessary, but it's best to avoid parroting the lead.
- Emphasize the positive unless the story demands the negative.
- Include a subject and predicate (expressed or implied).
- Try to use a verb in the top line.
- Maintain neutrality.
- Remember the rules of grammar and observe them.
- Eliminate articles and most adjectives and adverbs—except in feature headlines.
- Try to arouse the reader's interest. Remember that one function of the headline is to attract the reader to the story.
- Try to capture the flavor of the story.
- Make certain your headline is easy to read.
- Abbreviate sparingly.
- Verify the accuracy of your headline and be certain it has no double meaning.
- Verify that you've written the headline required.
- Verify the count of each line.

- Use short, simple words, but avoid such overworked and misused words as *flay, rap, eye* (as a verb) and *probe.* The use of such words is known as *headlinese.*
- Make the headline definite. Tell specifically what happened.
- If the main element is a characterization, the subsidiary element should support that characterization.
- Never exaggerate. Build the headline on facts in the story. If a statement is qualified in the story, it must be qualified in the headline as well.
- Make the headline complete in itself. Each headline must have a subject and a predicate. The implied verb *is* may be used. (**Note:** In feature headlines, this rule is sometimes violated. You may want to use a headline that is similar in construction to a book or magazine, suggestive of the story rather than a synopsis of it.)
- Never write a headline that begins with a verb and has no subject, such as **VOTE AGAINST COMPENSATION BILL.** The result is a headline that commands the reader to do something, a meaning entirely different than the one intended.
- Phrase your headline in the present tense if possible, even though the event has been completed. This form is known as the historical present. Do not, however, link the present tense with a past date, such as **JOHN DOE DIES YESTERDAY.** If a past date is so important that it must be used, the verb should be in the past tense. In most cases, the date can be omitted: **JOHN DOE DIES OF PNEUMONIA.**
- Do not use the present tense to indicate future events unless some word is included to clarify the meaning, as **CITY COUNCIL MEETS TONIGHT.**
- Don't write a headline after a single reading of the story.
- Don't be afraid to ask questions if you fail to understand the story.
- Don't just fill a line. Say something.
- Don't violate the rules of headline writing just to make a line fit.
- Don't use common last names *(Smith, Jones,* etc.) or other names that are not easily recognized.
- Don't use the speaker's name in the top line. What he or she said is more important.
- Don't use *is* and *are* in headlines, except feature headlines.
- Don't mislead the reader.
- Don't split nouns and their modifiers, verb forms and prepositional phrases over two lines.
- Don't use *said* when you mean *said to be.*
- Don't use *feel* for *believes* or *thinks.*

- Don't pad headlines with unnecessary words.
- Don't use double quotation marks in headlines or in other devices using headline type.
- Don't use slang.
- Don't write question heads.

Some of these rules, and a few this editor missed, merit further explanation.

## Use the Present Tense

Most news concerns past events. But to give the effect of immediacy, the headline uses the present tense for past events:

**BRITISH DOCTORS VOUCH FOR BIRTH CONTROL BILL**

**CITY HAS BUMPER CROP OF JUNK CARS**

**JONATHAN DOE DIES AT 65**

If, however, the headline announces a future event in the present tense, the reader won't know whether the event has occurred or will occur. **POWELL'S WIFE TELLS EVERYTHING** means that the wife has testified. But if the reader learns from the story that the testimony will not be given until the following day, the reader knows the head was misleading. The head should have read **POWELL'S WIFE TO TELL EVERYTHING.** The present tense can never be used if the date is included on a past event: **JONATHAN DOE DIES WEDNESDAY.**

On future events the headline may use the future, *will be;* the infinitive, *to be;* or the present: **TRAFFIC PARLEY OPENS MONDAY.**

## Punctuate Headlines Correctly

The period is never used in headlines except after abbreviations. Instead of full quotation marks, use single quotation marks, which take less space and may be more appealing typographically in headline-size type. The comma may replace *and,* and a semicolon may even indicate a complete break:

**TUMBLING SPACECRAFT TANGLES CHUTE;**
**COSMONAUT PLUMMETS TO DEATH**

The semicolon also indicates a full stop. Unless it is a last resort, neither the dash nor the colon should be used as a substitute for *says.* When used, the colon comes after the speaker, and the dash after what was said:

**MCCOY: DUAL ROLE TOO BIG**

**DUAL ROLE TOO BIG–MCCOY**

## Use Abbreviations Sparingly

Few beginning headline writers have escaped the abbreviation addiction. It occurs when the writer tries to cram too much into the head. The story said a woman under hypnosis had imagined herself as a reincarnation of an 18th-century Bridey Murphy in Ireland. The theory was discounted by a professor of psychology at a state university. This is how a student headlined the story: **CU PSYCH. PROF. DOUBTS B.M. STORY.** A simple head such as **PROFESSOR DOUBTS 'BRIDEY' CLAIMS** would have given readers enough information to lead them into the story.

Abbreviations clutter the headline: **MO. VILLAGE U.S. CHOICE FOR PAN-AM.** It could have been written as **MISSOURI FAVORED AS PAN-AM SITE.**

An abbreviation that has more than one meaning leads to confusion, especially when the headline writer is also guilty of poor phrasing: **TEN GIRLS ARE ADDED TO ST. VINCENT CANDY STRIPER UNIT** or **ILL. MAN ASKS PA. TO JOIN MISS. IN MASS. PROTEST** or **NEBRON NEW L.A. MUNI ASST. P.J.** (Nebron was elected assistant presiding judge of the Los Angeles Municipal Court).

Headline writers frequently overestimate the ability of readers to understand the initials used in the headlines. Some are easily recognized, such as FBI, CIA, NBC. Others aren't, such as AAUW, NAACP, ICBM. On many newspapers the style calls for abbreviations and acronyms without periods in headlines. Other newspapers use periods for two-letter abbreviations, including U.S. and U.N., but TV is a frequent exception.

Some contractions are acceptable in heads; others aren't. *Won't, don't* and *shouldn't* give no trouble, but *she'll, he'll, who're* and the "s" and "d" contractions do: **TRIPLETS 'FINE'; SO'S THE MOTHER OF 22** or **MOTHER'D RATHER SWITCH THAN FIGHT.** Try to read this aloud: **ANY MORE SERVICE, THE TOWN HALL'LL COLLAPSE.**

## Use Correct Grammar

Although headline writers must constantly compress statements, they have no license to abuse the language. A grammatical error emblazoned in 48-point type may be more embarrassing than a half-dozen language errors buried in body type. The writer normally would say **RUSSIAN GIRLS URGED TO STOP COPYING PARIS.** But the second line was too long so the writer settled for **RUSSIAN GIRLS URGED 'STOP COPYING PARIS'.** A comma should have been used to introduce the quoted clause. Transposing the lines would have produced a better head. A headline read **JONES BEST OF 2 CHOICES, SENATOR SAYS.** He can't be the best if there are only two choices.

One headline read: **WOMAN REPORTS SHE IS ROBBED BY MAN POSING AS INSPECTOR.** The present "reports" is correct. However, the second verb should be "was" to show that she is reporting a previous event.

A standing rule on use of proper names in headlines is that the names should be instantly recognizable to most readers. **UNION UNDER INVESTIGATION GAVE TO TIER-**

**NAN CAMPAIGN.** Unless this headline is for a Rhode Island paper, most readers won't recognize the name of a former Rhode Island congressman.

Copy editors who can't catch spelling errors have no place on the copy desk. They are a menace if they repeat the errors in the headline, as in these:

**RODEO PARADE HAS GOVERNOR AS MARSHALL**

**KIDNAP VICTIM TRYS TO IDENTIFY CAPTORS**

## Avoid Sports Clichés

Sports pages report on contests involving action and drama, and headlines over such stories should be the easiest in the paper to write. Yet, because of the jargon used by sports writers and the numerous synonyms signaling a victory of one team over another, the sports story headline has become a jumble.

A reader says to his companion, "I see that the Jayhawks crunched the Cornhuskers." "Yeah?" asks a University of Nebraska basketball fan. "By how much?" "Ninety-eight to 94," replies the reader. If that's a crunching, what verb describes a 98-38 victory?

The struggle to find substitutes for *wins, beats* and *defeats* produces verbs such as *bests, downs, smears* and *swamps*. Presumably, the reader reads a sports page to find out who wins in what races. What does it matter, then, if simple words like *wins* or *defeats* are used over and over? Certainly they are better than editorialized counterparts such as *clobbers, wallops, flattens* and *trounces*.

## Avoid Slang

A straight head that tells the reader precisely what happened is always better than one in which the writer resorts to slang. Slang in heads, as well as in copy, lowers the tone of the paper and consequently lowers the readers' estimation of the paper. The headline, no less than the copy, should speak to the general reader, not to reporters and other specialists.

Here is an object lesson from a San Diego newspaper. The second edition carried a six-column head in the lifestyle section: **KIDS GOING TO POT ARE AIDED.** Under the head was a three-column photo showing a girl sitting on a bed in a holding room at Juvenile Hall. The picture also revealed a toilet stool in one corner of the room. To the relief of the embarrassed editor, the top line in the third edition was revised to **TEEN NARCOTIC USERS AIDED.**

## Avoid Word Repetition

Another restriction on headline writing is that major words in the headline cannot be repeated except for effect in a feature head. The rule is intended to prevent obvious padding, such as **CAMPUS TO LAUNCH CAMPUS CHEST DRIVE** and **WIND-LASHED BLIZZARD LASHES PLAINS STATES.**

If the main head contains a word like *fire,* the secondary headline could easily include a synonym. *Blaze* would be acceptable; *inferno* would not.

Repetition is sometimes used deliberately to heighten a feature:

**THINKERS FAILURES, PROFESSOR THINKS**

**NEW LOOK? NEVER! OLD LOOK'S BETTER**

**POKEY DRIVING SENDS THREE BACK TO POKEY**

## Eliminate Headlinese

Faced with the problem of making a statement of a nine-unit line, headline writers have to grab the shortest nouns and verbs possible. They are tempted to use overworked words such as *hits, nabs, chief* or *set* because such words help to make the headline fit. Or they may reach for words with symbolic meanings, such as *flays, slaps, grills, hop* or *probe.* Nothing is *approved;* it is *OK'd* or *given nod.* All are headline clichés and have no place in today's newspaper.

The headline should not falsify the story. Many of the headlinese words do, at least by implication. If the story tells about the mayor mildly rebuking the council, the headline writer lies when he or she uses verbs like *hits, slaps, scores, raps, rips* or *flays.* An *investigation* or a *questioning* is not necessarily a *grilling;* a *dispute* is not always a *row* or a *clash. Cops* went out with prohibition; today's word is *police.*

Others that should be shunned are "quiz" for *question,* "hop" for *voyage,* "talks" for *conference,* "aide" for *assistant,* "chief" for *president* or *chairman,* "solon" for *legislator* or *congressman,* "probe" for *inquiry,* "nabs" for *arrests,* "meet" for *meeting,* "bests" for *defeats,* "guts" for *destroys,* "snag" for *problem,* "stirs" for *incites* and "hike" for *increase.*

## Watch for Libel

Because of the strong impression a headline may make on a reader, courts have ruled that a headline may be actionable even though the story under the head is free of libel. Here are a few examples:

**SHUBERTS GOUGE $1,000 FROM KLEIN BROTHERS**

**'YOU WERE RIGHT,' FATHER TELLS COP WHO SHOT HIS SON**

**McLANE BARES OLD HICKORY FRAUD CHARGES**

**DOCTOR KILLS CHILD**

**A MISSING HOTEL MAID BEING PURSUED BY AN IRATE PARENT**

**JOHN R. BRINKLEY–QUACK**

A wrong name in a headline over a crime story is one way to involve the paper in a libel action.

The headline writer, no less than the reporter, must understand that under the U.S. justice system a person is presumed innocent of any crime charged until proved guilty by a jury. Heads that proclaim **KIDNAPPER CAUGHT, BLACKMAILER EXPOSED, ROBBER ARRESTED** or **SPY CAUGHT** have the effect of convicting the suspects (even the innocent) before they have been tried.

If unnamed masked gunmen hold up a liquor store owner and escape with $1,000 in cash, the head may refer to them as "robbers" or "gunmen." Later, if two men are arrested in connection with the robbery as suspects or are actually charged with the crime, the head cannot refer to them as "robbers" but must use a qualifier: **POLICE QUESTION ROBBERY SUSPECTS.** For the story on the arrest the headline should say **TWO ARRESTED IN ROBBERY,** not **TWO ARRESTED FOR ROBBERY.** The first is a shortened form of "in connection with"; the second makes them guilty. Even "in" may cause trouble. **THREE WOMEN ARRESTED IN PROSTITUTION** should be changed to **THREE WOMEN CHARGED WITH PROSTITUTION.**

The lesson should be elementary to anyone in the publishing business, but even the more carefully edited papers are sometimes guilty of printing heads that jump to conclusions. This was illustrated in the stories concerning the assassination of President John F. Kennedy. Lee Harvey Oswald was branded the assassin even though, technically, he was merely arrested on a charge of murder. In a statement of apology, the managing editor of *The New York Times* said his paper should not have labeled Oswald an assassin.

In their worst days, newspapers encouraged headline words that defiled: **FANGED FIEND, SEX MANIAC, MAD-DOG KILLER.** Even today some newspapers permit both reporter and copy editor to use a label that will forever brand the victim. When a 17-year-old boy was convicted of rape and sentenced to 25 to 40 years in the state penitentiary, one newspaper immediately branded him *Denver's daylight rapist.* Another paper glorified him as *The phantom rapist.* Suppose an appeal reverses the conviction? What erases the stigma put on the youth by the newspaper?

The copy editor who put quotation marks around **HONEST COUNT** in an election story learned to his sorrow that he had committed libel for his paper. The implication of the quotes is that the newspaper doesn't believe what is being said.

## Some Miscellaneous Rules

Emphasis in the headline should be on the positive rather than the negative when possible. If the rodeo parade fails to come off as scheduled because of rain, the head makes a positive statement: **RAIN CANCELS RODEO PARADE,** not **NO RODEO PARADE BECAUSE OF RAIN.** The news value is lacking in the headline that says **NO ONE HURT AS PLANE CRASHES.** The positive statement would be **90 PASSENGERS ESCAPE INJURY AS**

**PLANE CRASHES INTO MOUNTAIN.** Here are three negating words in a headline: **TAX WRITERS VETO LIDS ON OIL WRITE-OFF.** Better: **TAX WRITERS LEAVE OIL TAX AS IT STANDS.**

The negative is illustrated in this headline from an English paper: **ONLY SMALL EARTHQUAKE; BUT NOT MANY KILLED.**

This admonition does not apply to feature heads in which the negative helps make the feature:

**NO LAWS AGAINST DROWNING, BUT IT'S UNHEALTHY**

**NOT-SO-GAY NINETIES** (on weather story)

**LAUNDRY GIVES NO QUARTER UNTIL SUIT IS PRESSED**

The question head, except on features, is suspect for two reasons: It tends to editorialize, and newspaper heads are supposed to supply answers, not ask questions. If the headline asks the reader a question, the answer, obviously, should be in the story. If the answer is buried deep in the story, the question headline should be shunned. A five-column head asked, **DID ANASTASIA MURDER HELP KILL BARBER SHAVES?** The lead repeated the same question, but the reader was compelled to look through a dozen paragraphs only to learn that the question referred to a frivolous remark that should have been used only to color the story.

## COMMON HEADLINE PROBLEMS

The headline must be as accurate as the story itself. Like inaccurate stories, inaccurate headlines invite libel suits and destroy one of the newspaper's most valuable commodities—its credibility. Inaccuracy, though, is merely one of many pitfalls in headline writing. Other common problems include overstating, missing the point of the story, taking statements out of context, confusing the reader, attempting poor or inappropriate puns, showing poor taste, using words with double meanings and giving away the punch line of a suspended-interest story.

### Inaccurate Headlines

The key to ensuring accuracy is close and careful reading of the story. Erroneous headlines result when the copy editor doesn't understand the story, infers something that is not in the story, fails to portray the full dimension of the story or fails to shift gears before moving from one story to the next. Some examples:

**MINISTER BURIED IN HORSE-DRAWN HEARSE**
(The hearse and horse participated in the funeral procession, but they were not buried with him.)

**COWBOYS NIP JAYHAWKS 68-66 ON BUZZER SHOT**
(The lead said that Kansas [the Jayhawks] beat Oklahoma State [the Cowboys] by two points in a Big Eight conference basketball game.)

## *Headline Writing Is Fun*

It's fun to write headlines because headline writing is a creative activity. Copy editors have the satisfaction of knowing that their headlines will be read. They would like to think that the head is intriguing enough to invite the reader to read the story. When they write a head that capsules the story, they get a smile from the executive in the slot and, sometimes, some praise.

Somerset Maugham said you cannot write well unless you write much. Similarly, you can't write good heads until you have written many. After copy editors have been on the desk for a while, they begin to think in headline phrases. When they read a story, they automatically reconstruct the headline the way they would have written it. A good headline inspires them to write good ones, too.

They may dash off a head in less time than it took them to edit the copy. Then they may get stuck on a small story. They may write a dozen versions, read and reread the story and then try again. As a last resort, they may ask the desk chief for an angle. The longer they are on the desk the more adept they become at shifting gears for headline ideas. They try not to admit that any head is impossible to write. If a synonym eludes them, they search the dictionary or a thesaurus until they find the right one.

If they have a flair for rhyme, they apply it to a brightener: **NUDES IN A POOL PLAY IT COOL AS ONLOOKERS DROOL.**

Every story is a challenge. After the writer has refined the story, it almost becomes the copy editor's story. The enthusiasm of copy editors is reflected in a newspaper's headlines. Good copy editors seek to put all the drama, the pathos or the humor of the story into the headline. The clever ones, or the "heady heads," as one columnist calls them, may show up later in office critiques or in trade journals:

**COUNCIL MAKES SHORT WORK OF LONG AGENDA**

**HEN'S WHOPPER NOW A WHOOPER**

**STOP THE CLOCK; DAYLIGHT TIME IS GETTING OFF**

**LAKE CARRIERS CLEAR DECKS FOR BATTLE WITH RAILROADS**

**'DOLLY' SAYS 'GOLLY' AFTER HELLOFUL YEAR**

**TICKETS CRICKET, LEGISLATORS TOLD**

**QUINTS HAVE A HAPPY, HAPPY, HAPPY, HAPPY, HAPPY BIRTHDAY**
(First birthday party for quintuplets)

**PADUCAH'S BONDING LAW SAID HAZY**
(The details were hazy; the subject of the story was hazy about the details.)

**3 IN FAMILY FACE CHARGES OF FRAUD**
(They were arrested in a fraud investigation, but the charges were perjury.)

**BLACK CHILD'S ADOPTED MOTHER FIGHTS ON**
(The child didn't select the mother; it was the other way around. Make it "adoptive.")

**DO-NOTHING CONGRESS IRKS U.S. ENERGY CHIEF**
(The spokesman criticized Democrats, not Congress as a whole, and the "do-nothing" charge was limited to oil imports.)

**FOUR HELD IN ROBBERY OF PIGGY BANK**
(The bank was the loot; the robbery was at a girls' home.)

**WHITE HOUSE HINTS AT CEILING ON OIL SPENDING**
(The subject of the story was oil imports.)

**GREEK PLEBISCITE SET DEC. 8**
(The vote was set for Dec. 8. The setting didn't come on that date.)

**IF THAT POOCH BITES YOU CAN COLLECT $200**
(The story said the animal had to have rabies for you to collect up to $200.)

**DIDN'T LIKE HER FACE, SHOOTS TV ANNOUNCER**
(A man fired a gun at her but missed.)

**GRAHAM BACKS STERILIZATION VIEW**
(Although he backed one viewpoint, he criticized the program generally.)

**BISHOP SAYS SEGREGATION IS JUSTIFIED**
(The lead quoted the bishop as saying that Christians are not morally justified in aiding segregation.)

**SCHOOLS HELP PAY EXPENSES FOR FORSYTH COUNTY'S DOGS**
(The story said the dog-tax money helps support the schools.)

**CIVIC BALLET AUDITIONS SCHEDULED SATURDAY**
(One listens to auditions; one looks at tryouts or trials.)

**YOUTH, 19, BREAKS PAROLE, SUBJECT TO WHIPPING POST**
(The story correctly said he had been on probation.)

**CIRCUS CLOWN'S DAUGHTER DIES FROM HIGH-WIRE FALL**
(Second paragraph of story: "The 19-year-old aerialist was reported in good condition at Paterson General Hospital with fractures of the pelvis, both wrists and collarbone.")

**INDIA-CHINA RELATIONS WORSEN**
(The story was datelined Jakarta and had nothing to do with India.)

**CHICAGO TO SEE OUTSTANDING MOON ECLIPSE**
(The story wasn't about a moon eclipse; it was about an eclipse of the sun. And "outstanding" is hardly an appropriate way to describe an eclipse.)

**BRILLIANT MODERN STUDY OF HOLY ROMAN EMPEROR**
(The story was a review of a book on Emperor Hadrian, who was born A.D. 76 and died in 138. The Holy Roman Empire lasted from A.D. 800 to 1806.)

**ISRAELIS RELEASE ARABS TO REGAIN HEROES' BODIES**
(Nowhere does the story call them heroes. The two were executed during World War II for assassinating a British official.)

## Overstating

Akin to the inaccurate headline is one that goes beyond the story, fails to give the qualifications contained in the story or confuses facts with speculations.

Examples:

**WEST LOUISVILLE STUDENTS AT UL TO GET MORE AID**
(The lead said they may get it.)

**INTEGRATION MAY IMPROVE LEARNING, STUDY INDICATES**
(But the lead was two-sided, indicating an improvement in one area and a loss in another. The headline should reflect divided results or views, especially in highly emotional news subjects.)

**PAKISTAN, U.S. DISCUSS LIFTING OF EMBARGO ON LETHAL WEAPONS**
(The story said, correctly, that the embargo may be eased. And aren't all weapons lethal?)

**ARABS VOTE TO SUPPORT PLO CLAIM TO WEST BANK**
(The story said that Arab foreign ministers voted to recommend such action to their heads of state. The head implies final action.)

**SCHOOLS GET 60% OF LOCAL PROPERTY TAX**
(This reflects fairly what the lead said but fails to reveal an explanation, later in the story, that the schools get a proportion of the contributions of various levels of government —federal, state and local. Although the local property tax contributes 60 percent, the amount is far less than 60 percent of the total local property tax.)

Here's what happens when a copy editor gets carried away:

**BACON ENLIVENS BUTTERMILK MUFFINS**
(The genius who thought of enlivening buttermilk muffins must be assumed to have done so with his tongue in his cheek and a piece of paper in his mouth.)

**COUNTY HEAVENS TO EXPLODE WITH COLOR**
(Perhaps a booster head for a fireworks display, but slightly exaggerated.)

## Commanding

Headlines that begin with verbs can be read as commands to the reader and should be avoided. A New York City newspaper splashed a 144-point headline over the story of the shooting of Medgar Evers, a civil rights advocate. The head: **SLAY NAACP LEADER!** Another head may have given the impression that a murder was being planned: **SLAYING OF GIRL IN HOME CONSIDERED.** In reality, police were trying to determine whether the murder took place in the girl's house or in a nearby field.

Here are some examples of heads that command:

**SAVE EIGHT**
**FROM FIRE**

**ARREST 50 PICKETS**
**IN RUBBER STRIKE**

**ASSASSINATE U.S. ENVOY**

**BUY ANOTHER**
**SCHOOL SITE**

**FIND 2 BODIES,**
**NAB SUSPECT**

## Editorializing

The reporter has ample space to attribute, qualify and provide full description. The copy editor, however, has a limited amount of space in the headline to convey the meaning of the story. As a result, there is a tendency to eliminate necessary attribution or qualification and to use loaded terms such as *thugs, cops, pinkos, yippies* and *deadbeats* to describe the participants. The result is an editorialized headline.

Every word in a headline should be justified by a specific statement within the story. Was the sergeant who led a Marine platoon into a creek, drowning six recruits, drunk? Most headlines said he was, but the story carried the qualification "under the influence of alcohol to an unknown degree."

Consider this common construction: **AUTHOR SCORES FEDERAL MISUSE OF STRIP STUDIES.** The head states as a fact that the federal government is misusing studies on strip mining. The story credited that view only to the author. In other words, the headline states an opinion as a fact.

Even though the headline reports in essence what the story says, one loaded term will distort the story. If Syria, for reasons it can justify, turns down a compromise plan offered by the United States concerning the Golan Heights problem, the head creates a negative attitude among readers when it proclaims **SYRIA SPURNS U.S. COMPROMISE.**

It is often difficult to put qualification in heads because of count limitations. But if the lack of qualifications distorts the head, trouble arises. A story explained that a company that was expected to bid on a project to build a fair exhibit was bowing out of the project because the exhibit's design was not structurally sound. The headline, without qualification, went too far and brought a sharp protest from the construction company's president: **BUILDER QUITS, CALLS STATE WORLD'S FAIR EXHIBIT 'UNSOUND'.**

## Sensationalizing

Another temptation of the headline writer is to spot a minor, sensational element in the story and use it in the head. A story had to do with the policy of banks in honoring outdated checks. It quoted a bank president as saying, "The bank will take the checks." In intervening paragraphs several persons were quoted as having had no trouble cashing their checks. Then in the 11th paragraph was the statement: "A Claymont teacher, who refused to give her name, said she had tried to cash her check last night, and it had been refused." She was the only person mentioned in the story as having had any difficulty. Yet the headline writer grabbed this element and produced a head that did not reflect the story:

**STATE PAYCHECKS DATED 1990**
**Can't cash it,**
**teacher says**

## Missing the Point

The process of headline writing starts as soon as a copy editor starts reading the story. If the lead can't suggest a headline, chances are the lead is weak. If a stronger element appears later in the story, it should be moved closer to the lead. The point is so elementary that every reporter should be required to take a turn on the copy desk if for no other reason than to learn how to visualize the headline before beginning the story.

Although the headline ideally emerges from the lead, and generally occupies the top line (with succeeding lines offering qualifications or other dimensions of the story), it frequently has to go beyond the lead to portray the full dimensions of the story. When that occurs, the qualifying paragraphs should be moved to a higher position in the story. Example: **U.S. COMPANY TO DESIGN SPYING SYSTEM FOR ISRAEL.** The lead was qualified. Not until the 15th paragraph was the truth of the head supported. That paragraph should have been moved far up in the story.

The head usually avoids the exact words of the lead. Once is enough for most readers. Lead: "Despite record prices, Americans today are burning more gasoline than ever before, and that casts some doubt on the administration's policy of using higher prices to deter use." Headline: **DESPITE RECORD GASOLINE PRICES, AMERICANS ARE BURNING MORE FUEL.** A paraphrase would avoid the repetition: **DRIVERS WON'T LET RECORD GAS PRICES STOP THEM FROM BURNING UP FUEL.** Since the story tended to be interpretive, the head could reflect the mood: **HANG THE HIGH PRICE OF GASOLINE, JUST FILL 'ER UP AND LET 'ER ROAR.** Most copy editors try to avoid duplicating the lead, but if doing so provides the clearest possible head, it is a mistake to obfuscate.

## Boring the Reader

A headline that gives no more information than the label on a vegetable can is aptly known as a label head. Generally these are the standing heads for columns that appear day by day or week by week, like **SOCIAL NOTES** or **NIWOT NEWS.** They tell the reader nothing. They defy the purpose of a display line, which is to lure the reader.

Almost as bad are the yawny, ho-hum heads that make the reader ask, "So what else is new?" The writer who grabs a generality rather than a specific for the head is more than likely to produce a say-nothing head. Such writers prefer **MANY PERSONS KILLED** to **1,000 PERSONS KILLED.** *Factors* is about the dullest headline word imaginable. **FACTORS SLOW CAR INPUT.** Something is responsible for either output or input. In this case, a parts shortage was the main cause.

Notice how little information is provided in the following samples:

**FINANCIAL PROGRAM<br>EXPLAINED**

**DEVELOPMENT PLANS<br>DESCRIBED**

**NEWARK ROTARY TOLD OF PLANNING**

**CLASS NIGHT TO BE TODAY**

**PAN AM JET LANDS SAFELY**

**WADSWORTH DERAILMENT PUTS 13 CARS OFF RAILS**

**CARS COLLIDE AT INTERSECTION**

**CORONER SEEKS CAUSE OF WHY DRIVER DIED**

**AUTOPSY SCHEDULED FOR DEAD AKRONITE**

**BROYHILL SPEAKS AT BIG PICNIC**

**ROTARIANS HEAR KOREAN BISHOP**

**MEETING TO BE HELD**

## Stating the Obvious

Readers read newspapers to get the news. If the headline tells them the obvious, they have been short-changed. Here are examples of the obvious statement:

**FALL SHIRTS OFFER NEW INNOVATIONS**
(Not to be confused with those old innovations.)

**CORN FIELD SELECTED AS PLACE FOR ANNUAL COUNTY HUSKING BEE**

**WARM HOUSE BEST IN COLD CLIMATE**

**TURKISH SHIP SINKS IN WATER NEAR CYPRUS**

**CALIFORNIAN, 20, DROWNS IN WATER OF LAKE MOHAVE**

## Rehashing Old News

Some stories, like announcements, offer little or no news to invite fresh headlines. Yet even if the second-day story offers nothing new, the headlines cannot be a repetition of the first-day story lead.

Suppose on Monday the story says that Coach Mason will speak at the high school awards dinner. If Mason is prominent, his name can be in the head: **MASON TO SPEAK AT AWARDS DINNER.** On Thursday comes a follow-up story, again saying that Coach Mason will be the awards dinner speaker. If the headline writer repeats the Monday headline, readers will wonder if they are reading today's paper. The desk editor will wonder why copy editors won't keep up with the news. The problem is to find a new element, even a minor one, like this: **TICKETS AVAILABLE FOR AWARDS DINNER.** So the dinner comes off on Friday, as scheduled. If the Saturday headline says **MASON SPEAKS AT AWARDS DINNER,** readers learn nothing new. The action is what he said: **MASON DENOUNCES 'CRY-BABY' ATHLETES.** Or if the story lacks newsworthy quotes, another facet of the affair goes into the head: **30 ATHLETES GET AWARDS.**

## Missing a Dimension

The headline must portray the story in context. That is, the headline should not repeat what was said yesterday or the day before that or a week ago. It may be a second-day story with a fresh angle. To judge the news fairly and accurately, the copy editor must keep up with the news through daily reading of a newspaper.

There is no trick in writing a vague, generalized headline statement. The real art of headline writing comes in analyzing the story for the how, the why or the consequences. If the story has more than one dimension, the head should reflect the full story, not part of it. Take this horrible example:

**D.J. COMPANY UNION (IAPE) WITHDRAWS COLA PROVISION**

That's three too many abbreviations. Worse, the head fails to capture the real meaning of the story, which was that the union (Independent Association of Publishers Employees Inc.) reached an agreement with Dow Jones & Co. providing for across-the-board pay increases. In return, the union agreed to withdraw its demand for a cost-of-living adjustment (COLA). And IAPE is an independent union, not a company union.

Other examples:

**DOCTORS OFFER SNIPER REWARD**
(The reward was not offered to him, it was offered for his capture.)

**REPORT SAYS SCHOOL 'INTEGRATION' STILL NOT ACHIEVED**
(The report was already several days old. This was a columnist's view of the matter, and it called for something more than a straight news head.)

**CARROLL TO USE FUNDS FOR COAL-ROAD REPAIR**
(This was not the news. The governor had said previously that he might use general fund money. The head should have reflected the fact that the governor now plans to use road fund money.)

**DOCUMENTS LINK ROYAL FAMILY TO SEX SCANDAL**
(Somehow the head should have told readers the story did not describe scandals in Britain's present royal family but concerned an unlordly lord in Queen Victoria's court a century ago.)

**CORONER SAYS TEETH AID IN IDENTIFYING DEAD**
(That was news decades ago. The news, as explained in the story, was about the number of methods used in medical detective work.)

**POLICE CHASE, CAPTURE 2 ALABAMA VAN THIEVES**
(Unless the item is for an Alabama paper, who cares about a run-of-the-mill crime? It turns out the thieves were 12 years old, a fact that should have been in the headline to justify the story.)

If only one element emerges in the headline, the head fails to do justice to the story. This headline is weak: **MAN INJURED IN ACCIDENT.** At least one person in a community is injured in an accident nearly every day. The word *man* is a faraway word. *Driver* is closer. *Injured in accident* can be shortened to *injured* if the word *driver* replaces *man*. Now the top line can read: **DRIVER INJURED.** A second element in the story shows that he was wearing a seat belt.

Marrying the two ideas produces a head like this: **INJURED DRIVER WORE SEAT BELT.** The original head is passable but weak. The revised head gives more information and is an attention-getter.

Notice how a good copy editor can make a pedestrian head come alive and have more meaning:

| ORIGINAL | REVISED |
|---|---|
| **6 PRIESTS LOSE DUTIES AFTER RAP OF BISHOP** | **PRIESTS FIRED AFTER CALLING BISHOP CALLOUS** |
| **MAN SAYS ROBBER RETURNED TO HOUSE A SECOND TIME** | **INTRUDER HITS, ROBS MAN IN HOME, RETURNS LATER FOR LOST JACKET** |
| **GARDENING IDEA FROM MEXICO HELPS INCREASE TOMATO OUTPUT** | **TEXAN DOUBLES TOMATO CROP WITH MEXICAN WATER TIP** |
| **FOUR-CAR ACCIDENT FRIDAY** | **6 INJURED AS 4 CARS COLLIDE** |
| **GARY PROJECT FUNDED BY U.S. CLOSES DOWN** | **SOUTH INC. RUNS OUT OF MONEY, CLOSES IN GARY** |
| **MOTORIST, 59, DIES OF APPARENT CORONARY** | **MOTORIST DIES AT WHEEL; CAR HITS TELEPHONE POLE** |
| **N.U. TRUSTEES O.K. NEW HEARING SYSTEM** | **N.U. STUDENTS GET BIGGER JUDICIAL ROLE** |
| **SENATE AMENDS BILL TO CUT PAY INCREASE FOR TOP JUDGES** | **SENATE AMENDS BILL TO GIVE TOP JUDGES 15% PAY BOOST** |
| **STATISTICS RELEASED ON TEST** | **HERE'S THAT TEST; CAN YOU PASS IT?** |
| **STATE APPROVES HOUSING PLAN** | **STATE APPROVES ANOTHER DORM** |
| **10¢ FARE IS URGED FOR THE ELDERLY** | **10¢ FARE IS URGED FOR THOSE OVER 65** |

## Muddling the Head

Because many words can be either verbs or nouns, the headline writer should make sure that such words can't be taken both ways. The reader will likely ascribe the wrong meaning:

**POPULATION GROWTH: DOOM WRITERS' FIELD DAY**
(Cue: "doom" is intended as a noun. )
Better: **DOOM WRITERS CAPITALIZE ON POPULATION GROWTH**

**STUDY HERALDS COP SELECTIVITY PROFILE**
(Cue: "heralds" is a verb, "cop" a noun.)
Better: **STUDY DRAWS PROFILE OF AN IDEAL COP**

**FLOURISH FLOORS DRABNESS**
(Cue: The second word is the verb.)

**RESORT WEAR SHOWING CUES ORCHESTRA WOMEN'S BENEFIT**
(It's anyone's guess what this means with so many words that double as nouns or verbs.)

**FLEXIBLE CAN NOT DANGEROUS**
(Can is a noun here, not a verb.)

**PROJECT JOB NEEDS WITH NEW METHOD**
("Project" is intended as a verb. Perhaps "foresee" would do the trick.)

The lack of a verb may force the reader to reread the head:

**4 CHILDREN DIE IN FIRE WHILE MOTHER AWAY**
(If the count won't permit "mother is away," the head should be recast: )
**MOTHER AWAY; 4 CHILDREN DIE IN HOME FIRE**

**PHYSICIAN SAYS PRESIDENT WELL**
(But "president" appears to be the object of "says," and the head literally says the physician is a capable elocutionist. When the lines are transposed, president is clearly the subject of the implied verb *is:* **PRESIDENT WELL, PHYSICIAN SAYS.)**

Some desk chiefs have an aversion to using auxiliary verbs in heads. Others insist that if the verb is needed to make the head clear, it should be used or the head should be recast. Often the lack of a vital auxiliary verb produces gibberish:

**ROOKIE ADMITS PRISONERS STRUCK**

**THAI PLEAS SAID LESS IMPORTANT THAN U.S. LIVES**

**EAST-WEST RAIL SERVICE SAID UNFEASIBLE**

**EX-CONVICT FATALLY SHOT FLEEING COP**
(Here is the copy editor's thought: "Ex-convict (is) fatally shot (while) fleeing (from) cop." But many readers will follow the normal order of subject, then predicate, so they read "(An) ex-convict fatally shot (a) fleeing cop." Not good, but perhaps passable: **EX-CONVICT KILLED WHILE FLEEING POLICE.**)

**TROOPER KILLS MAN WHO HAD SLAIN WIFE**
(Does this mean the trooper killed the man who had slain his (the trooper's) wife? Or that the trooper killed a wife slayer? Or that he killed a man whose wife had been slain by someone else?)

**OFFICIAL SAYS CIA, FBI MAY HAVE 'DESTROYED' FILES**
(To most readers this means the two agencies possibly destroyed some secret files. The quotes around "destroyed" suggest that perhaps they didn't destroy the files. The story, though, said that files previously reported as having been

## *The Magazine Title*

Most of the headline-writing techniques discussed in this chapter are those that have evolved at newspapers. But newspapers increasingly are adopting techniques pioneered at magazines. These include a *title*, as opposed to a headline, with a *conversational deck*, a secondary headline written more like a sentence than a headline. *Sports Illustrated*, whose editors are masters of writing titles with conversational decks, yields some excellent examples:

**WHAT PARITY?**
With 11 Super Bowl losses in a row,
the AFC has a way to go to beat NFC powers Dallas and San Francisco

**PRIME MIME**
Mitch Richmond of the Kings can mimic anyone,
but on the floor he's best playing himself

**RUN FOR THE ROSES**
After Michigan stunned previously unbeaten
Ohio State, joy bloomed at Northwestern

**LOVE STORY**
The death of 28-year-old Sergei Grinkov was the final chapter in one of sport's great romances

Newspapers sometimes call this convention a hammer head, but its marriage with conversational decks rather than a secondary headline was popularized in magazines. The idea is to catch the reader's eye with a bold title, then follow with an explanation that gives more detail.

destroyed may still be in the hands of the CIA and FBI. A clearer head: **OFFICIAL SAYS CIA, FBI MAY HOLD 'DESTROYED' FILES.**

An easy way to get tortured prose in headlines is to clothe nouns in human garb:

**FEAR DRIVES GUARDSMEN TO PANIC**

**SPAN CRASHES HURT FOUR, NONE SERIOUS**
(Some writers apparently think a headline provides a license for changing an adverb to an adjective.)

**SMOKE BRINGS FIREFIGHTERS TO GALT HOUSE**
(If smoke can impel action, it would "send" the firefighters.)

**RAINS FORCE ROADS TO CLOSE; FEW FAMILIES TO EVACUATE**
(Rains can't force roads; they can force somebody to close the roads.)

Again, a cardinal rule in headline writing is that the headline, standing alone, must be instantly clear to readers. If the headline puzzles them, they assume the story also will be puzzling and will turn to something they can understand. Examples:

**SCARCITY SANDWICHES JIM BETWEEN SIGNS OF THE TIMES**
(This makes sense only if readers know that during an inauguration ceremony a student named Jim paraded through the audience wearing a sandwich board sign reading, "I need a job.")

**NEXT SURGE IN FOOD COSTS TO BE MILD**
(If it's mild, how can it be a surge?)

**DOCTOR URGES SEX, ABORTION RULE SHIFTS**
(What the good doctor said was that sex taboos are out of date.)

**OUT-OF-TOWN BUSCH STRIKE IS FELT HERE**
(Busch refers to a beer.)

**WATER FALLS; CALLS BUILD TO NIAGARA**
(Want to know what idea this head is trying to convey? The water pressure in Penns Grove had fallen and the water company had received many complaints.)

**5 FROM MT. PLEASANT WIN HANDICAPPED ESSAY PRIZES**
(What's a handicapped essay? Or were the prizes handicapped?)

**SUIT CURBING SHED, FENCE SITES LOSES**
(If readers take long enough, eventually they will understand that someone lost a lawsuit that would have restricted sites for sheds and fences.)

**PARKED CAR COLLIDES WITH CHURCH**
(A church on wheels, perhaps?)

**ELM DISEASE IS THRIVING THROUGHOUT LONGMONT**
(Somehow one doesn't normally think of a disease as thriving. Make it "spreading.")

## Writing Bad Puns

The rule is that a pun in a head must be a good one or the impulse to commit it to print should be suppressed. When they're bad they're awful:

**UNBREAKABLE WINDOW**
**SOLVES A BIG PANE**

**BATTLE OF BUICKS SATURDAY**
**You buffs 'auto' be told;**
**'Little Indy' revving up**

This story illustrates the danger of trying to be too cute:

The president of the American Foundrymen's Society said in an interview that small, specialized foundries are shrinking in number because of problems concerned with scrap shortages, high material costs and the high price of conforming to anti-pollution laws and other requirements. Although the number of such small foundries is not great, the shrinkage is causing problems for the Defense Department, designers, manufacturers and investors. Overall, the statement said, the foundry business is flourishing as the nation's sixth largest industry in terms of value added.

Some headline writers could not resist the foundry-foundering pun and thus distorted the story's meaning:

**AMERICA'S FOUNDRY BUSINESS IS FOUNDERING**

**FOUNDRY BUSINESS FLOUNDERS**
(This proves that the writer couldn't distinguish between *founder* and *flounder*, a species of fish.)

Other headline writers stressed only one dimension of the story:

**FOUNDRIES DWINDLING**

**FOUNDRY FAILURES ARE ACCELERATING**

**FOUNDRY INDUSTRY IN BIG TROUBLE**

**FOUNDRY INDUSTRY SUFFERING SHRINKAGE**

A spot survey of papers using the story showed that slightly more than half (58 percent) of the headlines were reasonably fair and accurate:

**SMALL FOUNDRIES ARE DISAPPEARING**

**SMALL FOUNDRIES HAVING BIG FINANCIAL WOES**

**FOUNDRY FIRMS HURT BY MATERIAL SCARCITY**

A few headlines got both dimensions:

**INDUSTRY, MILITARY HURT BY FOUNDRY ILLS**

A few were incredibly inept:

**FOUNDRY INSTITUTE SHRINKAGE EYED**

**AMATEUR BOAT BUILDERS FACE PROBLEMS**

**INDUSTRY NEEDS HELP, FERRET OUT FOUNDRY**

## Avoiding Bad Taste

Newspapers must, of necessity, reveal human sorrows as well as joys, afflictions as well as strengths. No story or headline should mock those who have misfortunes. The newspaper belongs in the parlor where good taste is observed.

A story related that a Johannesburg motorist, whose car stalled on railroad tracks, died under the wheels of a train when he was unable to release his jammed seat belt. The victim may have been unknown to readers in an American community, but death is a common tragedy and should be treated with respect, something the copy editor forgot: **BELTED TO DEATH.**

A minor story told about a man digging his own grave, starving while lying in it for 21 days and dying two hours after being found. The headline: **DOWN ... AND THEN OUT.**

Another story related how a woman survived a 200-foot leap from a bridge, suffering only minor injuries. Investigators said the woman landed in about four feet of water near shore. The item was insensitively headed: **HIGHER BRIDGE NEEDED.**

## Avoiding Double Meanings

A headline is unclear if it can imply more than one meaning. Some readers may grasp the meaning intended; others won't. An ad writer for a coffee company created a double meaning in this slogan: "The reason so many people buy Red & White Coffee is that 'They Know No Better.'" Here are some headline examples:

**HILLARY CLINTON ON WELFARE**
(She was speaking about it, not accepting it.)

**BOY STRUCK BY AUTO IN BETTER CONDITION**
(Than if he had been hit by a truck?)

**RECTOR SEES SEX AS GOURMET MEAL**
(He said people would be healthier if they looked on sex as a gourmet meal rather than something distasteful.)

**RAPE CLASSES PLANNED**
(Rape prevention will be the subject of the classes.)

**YWCA OPENS PUBLIC SERIES WITH ABORTION**
(Abortion will be discussed at the first class meeting.)

**FLINT MOTHER-IN-LAW WOUNDED IN ARGUMENT**
(That's better than being shot in the head.)

Place names like Virgin, Utah; Fertile, Minn.; and Bloomer, Wis., inevitably invite a two-faced headline if the town is used:

**VIRGIN WOMAN**
**GIVES BIRTH**
**TO TWINS**

**MAN LOSES HAND**
**IN BLOOMER**

Other geographical terms:

**BOOK IN POCKET**
**SAVES MAN SHOT**
**IN SOUTH END**

**THREE BOSTON**
**WAITRESSES SHOT**
**IN NORTH END**

Unusual family names of officials—Love, Fortune, Dies, Church, Oyster—also invite two-faced headlines:

**OYSTER PROBES**
**UNKNOWN JAM**

**WALLACE ATTACKS**
**U.S. GRANT**

**BILLY HOOKS**
**PATIENT IN DURHAM**

**PICKS FORTUNE**
**FOR INDIANA**
**REVENUE CHIEF**

**FINK HEADS BRIDGE**
**CHARITY UNIT**

**SLAUGHTER RE-CREATES**
**CONSTANTINE'S ROME**

Presidents and presidential candidates have been victims of two-faced heads:

**CLINTON VISITS**
**HURT SOLDIERS**

**JOHNSON PUTTING RUSTY**
**ON WHITE HOUSE GREEN**

**FORD, REAGAN NECK IN PRESIDENTIAL PRIMARY**

**LBJ GIVING BULL TO MEXICAN PEOPLE**

**ROBERT KENNEDY STONED**

**GOOSE GIVEN TO EISENHOWER**

*Case* and *chest* produce these headlines:

**ORD PHILLIPS GETS TWO YEARS IN CIGARETTE CASE**

**CHEST PLEAS ISSUED FOR MOTHER'S MILK BANK**

The worst possible headline verb is *eyes:*

**FREAR RESTING, EYES RETURN**

**GREEN EYES MAJOR TITLE**

**SIDEWALKS TO BE EYED IN ELSMERE**

**ALLEGED RHODES PERJURY SAID EYED**

More double-takes in headlines:

**BEHEADING CAN CAUSE KIDS STRESS**

**SUSPECT'S COUNSEL SAYS: WINSETT QUIZZED IN NUDE**

**SWINE HOUSING TO BE AIRED**

**TOP SWINE PRIZE TO COUNTY YOUTH**

**PUBLISHER SAYS BAR ENDANGERS PRESS FREEDOM**

**TV NETWORKS AGREE TO POLICE VIOLENCE**

**RELATIVES SERVED AT FAMILY DINNER**

**MAN WITH TWO BROKEN LEGS SAVES ONE FROM DROWNING**

**BED AFLAME, JUMPS FROM FOURTH FLOOR**

**TAHOE MAN SENTENCED TO 28 YEARS IN CALIFORNIA**

**HILLARY'S VISIT TO STRESS WOMEN**

**POLICE SLAY SUSPECT BOUND OVER FOR TRIAL**

**LOCAL OPTION FAST TIME OFFERED TO SKIRT PROBLEM**

**U.S. TO FIRE EUROPE INTO STATIONARY ORBIT**

**CIA AGENT DONATED TO DEMS**

**ANDALUSIA GIRL IMPROVED AFTER DRINKING POISON**

**MANY WHO MOVED TO FLORIDA LEAVE AFTER DEATH**

**BOY CHASING FOX FOUND RABID**

**FRANKLIN PAIR IS IMPROVED AFTER SHOOTING**

**CHEMOTHERAPY FEARS INCREASE AFTER DEATH**

**EXPECTANT MOTHER, 23, IS ANXIOUS FOR FACTS**

**SEX EDUCATOR SAYS KINDERGARTEN'S THE TIME**

**PRESIDENT SAYS WOMEN RESPONDING ADEQUATELY**

**NEW RESTROOMS BIG ASSETS FOR SHOPPERS**

**OKLAHOMAN HIT BY AUTO RIDING ON MOTORCYCLE**

**ENGINEERS TO HEAR GROUND WATER TALK**

**TWO ACCUSED OF KIDNAPPING SLAIN MAN**

**STATE DINNER FEATURED CAT, AMERICAN FOOD**

**RAFFLE DRAWING FOR BABY AT PUB ON JAN. 8**

**ADMIRAL LIKES TO MAKE WAVES**

**BRADLEY SCHOOL GETTING NEW HEAD**

**MAN ON WAY TO ITALY TO SEE FAMILY KILLED**

**GLACIER LAKE STILL UP IN THE AIR**

**CHIEF SAYS U.S. COURTS ULCERS**

**PENNSAUKEN'S SAFETY CHIEF QUITS BLAMING POLITICS**

**YOUTH HANGS SELF IN CELL AFTER UNCLE TRIES TO HELP**

**HEROIN BUSTS UP**

**MAN WHO SHOT HIMSELF ACCIDENTALLY DIES**

**STORES ON 4TH BETWEEN WALNUT AND CHESTNUT TO DRESS UP REARS**

Even the lifestyle pages contain two-faced headlines:

**ITALIAN COOKIES EASY TO MAKE**

**FRESH DATES ARE GREAT**

**O'BRIEN PEAS IN SQUASH**

**MALE UNDERWEAR WILL REVEAL NEW COLORFUL SIGHTS**

**GUIDE, DON'T PUSH CHILD**

**NURSES AWARDED FOR POSTER ART**

**STRONG ATTIRE RIGHT FOR BILL'S DEATH**

**CARRIES ON FOR HUSBAND**

**DESIGNER'S DEATH BLOW TO THEATER**

## Giving Away the Punch Line

Some stories are constructed so that the punch line comes at the end, rather than at the beginning. Obviously, if the point of the story is revealed in the headline, the story loses its effectiveness. The following story calls for a head that keeps the reader in suspense until the end:

> One Saturday afternoon not long ago a night watchman named Stan Mikalowsky was window-shopping with his 5-year-old daughter, Wanda, and as they passed a toy shop the child pointed excitedly to a doll nearly as big as she was.
>
> The price tag was only $1 less than the watchman's weekly pay check, and his first impulse was to walk away, but when the youngster refused to budge he shrugged and led her into the store.

When Stan got home and unwrapped the doll his wife was furious.

"We owe the butcher for three weeks, and we're $10 short on the room rent," she said. "So you got to blow a week's pay for a toy."

"What's the difference?" said the night watchman. "Doll or no doll, we're always behind. For once let the kid have something she wants."

One word led to many others, and finally Stan put on his hat and stomped out of the house.

Mrs. Mikalowsky fed the child and put her to bed with the doll next to her and then, worried about Stan, decided to go looking for him at the corner bar and make up with him. To keep his supper warm, she left the gas stove on, and in her haste threw her apron over the back of the chair in such a way that one of the strings landed close to a burner.

Fifteen minutes later when the Mikalowskys came rushing out of the bar, their frame house was in flames and firefighters had to restrain the father from rushing in to save his daughter.

"You wouldn't be any use in there," a police officer told him. "Don't worry, they'll get her out." Firefighter Joe Miller, himself a father, climbed a ladder to the bedroom window, and the crowd hushed as he disappeared into the smoke. A few minutes later, coughing and blinking, he climbed down, a blanket-wrapped bundle in his arms.

The local newspaper headlined its story with the line that should have been saved for the finish:

**FIREMAN RESCUES LIFE-SIZE DOLL**
**AS CHILD DIES IN FLAMES**

# CHAPTER 9

# EDITING WIRE NEWS FOR PUBLICATION

News from the wire services plays different roles in the various media. For radio, television and the new on-line media, breaking news provided by the wires remains a staple of the daily news report. For newspapers, which usually get second crack at wire stories after the broadcast media have used them, interpretation has become more important.

To be sure, newspapers still contain plenty of inverted pyramid stories provided by the wires. But, increasingly, those are short summaries confined to roundup columns. The wire stories favored by newspapers today are those that expand on the bare-bones reports provided by radio and television. The mission of newspapers is to interpret, to explain and to amplify. That's true of local news, too, but the trend is particularly noticeable in the wire report.

Today, most newspapers prefer what editors call second-day leads for wire stories. For example, a broadcast report of an airplane crash might be written like this for radio, television and the on-line media:

> BOGOTA, Colombia (Reuters)—All 164 people aboard an American Airlines passenger jet en route from Miami died late Wednesday when it slammed into a mountain in southwest Colombia and burst into flames, authorities said Thursday.
>
> "All we know is that the plane was torn to pieces," said Alvaro Cala, the head of Colombia's Civil Aviation department.
>
> The accident involving a Boeing 757 was the most deadly involving a U.S. carrier since the 1988 bombing of a Pan Am flight over Lockerbie, Scotland, that killed 270 people. It was also the worst in Colombian history.
>
> The crash of American Flight 965, en route from Miami to Colombia's southwest city of Cali, took place within hours of the Dec. 21, 1988, anniversary of the Lockerbie crash.
>
> There is no evidence so far that the disasters are in any way related, but American Airlines spokesman Al Comeaux in Fort Worth, Texas, said the pilot, who was scheduled to

touch down in Cali at 9:54 p.m. EST, apparently was unable to radio any distress call to local air traffic controllers.

That inverted pyramid account serves the broadcast and on-line media well, but by the time the local newspaper is published tomorrow morning, a fresh approach is required:

**From AP and Reuter Reports**

BOGOTA, Colombia–U.S. government agencies were working feverishly Thursday to determine if the crash of an American Airlines 757 in Colombia was caused by terrorists. The crash came on the anniversary of the worst disaster in U.S. aviation history, the 1988 terrorist explosion of a Pan Am flight over Lockerbie, Scotland.

The crash, which killed all 164 people bound for Cali from Miami aboard Flight 965, occurred when the plane slammed into a mountain in southwest Colombia and burst into flames, authorities said. It was the first-ever crash of a Boeing 757, and it came in good weather conditions. The pilot had issued no radio warning of mechanical difficulty, and all navigation aids in the area were working properly, Colombian officials said.

Newspapers, then, try to answer the question: Why did this happen? Seldom do they have the opportunity to beat the electronic media to a wire story. That reality leads to a different approach to writing the story. It also dictates a different approach to the headline. Headline writers are admonished to emphasize the new angle, not the old, in the headline. For the plane crash story, tomorrow morning's newspaper headline might read like this:

**U.S. SUSPECTS TERRORIST LINK TO AIR CRASH**

Taking that approach is tacit recognition of the fact that the various media have different strengths. The broadcast and on-line media excel at delivering the news with speed; newspapers have the time and space to provide analysis and insight.

## SOURCES OF WIRE NEWS

The dominant wire service in the United States is The Associated Press, a cooperative owned by member newspapers and broadcast stations. Privately owned United Press International, once a major competitor, has shriveled into a minor player after going through a series of bankruptcies and ownership changes during the past 30 years. UPI's main presence today is in radio newsrooms, where it provides a bare-bones news service at a relatively low price. UPI's news report is no longer adequate for newspapers to use as their only source of wire news, and few newspapers in today's tough market can afford two mainstream services.

The demise of UPI has opened the door to the U.S. market for British-owned Reuters, French-owned Agence France Presse and other foreign-based services. Major U.S. newspapers often subscribe to such services, which provide excellent alternatives to AP for international news at affordable rates. Increasingly, the major foreign services cover U.S. news as well.

Another source of wire news is the so-called supplemental wire services—syndicates formed by major metropolitan newspapers, alliances of such papers or a newspaper group. Such services make it possible for a newspaper in Danville, Ill., to carry a major investigative piece from *The New York Times* on the same day the *Times* itself carries the story. Thus, even small newspapers have the opportunity to provide investigative accounts and depth reporting that radio and television rarely offer. Among the supplemental services are *The New York Times* News Service, the *Washington Post–Los Angeles Times* Syndicate, and the Copley, Gannett and Scripps-Howard news services. By syndicating their news, publishers participating in the supplemental services are able to recoup some of the costs of news gathering and, in fact, have been able to expand news coverage.

The services mentioned are among more than 200 syndicates offering news, features, pictures and special services. In addition to giving spot and secondary news, these services provide news of sports, food and fashion, and bylined columns and features that cover everything from personal computing to zoo animals. Some of that material is sent to newspapers by mail, but increasingly it is transmitted into the newspaper's computer system through the AP's satellite system. AP charges a small fee for the service.

News is visual as well as written, so the wire services and syndicates often handle pictures and graphics in addition to text. AP has special networks for delivering both photos and graphics. Photos are transmitted into a computer known as a *Leaf desk*, named for the company that designed it. Today, almost all wire photos are processed in digital form rather than with conventional photo processing techniques. Wire photos are edited for brightness and contrast right on the Leaf desk or on a personal computer. Information graphics—maps, charts and graphs—are delivered directly to Macintosh computers.

With UPI's demise, that service's once-powerful picture service also suffered. UPI then linked with the French news agency, AFP, to retransmit its photos in the United States. Reuters also has entered the picture business here, as have several syndicates.

## HOW THE WIRES OPERATE

Stories delivered by the wire services come from several sources:

- Copy developed by the agencies' own large staffs of reporters, feature writers, analysts, columnists and photographers.
- Rewrites of stories developed by subscribers. Newspapers or broadcast stations contracting with a wire service agree to make their own news files available to the service, either by providing proofs or computerized versions of stories.

Other wire service staffers rewrite from any source available—smaller papers, research reports and other publications.

- Stringers or correspondents in communities where there is no bureau. Such stringers frequently are newspaper reporters and are called stringers because of the old practice of paying correspondents by their strings of stories represented in column inches.
- Exchanges with other news agencies, such as foreign agencies.

A reporter telephones a story to the state bureau of AP. If the story has statewide interest, AP files the story on its state wire. If the story has regional interest, the state bureau offers it to a regional bureau, or, in some cases, the state office may offer the story directly to the national desk.

The national desk thus becomes the nerve center for the entire operation of the news agency. That desk collects news from all the state, regional and foreign bureaus, culls the material, then returns it to the regional and state bureaus or to subscribers directly.

The operation sometimes is referred to as a gatekeeping system. A Dutch story, for example, would have to get by the Amsterdam office before it could be disseminated in Holland. The same story would have to clear the London bureau before being relayed to New York and the national desk. That desk then would decide whether the story should be distributed nationally. The desk could send the story directly to newspapers or route it through regional and state bureaus. In the latter cases, the regional bureau would judge whether to transmit the story to a state bureau, and a state bureau would have the option of relaying the story to subscribers. A wire editor then would accept or reject the story. Finally, the reader would become the ultimate gatekeeper by deciding which stories to read and which to ignore.

Traditionally, the wire service has opened the news cycle with a news *budget* or summary that indicates to editors the dozen or more top national and international stories that were in hand or were developing. Today's wire editors may get a four- or five-line abstract of the complete offering—foreign, national, regional and state—transmitted directly to the newspaper's computer. From these abstracts or from computer directories, wire editors select the stories in which they think their readers are interested. Then they retrieve those stories directly from the newspaper's computer.

## BUDGETS AND PRIORITIES

The wire services operate on 12-hour time cycles—PMs for afternoon papers, AMs for morning papers. The broadcast wire is separate (see Chapter 15), but major broadcast stations also take the newspaper wire, which is more complete. The cycles often overlap so that stories breaking near the cycle change are offered to both cycles, or stories early in one cycle are picked up on stories late in the other cycle.

For a morning paper, the wire news day begins about noon with the following:

```
a200
d lbylczzcczzc
Starting AMs Report, a201 Next
1202pED 03-07
```

Then follows the news digest (or budget), notifying editors of the dozen or more top news stories in sight (Figure 9–1). An addition to the first budget may come a short time later. The budget is used to give wire editors a glimpse of the major stories forthcoming and aids them in planning space allocations.

Major stories breaking after the budget has been delivered are designated in degree of importance with a priority system:

FLASH—a two- or three-word statement alerting editors to a story of unusual importance: President shot. On the old teleprinter machines a warning bell signifying a flash brought editors running to the machine. Today a flash is seldom used because a bulletin serves the same purpose and is almost as fast.

BULLETIN—a short summary, usually no more than a lead, of a major story. Again, it is used primarily to alert editors and is not intended for publication unless the bulletin arrives at deadline. A bulletin also may be used to signal corrections, such as a mandatory "kill" on a portion of a story that could be libelous.

BULLETIN MATTER—expands the bulletin with more, but brief, details. Unless the deadline is a factor, the wire editor holds up the story for a more detailed account.

URGENT—calls editors' attention to new material or to corrections of stories sent previously.

## SORTING THE PIECES

Wire editors have two considerations in selecting wire copy for publication—the significance of the stories and the space allotted for wire copy. If space is tight, fewer wire stories are used and heavier trims may be made on those that are used.

Budget stories usually, but not necessarily, get top priority. When stories listed on the budget arrive, they are so indicated by BUDGET, BJT or SKED, together with the length in words. If such stories are developing or are likely to have additional or

**A Typical Wire-Service Budget**

AP News Digest—Thursday, December 21, 1995—5 a.m. Update

Top stories at this hour from The Associated Press. The supervisor is Dan Freeman (212-621-1608). On the photo desk: Tim Black (212-621-5555). For reruns, please call your AP bureau or the service desk (212-621-5555).

ECONOMIC REPORTS:

EDITORS: This change reflects the impact of the partial government shutdown. Postponed: Labor Department weekly jobless claims and Commerce Department special combined personal income figures for October and November.

AIRPLANE CRASH: American Airlines Flight From Miami Crashes in Colombia.

BOGOTA, Colombia—An American Airlines plane carrying 159 people from Miami to Cali crashed in the Andes Mountains on its final approach to the southwestern Colombian city. The cause of the accident, which police said occurred in a rebel "hot zone," was not known.

Slug PM-Colombia-Airplane. p0245.

By Chris Torchia.

BUDGET TALKS: Presidential Telephone Call Restarts Budget Talks.

WASHINGTON—The on-again, off-again budget talks between the Clinton administration and Republican lawmakers were on again today as 260,000 furloughed federal workers stayed off the job for a sixth day.

Slug PM-Budget-Shutdown. p0216.

By Dave Skidmore. AP Photos WX101, Clinton at a White House briefing Wednesday; WX102, Kasich at a Capitol Hill briefing; WX103, GOP freshmen at a Capitol Hill news conference.

YUGOSLAVIA: U.S.-Led NATO Force Takes Over Troubled U.N. Mission.

MRKONJIC GRAD, Bosnia-Herzegovinia—Many of the foreign soldiers here aren't new to Bosnia. But their mission as NATO troops is fresh—police the peace instead of often watching, hands tied, as people are slaughtered.

Slug PM-Yugoslavia. p0233.

By Tony Smith.

---

***Figure 9–1** This is the top of a typical Associated Press budget for afternoon newspapers. Information listed here helps editors plan their wire coverage. This particular budget informs editors that AP is offering three photos to accompany the story on federal budget negotiations.*

new material, the editor places each story in a folder or computer holding queue and concentrates on stories that will stand.

Eventually the stories in the folder or holding queue have to be typeset. The Associated Press has added a feature called DataRecaps to aid wire editors in handling a breaking story. Previously, editors had to assemble the pieces from multiple leads, inserts, subs and adds or wait until space was cleared on the wire for a no-pickup lead. Now the service notifies wire editors a recap is coming, then delivers a complete story at high speed.

Starts and stops occur even on copy apparently wrapped up. A story arriving early in the morning describes a congressional appropriation of $5.9 *billion* to provide jobs for the unemployed. Fifty items later editors are informed that the figure should be changed to $6.4 *million*. Still later, New York sends the message that the original $5.9 *million* should stand. Eventually the service again corrects the figure to the original $5.9 *billion*. Such changes pose few problems for the editor if the story has not yet been typeset. The editor simply changes the copy on the computer screen.

## EDITING THE WIRE

On each news cycle, even the wire editors of large newspapers have more stories than they can use (see Figure 9–2). On larger dailies using all the wires from the AP and several supplementals, the flow of copy is monumental. One way to handle this spate of copy is to categorize the news—one queue for stories from Washington, D.C., another for New York and international, another for national, another for regional, another for area copy, and the like. Most computer systems do that automatically.

An advantage of the paper's receiving more than one news service is that the editor can use the story from one service to check facts against the same story from another service, such as casualty figures, proper names and spellings. If there is a serious discrepancy in facts, the editor asks the state or regional bureau for verification.

### Wires Make Mistakes

Two points should be kept in mind as you edit wire copy. The first is that no wire service tailors copy for a particular newspaper. Abundant details are included, but most stories are constructed so that papers may use the full account or trim sharply and still have the gist of the report.

The second point is that the wire isn't sacred. AP has a deserved reputation for accuracy, impartiality and speed of delivery. It also makes errors, sometimes colossal ones. Other services do, too.

A source who turned out to be unreliable caused United Press to release a premature armistice story during World War I. A confused signal from a New Jersey courthouse caused AP to give the wrong penalty for Bruno Richard Hauptmann, convicted in the kidnapping and slaying of the Lindbergh child. The state wire, more often than not, is poorly written and poorly edited. Even the wire executives admit they still have bonehead editing and some stories that don't make sense. Stories abound with partial quotations, despite repeated protests from subscribers.

The Associated Press has an advisory committee, composed of managing editors who monitor writing performance. Here is one example from the APME (Associated Press Managing Editors) writing committee:

***Figure 9–2 A wire editor is responsible for sorting through the massive amount of wire news and making decisions about what will go into the newspaper.***
Photo by Philip Holman.

DETROIT (AP)—Two Detroit factory workers were killed Wednesday evening on their jobs, Detroit police said today. (Clearly an industrial accident.)

The two men, employees of the Hercules Forging Co., were identified as James H–, 51, the father of nine children, and M.C. Mc–, 37, both of Detroit. Mc– may have been struck accidentally, police reported. (Oh, somebody struck him accidentally, but hit the other fellow on purpose.)

Police said they had a man in custody and added he would probably be charged in connection with the deaths. (Let's see, both of them were killed by the same man, but one death may have been an accident, but the suspect is charged with two murders?)

Police say they interviewed 15 workers at the plant who could give no reason for the shootings. (Now gunplay gets into this story.)

Witnesses told police the man confronted H– with a carbine and shot him when he tried to run. He fired a second shot after H– fell, they said.

The third shot, apparently aimed at another workman, hit Mc–, police said. (Now we are beginning to understand that what started out sounding like an industrial accident has become double murder.)

## Peculiarities of Wire Copy

At most newspapers, wire copy is edited the same as local copy, but a few peculiarities apply to the editing of wire news. First, wire news, unlike local news, usually carries a *dateline,* which indicates the city of the story's origin.

Despite the name of the device, most newspapers no longer include in the dateline the date the story was written:

OLD DATELINE

OVERLAND PARK, Kan., Jan. 12 (AP)—An apparent good Samaritan who helped start a woman's car talked about how dangerous it was to be out at night, then pulled a gun and took her purse, police said.

MODERN DATELINE

OVERLAND PARK, Kan. (AP)—An apparent good Samaritan who helped start a woman's car talked about how dangerous it was to be out at night, then pulled a gun and took her purse, police said.

Note that the city of origin is in capital letters, and the state or nation is in uppercase and lowercase letters. At most newspapers, style calls for the dateline to be followed by the wire service logotype in parentheses and a dash. The paragraph indention comes before the dateline, not at the start of the first paragraph.

Stories that contain material from more than one location are called *undated* stories. They carry no dateline but a credit line for the wire service:

**By The Associated Press**

Arab extremists said today they would blow up American installations throughout the Middle East unless all Americans leave Beirut immediately.

Stories compiled from accounts supplied by more than one wire service carry similar credit lines:

**From our wire services**

LA PAZ, Bolivia—Mountain climbers experienced in winter climbing tried today to reach the wreckage of a Ladeco Airlines Boeing 727, which crashed Monday while approaching La Paz in a snowstorm.

When editors combine stories, they must be sensitive to the fact that wire stories are copyrighted. If material in the story was supplied exclusively by one service, that service should be credited within the text. Combining the stories this way often provides a newspaper's readers with a better story than could be obtained from either of the major services alone. Good newspapers make a habit of doing this frequently.

Wire stories often use the word *here* to refer to the city included in the dateline. If the editor has removed the dateline during the editing process, the city must be inserted in the text. Otherwise, *here* is understood to be the city in which the newspaper is published.

The Associated Press uses Eastern time in its stories. Some newspapers outside the Eastern time zone prefer to convert those times by subtracting one hour for Central time, two for Mountain or three for Pacific. When this is done, if the dateline remains on the story, it will be necessary to use phrases such as *3 p.m. St. Louis time* or *3 p.m. PDT* in the text. The newspaper's local style will specify the form to be used.

## Localizing Wire Stories

Wire stories often become ideas for local stories. If Congress has reduced the amount of money it will provide for loans to college students, a newspaper in a college town may want to contact the college's financial aid officer to determine what effect the measure will have locally.

When the Soviet Union shot down a Korean Airlines plane, the *Columbia Missourian* learned that two people were aboard who had just completed their doctoral degrees at the University of Missouri. Their daughter was with them aboard the jet. Because members of the family had lived in Columbia for several years, many people knew them. As a result, a major international story became a good local story:

---

Recent University of Missouri graduates Somchai Pakaranodom and his wife, Wantanee, were on their way back to Thailand after living in Columbia for four years. They never made it.

The couple, along with their 7-year-old daughter Pom, were among the 269 killed when a Soviet jet shot down Korean Air Lines Flight 007 north of Japan.

"We had laughed about the flight schedule," education Professor Dorothy Watson recalled Friday, sitting in her living room where a week earlier she had thrown a going-away party for the Pakaranodoms.

The couple left Columbia Saturday. They were traveling first to Chicago, then on to New York to catch the flight to Seoul and eventually on to Bangkok, Thailand.

With the sound of news reports about the crash coming over the television from the next room, Mrs. Watson fought back the tears. "I kept thinking they couldn't possibly be on the plane," she said. "It is just such a waste, so stupid."

At first she didn't connect the crash with her friends.

On Friday, another faculty member called to break the news to her. She said even then she refused to believe it, until Korean Air Lines confirmed the family was on the flight.

Both Fulbright scholars, the Pakaranodoms received their doctoral degrees from the university Aug. 5. Somchai's degree was in agricultural engineering, Wantanee's in reading education.

---

Localizing wire stories can be done in several ways. The story can be rewritten to emphasize the local angle, a separate local sidebar to the wire story can be written or a simple insert in the wire story may be sufficient.

# CHAPTER 10

# EDITING MAGAZINES AND CORPORATE PUBLICATIONS

## AN INDUSTRY OF DIVERSITY

No one seems to know how many magazines there are in the United States. Perhaps that's because almost every company and almost every nonprofit organization publishes one. Add those to the staggering number of titles found on newsstands and you have an elaborate assortment of titles covering every imaginable topic.

There is even a group called the College Fraternity Editors Association, made up of the editors of magazines published by fraternities—social, honorary and professional. More than 100 titles are put out by representatives of that group alone, yet they represent a tiny portion of the journalists who write and edit for nonprofit organizations. Included are journalists for such diverse groups as the American Red Cross, the National Rifle Association, the Campfire Girls and Rotary International.

Jobs for the magazine editor also abound in corporate communications. There, journalists may split time between public relations duties and writing and editing the company magazine, which is usually directed to both employees of the company and customers. Or, at larger companies, journalists may be hired to do editing only.

Then there are the specialty magazines with titles that range from *PC Computing* to *Golf* and from *Sewing* to *Home Improvement.* For almost any possible hobby or field of endeavor, there is a magazine or two—sometimes more—to match. Government gets into the act, too, with titles that range from *Missouri Resources* to *Arizona Highways.*

Finally, there are the big national magazines. Included in that group, of course, are the news magazines—*Time, Newsweek* and *U.S. News and World Report.* They rank among the few remaining general-interest magazines, which include *Life, Good Housekeeping* and *TV Guide.* And, while general-interest magazines are disappearing in favor of those that target specific audiences that advertisers covet, the few that remain are extraordinarily profitable. Their audiences often are huge. Few journalism graduates land jobs at the big national magazines right out of school, but many aspire to get there.

What makes the magazine industry appealing to many are its advantages over newspapers—the ability to produce a more attractive product with better color on top-quality paper, and less-frequent deadlines, which make it easier to practice a consistently high-quality brand of journalism.

## MAGAZINE CHARACTERISTICS

Magazines differ from newspapers in many ways, including:

- Magazines use a better grade of paper, or stock. The cover paper may be heavier than the paper for inside pages.

Informal
Alexa

OUTRAGEOUS
Cinema Solid

OLD WEST
Birch

Inviting
Lydian Cursive

Fluency
Brush

Far East
Peking

SPONTANEOUS
Herculanum

PROPER
Blackfriar Normal

HIGH TECH
Modula Black

POSTER
Falstaff

***Figure 10–1** Mood typefaces.*

- Magazines use more color, not only in illustrations but in type and decoration as well.
- Illustrations often are more dominant in a magazine than in a newspaper. The illustrations may run (or bleed) off the page or extend into the fold.
- Magazines breathe; they use air or white space to emphasize text and illustrations much more often than newspapers.
- Magazines vary typefaces to help depict the mood, tone or pace of the story (see Figure 10–1). They use initial capital letters to help readers turn to the message or to break up columns of type (see Figure 10–2). They wrap type around illustrations (see Figure 10–3).
- Magazines may use reverse plates (white on black) or use display type over the illustration (overprinting) for display headings or even text.
- Magazines may vary the placement as well as the design of the headline (see Figure 10–4).

A WILD Alaskan wolf yawns broadly as he lolls in the warm sun of the arctic spring. Trumpeter swans dabble

There aren't any weekend beer parties on campus. No pep rallies. No homecoming queen, no dormitories and no Saturday night

*Figure 10–2 (Top) Two-line sunken initial cap. (Bottom) Upright initial cap.*

The path of time that will lead us from today through the 2lst century will be lined with many bright new mechanical and electronic marvels, with better energy sources, and new construction materials and methods. How we respond to

*Figure 10–3 Wraparound type.*

*Figure 10–4 Headline design*

## *What Makes Magazines Different*

Newspapers, as earlier chapters have described, contain more and more stories written and edited in magazine style. The narrative, the personalized approach and other writing styles in vogue at newspapers originated in magazines. So what makes magazine writing and editing different?

The primary difference is that magazines write and edit *most* of their stories that way, while the inverted pyramid still plays an important role, perhaps still a dominant one, in newspapers. Magazine stories are consistently longer, too. It's not unusual for a major article to extend for eight or more magazine pages. Without a doubt, the average magazine story is much longer than the average newspaper story.

Longer periods between deadlines also give magazine editors more time to do their work. Typically, magazine stories are edited several times more than the average newspaper article. *Time* magazine's editors are notorious for rewriting and rewriting, then editing and re-editing, sometimes to the point that a reporter has difficulty recognizing his or her contribution to the article. Such meticulous attention to detail makes much of magazine writing more polished and refined than the average newspaper story.

Those same high standards often don't exist, however, at small magazines. When one editor does everything, from writing to editing to design, there is little opportunity, if any, for another set of eyes to challenge the copy.

## EDITING THE MAGAZINE

An editor who does not understand art cannot hope to produce a superior magazine. Magazine editing is essentially a joint endeavor; the editor provides editorial excellence, and the artist creates the visual image.

Front pages of daily newspapers often look alike. But the magazine comes in a distinctive wrapper or cover that reflects the nature of the publication, stresses a seasonal activity or merely directs readers to the "goodies" inside the magazine.

News, as we have seen, may be presented in many styles, but in newspapers the inverted pyramid usually prevails. In a magazine the space is likewise limited, but the writing style is more relaxed, more narrative and more personal. The pace of the magazine piece may be slower—but certainly no less dramatic—than that of the newspaper story. Here is the beginning of a magazine feature:

> In Chicago some middle-aged businessmen plan a skiing trip to Colorado. In Miami a middle-aged woman with high blood pressure seeks medical advice before leaving for Denver to visit her daughter. In Baltimore a family is cautioned against vacationing in Colorado because one of the children has a lung ailment.
>
> All three of these examples involve a change from low to high altitude, and owing to air travel, making the change in a relatively short period.
>
> Coupled with exertion, cold temperatures and high altitude, won't the businessmen who have been sedentary for months be risking heart attacks? Won't the visiting mother experience even higher blood pressure? And won't high altitude aggravate the child's illness?

Not until later in the story does the angle that normally would be in the lead of a news story appear:

> Contrary to the popular belief that reduced oxygen pressure at high altitude has an adverse effect on the coronary artery system, research indicates high altitude may be beneficial and even afford a degree of protection against coronary artery disease.

So rigid are the style requirements of some magazine editors that they lean heavily on staff writers, use staff writers to reshape free-lance material or buy only from free-lancers who demonstrate they are acquainted with the magazine's requirements.

## ARTICLE HEADINGS

Typically, a newspaper uses illustrations to focus the readers' attention on a page. It relies on the headline to lure readers into the story. But in a magazine the whole page—headline, pictures, placement—is designed to stop the readers in their tracks. They may get part of the story from a big dramatic picture before they ever see the head. This combination of elements must make readers say to themselves, "I wonder what this is all about."

The magazine editor is not confined to a few standardized typefaces for headings. Instead, the editor may select a face that will help depict the mood of the story. Nor is the editor required to put the heading over the story. It may be placed in the middle, at the bottom or on one side of the page.

The heading may occupy the whole page or only part of a page. It may be accented in a reverse plate or in some other manner. It may be overprinted on the illustration. More often it will be below the illustration rather than above it. Almost invariably it is short, not more than one line. Frequently it is a mere tag or teaser. A subtitle, then, gives the details:

> **Oil from the Heart Tree**
>
> An exotic plant from Old China produces
> a cash crop for the South
>
> **I Can HEAR Again!**
>
> This was the moment of joy, the rediscovering
> of sound: Whispers . . . rustle of a sheep . . .
> ticking of a clock
>
> **The Pleasure of Milking a Cow**
>
> Coming to grips with the task at hand
> can be a rewarding experience,
> especially on cold mornings

In magazines, only a few of the rigid rules that apply to newspaper headlines (see Chapter 8) remain in force. Rules of grammar and style are observed, but almost anything else goes. Magazine title writing is free-form in both style and content.

## MAGAZINE DESIGN

Type and illustrations, or gray and black blocks, are the dominant elements in newspaper page design. In magazine design a third block—white—is used more frequently. To the magazine art editor, white space is not wasted space but rather a means of emphasizing other elements.

The editor may use space generously around headings, between text and illustrations, and around illustrations. The editor deliberately plans to get white space on the outside of the pages. To gain extra space, illustrations may bleed off the page.

Some stories are told effectively in text alone; others are told dramatically in pictures. The ideal is a combination of text and pictures, and the emphasis depends on the quality of the illustrations or the significance of the text. A picture's value, says one editor, is best exploited when it sweeps the reader rhythmically into the text. Too often, the story is adequate but good pictures are lacking, thus robbing the story of its dramatic appeal and producing a dull page of straight text.

A magazine page usually has these elements:

- At least one dominant picture.
- A title, preferably with a subtitle.
- A block of text, usually beginning with a typographical device that will compel the reader's attention to the opening of the story.
- The opening device may be a dingbat, such as a black square, followed by a few words in all-capital letters. Or it may be an initial capital letter, either an inset initial (its top lined up with the top of the indented small letter) or an upright or stick-up initial (the bottom of the initial lined up with the bottom of the other letters in the line) (see Figure 10–2).

Simplicity is the keynote in effective page design. An easy, modular arrangement is more likely to attract readers than a tricky makeup with oddly shaped art and a variety of typefaces. Illustrations need not be in the same dimensions, but they should be in pleasing geometric proportions. Margins should be uniform or at least give the effect of uniformity. Usually the widest margin is at the bottom of the page, the next widest at the side, the third at the top and the narrowest at the inside or gutter. The content of the page is thus shoved slightly upward, emphasizing the eye level or optical center of a rectangle. The outside margin is larger than the gutter because the latter, in effect, is a double margin.

Kenneth B. Butler, author of a series of practical handbooks (published by Butler Typo-Design Research Center in Mendota, Ill.) treating the creative phases of magazine typography and design, advises design editors to touch each margin at least once, regardless of whether illustrations are used. He contends that the eye is so accustomed to the regular margin that even when the margin is touched only once, an imaginary margin is defined clearly in the reader's mind. If the illustration bleeds off the page, the margin on the bleed side may be widened to give more impact to the bleed device.

The art director must know the position of the page—whether left, right or double spread—and whether the page contains advertising. It also helps if the art director knows the content and appearance of the advertising on the page to avoid embarrassing juxtaposition. If the art director is working on a one-page design, he or she should know the content and appearance of the facing page.

The artist tries to visualize what the page is supposed to say. From experience the artist has developed a feel for the magazine page and knows how it will look on the finished page. The beginner may have to use trial and error to find an appropriate design. He or she may, for example, cut out pieces from construction paper to represent the black blocks, then juggle these blocks until they form a usable design.

Design is a means rather than an end. If the reader becomes aware of the design, the design is probably bad.

One danger most art directors seek to avoid is cluttering. This occurs when too many illustrations are placed on the same page, when the pages are crowded because of lack of spacing or uneven spacing, or when too many elements—dingbats, subtitles, boldface type—make the page appear busy. The primary goals of magazine design are to catch and direct the reader's attention and to make the pages easy to read.

## Copy Fitting

Widths of magazine columns may vary with the number, shape and size of the ads or the size and shape of the illustrations. It is not unusual for a magazine story to be strung over four pages in four different widths. The editor must be able to estimate whether the story will fit the space allocated. Computer terminals have simplified this process because they allow exact copy fitting even before the story is typeset. If computers are not available for copy fitting, the process is more difficult.

The most accurate manual method of determining copy length is by counting characters in the manuscript. These steps are used:

- Count the number of characters in the manuscript.
- Multiply the number of lines of copy by the number of characters to the average line. A line extending half the width of the line is counted as a full line.

## *The Importance of Transitions*

Magazine editors spend a lot of time worrying about transitions, or ways of leading the reader from one thought to the next within a story. Stories written in the inverted pyramid format often are completely lacking in transitions. Information in such stories is thrown together with little connectivity. In magazine stories, however, a premium is placed on transitions. In his book *Magazine Editing,* J.T.W. Hubbard quotes Leonard Robinson, a former editor of the *New Yorker* and *Esquire* on the subject:

> The ability to write a good transition is quite simply the difference between a $20,000-a-year editor and a $35,000-a-year editor.

Hubbard notes that a transition is only as effective as the article's underlying organization. "A good transition presupposes material that has already been laid into a clear organizational pattern; it also presupposes a sensitivity on the part of the editor to the complex web of relationships between the various parts of that article."

Good transitions are not the exclusive province of magazines. One editor learned that when he was regularly assigned to trim three syndicated columns into a fixed amount of space on the editorial page. On days when James J. Kilpatrick's column appeared, he made all the cuts in the other two.

"Kilpatrick writes so tightly, and his transitions are so well-integrated, that his stuff is almost impossible to cut," the editor said. "I'd rather spend the day at the dentist's office than try to cut a paragraph out of the middle of a Kilpatrick piece."

That's high praise, indeed, for one of the great newspaper writers of our time.

- Consult a type book to determine the CPP, characters per pica, for the body size and typeface. For example, 10-point Bodoni Book measures an average of 2.75 characters per pica. If the type line is to be 20 picas wide, 55 characters will fill one line of type.
- Divide the number of lines of type by the number of typeset lines per column inch. If the type is set in 10-point with 2-point spacing, the number of type lines per column inch is determined by dividing 72 (the number of points in an inch) by 12. The result will be the number of column inches the manuscript will occupy.

The same figure can be obtained by multiplying the number of typeset lines by the point size (including leading space or spacing between lines) of the typeface, then dividing the total points by 72 to find column inches. To convert into pica depth, divide the point total by 12 (points per pica) rather than by 72.

For fitting copy into a specified space, the method can be used in reverse. Suppose the space to be filled is 6 inches deep and 24 picas wide. The type is to be 12-point. The type chart shows 2.45 characters per pica or 59 characters to the 24-picas line. Twelve-point type set 6 inches deep requires 36 type lines. Multiplying 59 by 36 gives 2,124 characters. If the manuscript lines average 65 characters, 32 lines of the manuscript will be needed.

## Placement of Advertising

The usual newspaper practice is to pyramid the ads on the right of the page. In a magazine the ads generally go to the outside of the pages or may appear on both outside and inside, leaving the *well* for editorial copy. The ads need not restrict editorial display, especially if the well is on a double spread.

At magazines where the advertising manager determines ad placement, there is a give and take between ad manager and editor. The editor may want to start a story in a certain part of the magazine, but there is a two-column ad on the most likely page. The editor then asks the ad manager if the ad can be moved to another page. Unless the ad was sold with position guaranteed, the ad manager usually is able to comply.

## Imposition

Editors should know something about *imposition,* or the arrangement of pages for binding. Imposition refers to the way the pages are positioned on the reproduction proof and not the way they will appear in the magazine. The printer can give the editor the imposition pattern, or the editor can diagram the imposition if he or she knows whether pages in the form are upright or oblong.

For a 16-page form, the editor makes three right-angle folds and numbers the pages. This will show Page One opposite Page 8, 16 opposite 9, 13 opposite 12, 4 opposite 5. The remaining eight pages will be in this order—7 and 2, 10 and 15, 11 and 14, 6 and 3. For an oblong form, printed work and turn, the pattern is 1 and 16, 4 and 13, 5 and 12, 8 and 9, 15 and 2, 14 and 3, 11 and 6, 10 and 7. Again, the editor may determine the pattern by making three parallel or accordion folds and one right-angle fold. Or the editor may use the following formula: the size of the book plus one page. Thus, in a 32-page section, Page 4 is opposite 29 (33 – 4) (see Figure 10–5).

For a story spread over several pages, the editor tries to get the pages on the fewest forms possible to avoid tying up too many forms with one story. Understanding imposition also can serve as a guide in using color. If one page in a four-page form is in full color, four forms will be needed, one each for red, yellow, blue and black. The other three pages on the same form can accommodate color with little added expense.

# MAGAZINE PRODUCTION

A magazine editor relies on an artist to help achieve editorial excellence. By the same token the artist relies on a production expert, usually a printer, to help produce the best possible publication within the budget.

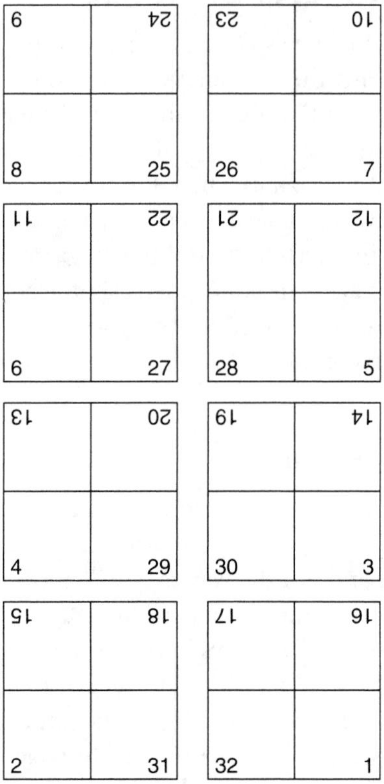

*Figure 10–5 Imposition pattern for a 32-page section in four-page forms. The pattern is obtained by gathering four quarter-sheets and making two right-angle folds. The facing pages total 32 (pages in the section) plus one.*

The editor is responsible for providing the printer with complete specifications of the magazine, not only the size of the publication, the number of pages and the press run but the use and placement of color, the number and size of the illustrations, the type area, size of type and any items that will require special handling.

The editor can save money, and make the printer happy, by giving clear and adequate instructions, editing the copy thoroughly rather than making changes in proofs, promptly reading and returning all proofs, meeting copy deadlines and giving the printer time to do top-quality work.

Today, magazines, like newspapers, are almost all written, edited and designed on computers. Copy fitting, title writing and design are made much simpler in the process. That, in turn, allows editors to focus on the editing and design of the magazine rather than wasting time on complex calculations for story lengths and caption fitting. Computers make it easier than ever to wrap type around illustrations and change the widths of columns—critical features of magazine design.

# CHAPTER 11

# USING PHOTOS AND GRAPHICS IN PRINT

## THE IMPORTANCE OF GRAPHIC ELEMENTS

At good newspapers, the days are gone when a city editor would say to a reporter, "That's a good story; now get me a photo to go with it." Today, good newspapers treat graphical elements—photos, charts, graphs and maps—as the editorial equivalents of stories. Accordingly, they are assigned with the same care, and often at the same time, as news and feature articles.

That development came about as newspapers of the 1980s and 1990s faced the realities of increasing competition for the reader's attention. Editors found that newspapers needed to be more attractive if they were to hold or increase readership. The result was an era in which newspaper design became almost as important to many editors as content. In general, the size and quality of newspaper photographs increased, and charts, graphs and maps proliferated. All of those elements were increasingly likely to be printed in color.

None of that should be surprising. We live in a visual society in which color and design play important roles. Visual appeal is used to market everything from television to cereal boxes, and newspapers are not exempt. That has necessitated a change in the way almost all newspapers look. The gray, vertical columns of the past have given way to modular design, color, more and larger pictures, and charts, graphs and maps, which editors call *information graphics.* To make those changes, newsrooms have been forced to change their internal structures.

Today, photo editors are key members of the newsroom management team, usually equal in rank to news, city or metropolitan editors. Design desks have been added in many newsrooms to relieve the copy desk of the chore of designing as well as editing. Many metros have gone so far as to appoint assistant managing editors for graphics. And graphic departments, which produce charts, graphs and maps, have sprung from nowhere, even at relatively small dailies.

Not surprisingly, technological advance has helped. Few newsrooms today are without personal computers with design and charting software. These machines allow editors and artists to create graphics that would have taken days to execute through traditional methods.

Now the art of photography itself is about to change as digital cameras threaten to make silver-based photography obsolete. Both local and wire service photos are being processed with computers and digital-imaging software (see Figure 11–1). Once again the newspaper industry is embracing technology to help it remain competitive. Most observers agree that the results to date are promising.

No fancy technology, however, can make a bad picture good or help an editor determine whether a graphic accomplishes what it purports to do. This chapter is designed not to make photographers or graphics artists of editors but to give those editors an appreciation of the role of photos and graphics in the appeal of newspapers and magazines. Visual literacy is critical to the competent editor.

Rewriting can turn a poorly written news story into an acceptable one. Little can be done to change the subject matter of clichéd photos—tree plantings, ribbon cuttings, proclamation signings and the passing of checks, certificates or awards

***Figure 11–1 Photos are now edited on computer workstations at most newspapers.***
Photo by Philip Holman.

from one person to another. Some newspapers and magazines use photographs of these situations simply because of the tradition that "chicken-dinner" stuff must be photographed. It is a tradition that should be scrapped, and most good publications have already have done so.

Only on a rare occasion would a city editor permit reporters to share their time with sports, Sunday supplements or the advertising department. Yet that is what happens on papers with a small staff of photographers.

One consequence is that too often good news and feature stories are less effective than they could be because they lack the additional information that accompanying pictures could provide. Another is that too often mediocre, space-wasting pictures from the wire services or syndicates are given more attention than they deserve.

A photo editor is almost as essential to a newspaper as a city editor. Some executive—preferably one with a background in photography—should be responsible for assigning photographers to news and feature events. Someone in authority should insist that all photos, including those from news agencies, be edited and that cutlines, or captions, be intelligently written.

If it is a good picture, it should get a good play, just as a top story gets a big headline. If pictures are a vital part of the story, editors should be willing to cut back on words, if necessary, to provide space for pictures. Some events can be told better in words than in pictures. Conversely, other events are essentially graphic, and editors need little or no text to get the message across.

Pictures can "dress up" a page. But if their only purpose is to break up the type, they are poorly used. The large number of pictures used, even on front pages, without an accompanying story suggests that the pictures are being used for their graphic value rather than for their storytelling value. Ideally, pictures and stories should work together.

Still pictures, even action shots, may not be able to compete with television in some ways, but still photography can add color to words and can capture moods. Originality starts with the picture. Its values are interest, composition and quality of reproduction.

## PHOTOGRAPHER-EDITOR RELATIONSHIPS

An encouraging development in recent years has been the trend toward making photographers full partners in the editorial process. Historically, photographers have been second-class citizens in the newspaper and magazine hierarchy. They have not enjoyed the prestige of the reporter or the copy editor, and with rare exception their opinions have been ignored or have not been solicited.

Many insightful editors have now realized that photographers possess an intangible quality known as visual literacy, a trait sometimes lacking in even the best of wordsmiths. Photographers should have a voice in how their pictures are displayed, and editors who have attempted to give them that voice invariably have been pleased with the results. The number of publications using pictures well is increasing each year, although leading picture editors agree that there is ample room for improvement. For every newspaper and magazine using pictures well, it is easy to find three still using the line-'em-up-and-shoot-'em approach.

A key to improvement in the quality of pictures is allowing the photographer to become involved in the story from the outset. If possible, the photographer should accompany the reporter as information for the story is gathered. If that is impossible, allowing the photographer to read the story—or to take time to talk with the reporter about the thrust of it—will help ensure a picture that complements the story. Publications throughout the country each day are filled with pictures that fail to convey a message because the reporter, photographer and editor failed to communicate.

An important part of this communications process involves writing the photo order, the document given to the photographer when an assignment is made. A photographer for the *Columbia Missourian* once received an order for a picture of an elementary school principal. The order instructed the photographer to meet the principal in his office at 3:15 p.m. and to take a picture of him at his desk. It mentioned that the principal would be unavailable until that time. The story focused on how the principal went out of his way to help frightened first-graders find the right bus during the first few weeks of school. Because the reporter failed to mention that when he wrote the photo order, the photographer followed his instructions exactly. He arrived at 3:15 and took the picture. Only after returning to the office did he learn of the thrust of the story and discover that the principal was "unavailable" 15 minutes earlier because he had been helping students find their buses. The best picture situation had been missed.

Words and pictures are most effective when they work together. For that to happen, reporters, photographers and editors must work together. Often the best approach is to let the photographer read the story and determine how to illustrate it. The photographer, after all, should be the expert in visual communication.

## EDITING DECISIONS

Most pictures, like most news stories, can be improved with editing. The photo editor, like the copy editor, must make decisions that affect the quality of the finished product. The photo editor must determine:

- Which photo or photos complement the written story or tell a story of their own.
- Whether cropping enhances the image.

- What size a photo must be to communicate effectively.
- Whether retouching is necessary.

## Selection

Picture selection is critical because valuable space is wasted if the picture does nothing more than depict a scene that could be described more efficiently with words. The adage that a picture is worth a thousand words is not necessarily true. If the picture adds nothing to the reader's understanding of a story, it should be rejected. Conversely, some pictures capture the emotion or flavor of a situation more vividly than words. In other situations, words and pictures provide perfect complements.

A talented photo editor, experienced in visual communication, can provide the guidance necessary for successful use of pictures. Smaller papers without the luxury of full-time photo editors can turn to their photographers for advice, but often the news editor or copy editor must make such decisions. When that is necessary, an appreciation for the importance of visual communication is essential for good results.

Internal procedures reflect a publication's picture selection philosophy. Some allow the photographer to make the decision; the pictures he or she submits to the desk are the only ones considered for publication. This procedure may ensure selection of the picture with the best technical quality, but that picture may not best complement the story. A photo editor, working closely with the photographer, the reporter, the city editor and copy editors, should have a better understanding of the story and be able to make the best selection. Contact prints, miniature proofs of the photographer's negatives, allow the photographer, reporter and editors to review all frames available so the best selection can be made (see Figure 11–2).

## Cropping

A photograph is a composition. The composition should help the reader grasp the picture's message clearly and immediately. If the picture is too cluttered, the reader's eyes scan the picture looking for a place to rest. But if the picture contains a strong focal point, the reader at least has a place to start. A prime job of a photo editor, therefore, is to help the photographer take out some unnecessary details to strengthen the overall view.

Certain elements within the picture could be stronger than the full picture. Some photo editors try to find these interest points and patterns by moving two L-shaped pieces of cardboard over the picture. This helps to guide the editor in cropping. The editor looks for a focal point, or chief spot of interest. If other points of interest are present, the photo editor tries to retain them. The editor searches for

***Figure 11–2** Contact print sheet. Photographers and editors often choose pictures from contact prints of the available negatives.*

patterns that can be strengthened by cropping. The pattern helps give the picture harmonious and balanced composition. Among these patterns are various letter shapes—L, U, S, Z, T, O—and geometric patterns such as a star, a circle, a cross or a combination of these.

Because most news and feature photos contain people, the editor strives to help the photographer depict them as dramatically as possible. The editor must decide how many persons to include in the picture, how much of a person to include and what background is essential.

Historically, publications have opted for the tightest possible cropping to conserve valuable space. Severe cropping, however, may damage a picture to the point that not printing it would have been preferable. Those who win awards for photo editing appreciate the fact that background is essential to some photographs. As a result, they tend to crop tightly less often than do editors with more traditional approaches to picture editing. In Figures 11–3A and 11–3B, tight cropping allows the reader to see interesting detail. But in Figures 11–4A and 11–4B, tight cropping eliminates the environment and damages the meaning of the picture. Those who can distinguish between these approaches are valuable members of newspaper and magazine staffs. They possess visual literacy.

## Sizing

The value of the picture, not the amount of space available, should determine the reproduction size of a photograph. Too often, newspaper editors try to reduce a photograph to fit a space and destroy the impact of the photo in the process. Common sense should dictate that a picture of 15 individuals will be ineffective if it appears as a two-column photo. More likely, such a photo will require three or even four columns of space.

Talented photo editors know that the greatest danger is making pictures too small. If the choice is between a two-column picture and a three-column picture, the wise photo editor opts for the larger size. Pictures can be too large, but more often they are damaged by making them too small. Another alternative may be available. Modern production techniques make it easy for the editor to publish a $2\frac{1}{2}$ column photo. Text to the side of it is simply set in a wider measure to fill the space.

Sizing of any photograph is an important decision, but sizing of pictures in multiphotograph packages is particularly important. In such packages, one photograph should be dominant. The use of multiple pictures allows the editor flexibility that may not exist in single-picture situations. If a picture editor selects a photo of a harried liquor store clerk who has just been robbed and a photo of the outside of the store where the robbery occurred, the editor has three choices:

- Devote equal space to the two pictures. This is the least desirable choice because neither picture would be dominant and, consequently, neither would have eye-catching impact.
- Make the outside shot dominant and the close-up of the clerk secondary. This would work, but the dominant picture, which serves merely as a locater, would have little impact. The impact of human emotion, evident in the clerk's face, would be diminished.
- Make the facial expression dominant with good sizing and make the outside shot as small as $1\frac{1}{2}$ columns. The outside shot, standing alone, would look

A

B

***Figures 11–3A and 11–3B** Footprint on the lunar soil. An example of how cropping (B) can bring out an interesting detail in a photograph (A). The close-up view was photographed with a lunar surface camera during the Apollo 11 lunar surface extravehicular activity.*

Photographs courtesy of the National Aeronautics and Space Administration.

A B

***Figure 11–4A** Tight cropping can occasionally destroy the impact of a picture.*
***Figure 11–4B** Here, tight cropping takes the farmer out of his environment by making it difficult or impossible for the reader to determine that the setting is a barn.*
*Columbia Missourian* photos by Manny Crisostomo.

ridiculous if used in that size. But used in conjunction with another, larger photo, it would work well.

Dramatic size contrast is an effective device in multipicture packages (Figures 11–5A and 11–5B). An editor trained in visual communication understands the usefulness of reversing normal sizing patterns for added impact.

***Figures 11–5A and 11–5B** Many editors would run the overall flooding shot larger than the picture of the farmer laying sandbags in place. The pairing, however, has more impact if the close shot of the farmer is run larger than the scene-setting overall picture.*

*Columbia Missourian* photos by Lee Meyer and Mike Asher.

### Retouching

Some pictures can be improved by *retouching,* the process of toning down or eliminating distractions within the frame. Minor imperfections in photos can be retouched with an airbrush, an instrument that applies liquid pigment to a photo surface by means of compressed air. Retouching also can be done by brushing on a retouching liquid or paste, or by using retouching pencils of various colors. Even regular pencils can be used to retouch imperfections in black-and-white photos.

Care must be exercised, however, to ensure that retouching does not change the meaning and content of the photo. Changing a photo to alter its meaning is as unethical as changing a direct quotation to alter a speaker's meaning.

This whole issue has taken on new importance in recent years with the arrival of computerized photo editing. An unscrupulous editor can move one of the Great Pyramids in relation to the Sphinx to improve a picture's composition or place the head of one person on the body of another. Both those things have happened in recent years as magazines and newspapers have grappled with the ethics of altering photographs.

Most good newspaper editors have taken the solid ethical position that nothing more than changing the brightness, contrast or sharpness of a picture should be tolerated. Anything more than that leads to deception, which is sure to destroy a publication's integrity, at least where photographs are concerned.

## PICTURES AS COPY

When the picture has been processed, someone—reporter or photographer—supplies the information for the caption, also known as the cutline. The picture and caption information then go to the appropriate department where the editor decides whether to use the picture and, if so, how to display it.

Before submitting a picture to the production department, the editor supplies enough information to get the correct picture in the correct place with the correct caption. A picture, like a story, generally carries an identifying slug. To ensure that the picture will match the screened mechanical reproduction, the cutline and, if need be, the story, the editor uses a slugline.

A slip of paper clipped on the picture or taped to the back normally contains information such as:

- The slug, or picture identification.
- The size of the desired screened reproduction.
- Special handling instructions.
- The department, edition and page.
- The date the photo is to appear.

- The date and time the picture was sent to the production department.
- Whether the picture stands alone or accompanies a story.

The picture is then routed directly to the camera room or indirectly through the art department. The caption goes to the composing room. Caption copy contains, in addition to the cutlines, essential directions to match text and picture. If the picture is to go with a story, the information is carried on both the cutline and the story copy. The reason is obvious. Unless properly slugged, the story may appear on Page 3 and the photo on Page 16.

Sometimes the photo may be separated from the story deliberately. A teaser picture may be used on Page One to entice readers to read the story on another page. If a long story has two illustrations, one illustration often is used on the page where the story begins and the other on the jump page. On major events, such as the death of a president, pictures may be scattered on several pages. In that case, readers are directed to these pages with a guideline such as "More pictures on pages 5, 7 and 16."

Sometimes the mechanical reproduction of the photo is made in reverse. The result can be ludicrous, particularly if the picture shows a sign or if the principals are wearing uniforms containing letters or numerals. The person responsible for checking page proofs makes sure the correct headline is over the correct story and the captions under pictures of a local politician and a jackass are not reversed.

## CHANGING PHOTO TECHNOLOGY

Technology is changing the way photographs are processed at newspapers and magazines. Today, many photographs are scanned into computers, where they are lightened or darkened, sharpened, cropped and sized using programs such as PhotoShop. More and more newspapers and magazines also are flowing photographs directly onto the page using personal computer programs such as PageMaker or Quark XPress. From there photographs are output to paper, to film or even directly to the printing plate along with the text.

Scanning of photos is necessary because most publications still prefer to use conventional photography to take pictures. Digital photography, in which photos are stored on miniature diskettes inside cameras rather than on film, is still in its infancy. Inexpensive digital cameras produce photos of lesser quality than conventional photography, and high-end digital cameras are still too expensive for widespread use.

Expect that to change quickly. Once it does, darkrooms and chemical photo processing will be a thing of the past. Digital processing saves time, which is of paramount importance for publications, particularly newspapers. Indeed, the move to digital photography is more likely to take place first at newspapers, where low-quality paper limits reproduction quality anyway. Magazines, which usually use glossy, high-quality paper, are more likely to remain with conventional photography until the quality of digital cameras improves.

Already, newspaper wire photos are transmitted and processed digitally, as described in Chapter 9; so, too, are many photos that appear in advertisements—photos of food items, small photos of houses in the classified section and similar items. Increasingly, newspapers are experimenting with digital photography on the sports and news pages, where timeliness and speed are important.

## THE ENLARGING-REDUCING FORMULA

Those newspapers and magazines still processing photos the old way typically start by making 8-by-10-inch prints. Photos then are enlarged or reduced in proportion to their width and depth to fit the assigned space on the page. A simple method of determining the proper proportion is to draw a diagonal line from the upper-left to the lower-right corner of the photograph as cropped. It is best to draw this line on the back of the picture to make sure the line is not accidentally reproduced in the newspaper. Then measure the reproduction width along the top of the picture and draw a vertical line at that point. The point where the vertical line intersects the diagonal line indicates the reproduction depth of the picture.

A more common procedure is to measure in picas or inches the width and depth of the photograph as cropped. Because the editor knows the width he or she wants the picture to be when it is reproduced, a proportion wheel (Figure 11–6) or calculator allows that editor to determine the reproduction depth quickly. The percentage of enlargement or reproduction is known as the SOR, or size of reproduction. If a photograph is being enlarged, the SOR will be greater than 100 percent. If the photograph is being reduced, the SOR will be less than 100 percent. Knowing this allows the editor to make a quick check to determine whether reproduction instructions to the engraver or camera room are correct. If the picture margins are uneven, the editor may place a sheet of tissue paper over the picture and draw the diagonal and connecting lines on the tissue paper.

## PICTURES CAN LIE

The picture editor makes the same kind of editorial judgment about a picture that the city editor and the wire editor make about a local story and a wire story. Does the picture tell the whole story or only part of it? Does it distort, editorialize, mislead? Does it omit important details or include details that create an erroneous impression? In other words, is the picture loaded?

The point was raised by James Russell Wiggins, former editor of *The Washington Post,* during a lecture at the University of North Dakota, "The camera," he said, "can be a notorious, compulsive, unashamed and mischievous liar."

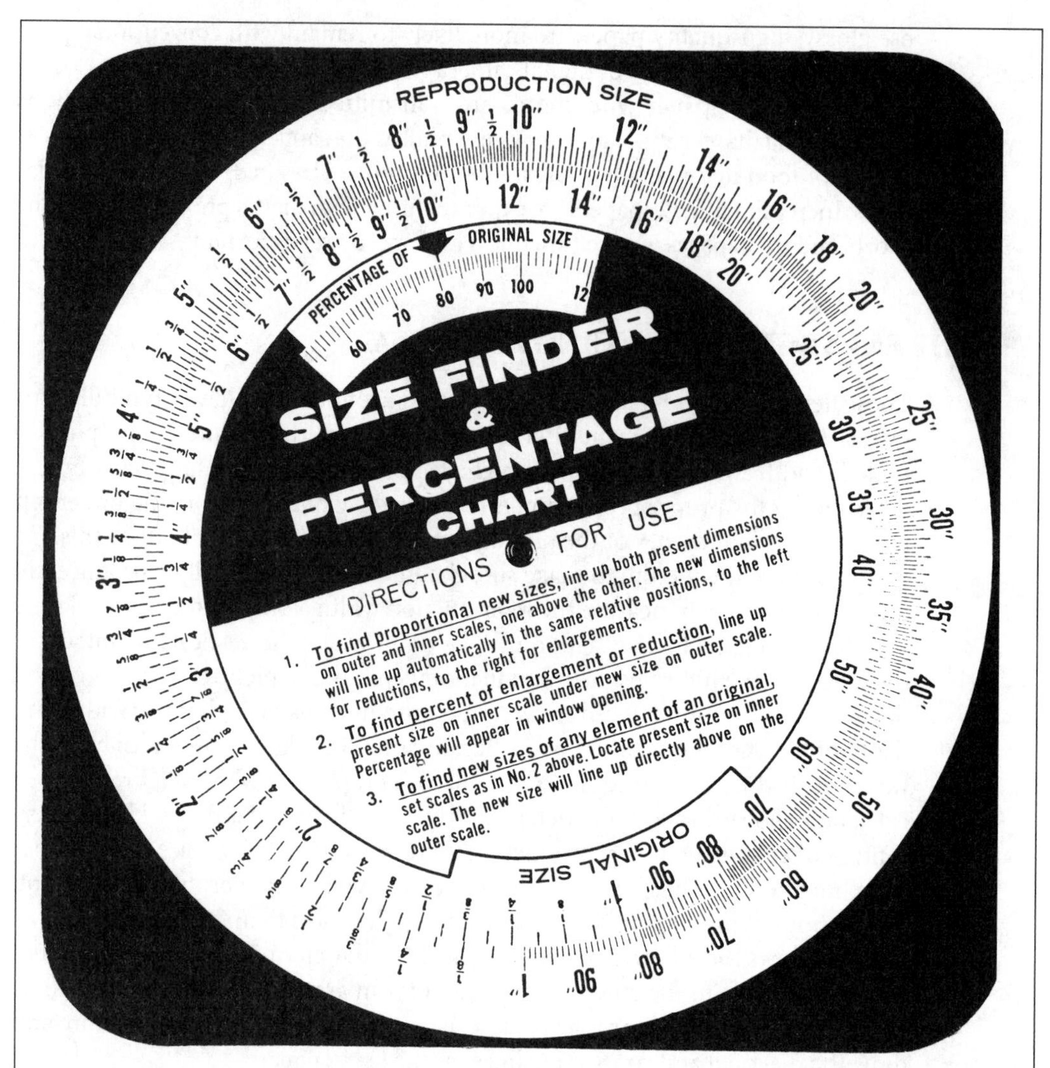


*Figure 11–6 Example of disc showing enlargement and reduction ratios. The finder is set at dimensions with $4\frac{1}{8}$" on the inner circle matching $3\frac{3}{8}$" on the outer circle. The figure $6\frac{5}{16}$" is opposite the inner figure of $7\frac{3}{4}$" and the SOR, or size of reproduction, is 81 percent. The numerals on the wheel, while marked as inches, also can represent picas if the newspaper uses that unit of measurement in sizing photographs.*

## *When Photos Lie*

"Photos never lie" is one of the oldest newspaper axioms. It's doubtful whether that was ever true, and it certainly isn't today. Computerized processing of photos makes it easier than ever to manipulate pictures in ways that are totally unethical.

*National Geographic* magazine moved a pyramid closer to the Sphinx to improve the composition of a photo, then listened to cries of outrage from photographers and others who objected to the practice. *TV Guide* put Oprah Winfrey's head on another person's body and suffered a similar fate.

One can't help but wonder, though, how many similar things have occurred without someone noticing. The fact is that tampering with the content of a photo is tantamount to printing a manufactured quotation and attributing it to a senator, the mayor or a police officer at the scene of a crime. Manufacturing untrue photos is just as wrong as manufacturing a quotation. It is unethical, and at many publications it is now grounds for dismissal.

Quality publications now limit computerized alteration of photographs to the equivalent of minor retouching, a process developed in the era of conventional photography. Editors and photographers are allowed to make a photo lighter or darker. They also are allowed to do electronic edge sharpening, the equivalent of improving the focus.

Most other forms of alteration are prohibited, although some publications allow the removal of distracting background elements. That might include removal of an electric wire dangling behind a subject in such a way as to appear the wire is emerging from the person's head. The best photographers eliminate such distractions the right way by making sure they aren't there in the original photo.

To illustrate, he said he once declined to print a photograph of President Harry Truman walking across the platform of Union Station before a backdrop formed by a row of caskets just shipped in from the Korean war. "What that camera said was that the Korean war was 'Truman's war,' just what thousands of the president's critics were saying."

He also commented on the distorted portrait of police officers during civil disorders. The pictures may have been representative of the action, but they failed to tell what really happened in perspective and why.

"The camera does not tell the truth," said Wiggins, "and because what it tells is not the whole truth, skepticism about the media rises in the minds of readers who know that police officers, whatever their undoubted faults, are not always wrong."

A picture may be striking and it may be narrative. But if it conveys a false or distorted impression, it would be better left unpublished. Picture editors often can show subjective judgment in the selection of pictures. Suppose an editor has four or five pictures of a public figure. Some editors will select the more favorable picture; others will pick the less favorable one. Many of the pictures used of former President Nixon, even before his resignation, were editorialized. Pictures of former Presidents Jimmy Carter and Lyndon Johnson were similarly criticized.

## TASTE IN PICTURE EDITING

It was a tragic fire in a metropolitan area. A woman and a child took refuge on an ironwork balcony. As firefighters tried to rescue them, the balcony collapsed, sending the woman to her death and the child to a miraculous survival. Photographers took sequence shots of the action (Figures 11–7A and 11–7B). Should a picture editor use the pictures?

Some readers will be incensed, accusing the papers of sensationalism, poor taste, invasion of privacy, insensitivity and a tasteless display of human tragedy to sell newspapers. Picture editors could reply that their duty is to present the news, whether of good things or bad, of the pleasant or the unpleasant. Defending the judgment to use the pictures on Page One, Watson Sims, then editor of the *Battle Creek* (Mich.) *Enquirer and News,* said, "The essential purpose of journalism is to help the reader understand what is happening in this world and thereby help him to appreciate those things he finds good and to try to correct those things he finds bad."

Of the flood of pictures depicting the war in Vietnam, surely among the most memorable were the Saigon chief of police executing a prisoner, terrified children fleeing a napalm attack, and the flaming suicide of a Buddhist monk. Such scenes were part of the war record and deserved to be shown.

Photos of fire deaths may tell more than the tragedy depicted in the burned and mangled bodies. Implicit could be the lessons of inadequate inspection, faulty construction, carelessness with matches, arson, antiquated firefighting equipment or the like.

Picture editors have few criteria to guide them. Their news judgment and their own consciences tell them whether to order a picture for Page One showing a man in Australia mauled to death by polar bears after he fell or dived into a pool in the bears' enclosure in a zoo. Of the hundreds of pictures available that day, surely a better one could have been found for Page One. If the scene causes an editor to turn away and say, "Here I don't belong," chances are the readers will have the same reaction. Not all of life's tragedies have to be depicted. The gauge is importance and newsworthiness.

## PICTURE PAGES

Some newspapers devote an entire page to pictures with a minimum of text. Some use part of the page for pictures, the rest for text matter. Some pages are made up of unrelated photos; some are devoted to related pictures. Some use part of the page for sequence pictures and leave the remainder for unrelated pictures or text matter. Increasingly, photo pages on a single subject are used to tell a story more forcefully than words can tell it (see Figure 11–8, page 230).

Here are a few pointers on picture pages:

- Three or four large pictures make a more appealing picture page than eight or 10 smaller ones.
- Let one picture, the best available, dominate the page.
- Emphasize the upper-left portion of the page either with a dominant picture or a large headline.
- If the content allows, crop some of the pictures severely to achieve either wide, shallow, horizontal ones or narrow, long, vertical ones.
- In a picture series or sequence, place a big picture in the bottom right corner of the page. It is the logical stopping point.
- Let the page breathe. White space makes both the pictures and the text stand out. But keep interior spacing standard and allow white space to bleed to the outside. Trapped interior white space is distracting.
- Don't align pictures with a T square. An off-alignment often provides extra white space or leaves room for a caption.
- If a picture page has to be made up in a hurry, pick the best picture, rough-sketch it on a dummy, slug and schedule the picture and get it to the production department. Then edit the other pictures. The cutlines can be written while the editor looks at photocopies of the pictures.
- Vary picture page patterns. Don't make today's picture page look like last Saturday's.
- Captions need not appear below the pictures. In fact, a narrow caption beside the picture may be easier to read than a wider one below it.
- In a sequence or series of pictures, don't repeat in one caption what was said in another.
- If all the pictures were taken by one photographer or provided by one wire service, a single credit line on the page will suffice. Too many credits give the page a bulletin-board effect.
- In a photo-essay page, keep the captions as brief as possible. Usually, the pictures tell most of the story, especially if the headline has established the theme.
- Headlines generally are more effective at the left or right of the page or under the main pictures. Occasionally the head may be overprinted on the main picture if the type does not rob the picture of important details.

***Figure 11–7A** One of the controversial sequence shots of a fire tragedy in a Boston apartment. Scores of readers protested the use of these widely distributed photos. Most editors defended the use of the pictures.*

Photos by Stanley Forman of the *Boston Herald-American*, distributed by UPI.

***Figure 11–7B** The second shot of the fire tragedy in Boston.*

COLUMBIA MISSOURIAN

# SHOWCASE

Sunday, November 13, 1994 — SECTION G

# Religious CONSTRUCTS

## The design of a house of worship reflects the foundation of its congregation's faith

The First Christian Church was dedicated in 1893.

Text by Todd J. Satter
Photos by Greg Lahann
Of the Missourian staff

Throughout history, religion has inspired some of the world's greatest architecture. From the ancient temples of the Near East to the Kaaba shrine in Mecca, from early synagogues to highly embellished Gothic cathedrals, much time and effort has gone into constructing places for worship and congregation.

"The best way to symbolize God's presence is for the community to have a place in which they can come together in worship," Father Michael Quinn of the Newman Center says.

A number of mid-Missouri structures serve as an important source of knowledge about these faiths, with the Christian church the most common type. While some churches cling to tradition and worship in conventional, elaborately designed buildings, others are moving toward a more modern style.

### NEWMAN CENTER

The Newman Center's new chapel is quite a departure from the traditional Catholic church. It has a simple, functional design constructed to fit its surroundings.

"This is so the environment is not so cluttered," Quinn explains. "It emphasizes the sense of community."

The Newman Center, 701 Maryland Ave., is characterized by a functional design with few embellishments. If not for the cross towering above the entryway, it easily could be mistaken for a more secular building. The most pervasive element of the exterior design is the window placement, most notably the wall of windows that round off the building's northwest corner.

"The windows are like the wings of the Holy Spirit reaching out, embracing you," Quinn says.

The stained-glass windows inside are one aspect that hearkens back to churches of old.

"Windows taught people who could not read traditionally," Quinn says. "Many of those early writings and images are still used in modern churches."

The Newman chapel's interior design might be the most important thing to parishioners. The Newman Center emphasizes the mean of hospitality, beauty and simplicity. Every step one takes upon entering the chapel symbolizes a much greater spiritual journey.

"As soon as you walk in, you come to the baptistry," Quinn says. "It's a reminder of committing ourselves more deeply to our baptismal covenant and union with Christ."

The positioning of the pews in a semicircle, as well as the altar and reading stand, enhance the sense of community promoted in the church.

### FIRST CHRISTIAN CHURCH

Other area congregations worship in older structures that permit a connection with history. First Christian Church, 101 N. Tenth St., is a prime example. In 1858, members of the church purchased the corner lot on which it is located. The first church on this site was constructed two years later, and the present church was dedicated in 1893, making it one of the oldest buildings in Columbia.

The Department of the Interior officially has recognized the sanctuary as a National Historic Place. According to First Christian secretary Linda Kersey, a building must be 100 years old and have maintained its original appearance to receive certification and registry as a National Historic Place.

"There's a lot of paperwork involved," Kersey says with a laugh.

The building was designed in the elaborate Richardson-Romanesque style, with stone walls and arched windows and entryways. The interior is equally impressive. Stained-glass windows, all of which are more than a century old, surround the sanctuary.

Virtually the only change that the sanctuary has undergone in the past 100 years is the rearrangement of the pews. Originally, the chancel was placed in a corner, with the pews in a fan-shaped arrangement. In 1929, it was changed to the current rectangular arrangement common to most Christian churches.

### ISLAMIC CENTER

The Islamic mosque has a different purpose for those who worship there.

"The mosque should be built as the focal point of the community," says Amor Cheriff, a member ofthe Islamic Center of Central Missouri, 201 S. Fifth St. Cheriff notes that with the mosque, appearance is not as important as function.

"Embellishments shouldn't annoy the person praying," he says. "You should be able to concentrate on your god. The mosque should just be some form that reflects Islamic architecture."

However, most mosques — including the Islamic Center — are built to resemble a mountaintop. This is a reminder of the early mosques that were constructed on cities' highest points so that everyone could hear the call to prayer.

Islam might seem different to the Western world because there is no

Please see MODERN, Page 2G

ABOVE: The use of light is important to the design of traditional Catholic churches such as Sacred Heart on Ninth Street.

LEFT: A portion of the Islamic Center is a mosque, which, in keeping with Islamic tradition, was designed to resemble a mountaintop.

BELOW: Modern synagogues such as the B'nai B'rith Hillel Foundation are designed to blend in with their surroundings.

*Figure 11–8 Newspaper picture pages require thoughtful planning, including varied picture sizes. Here, white space is allowed to bleed to the outside of the page and is not trapped within.*

*Columbia Missourian.*

## CAPTION GUIDELINES

Picture texts are known by many names—cutlines, captions, underlines (or overlines) and legends. *Caption* suggests a heading over a picture, but most editors now use the term to refer to the lines under the picture. *Legend* may refer either to the text or to the heading. If a heading or catchline is used, it should be under, not over the picture.

The editor "sells" the reporter's story by means of a compelling headline. By the same token, the picture editor can help control the photographic image with a caption message. The primary purpose of the caption is to get the reader to respond to the photo in the manner intended by the photographer and the picture editor.

Readers first concentrate on the focal point of the picture, then glance at the other parts. Then, presumably, most turn to the caption to confirm what they have seen in the picture. The caption provides the answers to questions of who, what, where, when, why and how, unless some of these are apparent in the picture.

The caption interprets and expands on what the picture says to the reader. It may point out the inconspicuous but significant. It may comment on the revealing or amusing parts of the picture if these are not self-evident. The caption helps explain ambiguities, comments on what is not made clear in the picture and mentions what the picture fails to show if that is necessary.

The ideal caption is direct, brief and sometimes bright. It is a concise statement, not a news story. It gets to the point immediately and avoids the "go back to the beginning" of the background situation.

If the picture accompanies a story, the caption doesn't duplicate the details readers can find in the story. It should, however, contain enough information to satisfy those who will not read the story. Ideally, the picture and the caption will induce readers to read the story. Normally the caption of a picture with a story is limited to two or three lines.

Even when the picture relates to the story, the caption should not go beyond what the picture reveals. Nor should the facts in the caption differ from those in the story.

Captions stand out in the publication's sea of words and strike the reader with peculiar force. Every word should be weighed, especially for impact, emotional tone, impartiality and adherence to rules of grammar and the accepted language.

Anyone who tries to write or rewrite a caption without seeing the picture risks errors. The writer should examine the cropped picture, not the original one. The caption has to confine itself to the portion of the picture the reader will see. If the caption says a woman is waving a handkerchief, the handkerchief must be in the picture. In a layout containing two or more pictures with a single caption, the writer should study the layout to make sure that left or right or top or bottom directions are correct.

## Writing the Caption

Here are some tips on caption writing:

- **Don't tell the obvious.** If the person in the picture is pretty or attractive, that fact will be obvious from the picture. The picture will tell whether a person is smiling. It may be necessary, however, to tell why he or she is smiling. An explanation need not go as far as the following: "Two women and a man stroll down the newly completed section of Rehoboth's boardwalk. They are, from left, Nancy Jackson, Dianne Johnson and Richard Bramble, all of West Chester." An editor remarked, "Even if some of the slower readers couldn't have figured out the sexes from the picture, the names are a dead giveaway."
- **Don't editorialize.** A writer doesn't know whether someone is happy, glum or troubled. The cutline that described the judge as "weary but ready" when he arrived at court on the opening day of trial must have made readers wonder how the writer knew the judge was weary.
- **Use specifics rather than generalities.** "A 10-pound book" is better than "a huge book." "A man, 70," is more descriptive than "an old man."
- **Omit references to the photo.** Because the readers know you are referring to the photograph, omit phrases such as "is pictured," "is shown" and "the picture above shows."
- **Use "from left" rather than "from left to right."** The first means as much as the second and is shorter. Neither *left* nor *right* should be overworked. If one of two boys in a picture is wearing a white jersey, use that fact to identify him. If the president is in a golf cart with a professional golfer, readers shouldn't have to be told which one is the president.
- **Avoid "looking on."** One of the worst things you can say about a person in a photo is that he or she is "looking on." If that is all the person is doing, the photo is superfluous. Perhaps something like this will help: "William McGoo, background, is campaign treasurer."
- **Don't kid the readers.** They will know whether this is a "recent photo." Give the date the photo was taken if it is an old photo. Also, let readers know where the picture was taken—but not how. Most readers don't care about all the sleet and snow the photographer had to go through to get the picture. Also, readers aren't stupid. If the caption says three persons in a Girl Scout picture are looking over a drawing of a new camp, readers aren't fooled if the picture shows two of the girls behind the drawing; they obviously can't be looking it over. If a special lens was used, resulting in a distortion of distance or size, the reader should be told what happened.
- **Write captions in the present tense.** This enhances the immediacy of the pictures they accompany. The past tense is used if the sentence contains the date or if it gives additional facts not described in the action in the picture. The caption may use both present and past tenses, but the past time-element should not be used in the same sentence with a present-tense verb describing the action.
- **Make sure the caption is accurate.** Double-check the spelling of names. The paper, not the photographer, gets the blame for inaccuracies. Caption errors occur because someone, the photographer or the reporter accompanying the photographer, failed to give the photo desk enough, or accurate, information from which to construct a caption. Apparently assuming that any big horn is a tuba, a caption writer wrote about a horn player with half his tuba missing. His editor was quick to reprimand, "Umpteen million high school kids, ex-bandsmen and musicians in general know better."
- **Double-check the photo with the caption identification.** The wrong person pictured as

"the most-wanted fugitive" is a sure way to invite libel.

- **Be careful.** Writing a caption requires as much care and skill as writing a story or a headline. The reader should not have to puzzle out the meaning of the description. Notice these jarring examples:

  Fearing new outbreaks of violence, the results of Sunday's election have been withheld.

  Also killed in the accident was the father of five children driving the other vehicle.

  Yum! Yum! A corn dog satisfies that ravishing fair appetite. (The word was *ravenous,* not *ravishing.)*

- **Don't hit the reader over the head with the obvious.** If the photo shows a firefighter dousing hot timbers after a warehouse fire and a firefighter already has been mentioned in the text, it is ridiculous to add in the caption that "firefighters were called."
- **Avoid last-line widows or hangers.** The last line of the caption should be a full line, or nearly so. When the lines are doubled (two two-columns for a four-column picture), the writer should write an even number of lines.
- **Captions should be humorous if warranted by the picture.** Biting humor and sarcasm have no place in captions.
- **The caption should describe the event as shown in the picture, not the event itself.** Viewers will be puzzled if the caption describes action they do not see. Sometimes, however, an explanation of what is not shown is justified. If the picture shows a football player leaping high to catch a pass for a touchdown, viewers might like to know who threw the pass.
- **Update the information.** Because there is a lapse between the time a picture of an event is taken and the time a viewer sees the picture in the newspaper, care should be taken to update the information in the caption. If the first report was that three bodies were found in the wreckage, but subsequently two more bodies were found, the caption should contain the latest figure.

  Or, for a picture taken in one season but presented in another, the caption should reflect the time difference. An example: "Big band singer Helen O'Connell, left, reminisces with Pat and Art Modell backstage at Blossom Music Center this summer." Reading that on Dec. 1, it is difficult to decide when it happened, especially with a present-tense verb and frost in the air.

- **Be exact.** In local pictures, the addresses of the persons shown may be helpful. If youngsters appear in the picture, they should be identified by names, ages, names of parents and addresses.
- **Credit the photographer.** If the picture is exceptional, credit may be given to the photographer in the caption, perhaps with a brief description of how he or she achieved the creation. On picture pages containing text matter, the photographer's credit should be displayed as prominently as the writer's. Photo credit lines seldom are used on one-column or half-column portraits.
- **Pictures without captions.** Although pictures normally carry captions, mood or special-occasion pictures sometimes appear without them if the message is obvious from the picture itself. Not all who look at pictures will also read the captions. In fact, the decline is severe enough to suggest that many readers satisfy their curiosity merely by looking at the picture.
- **Creating slugs.** In writing a series of captions for related shots, use only one picture slug, followed by a number—moon 1, moon 2 and so on.
- **Know your style.** Some papers use one style for captions with a story and another style for captions on pictures without a story (called *stand-alone* or *no-story).* A picture with a story might call for one, two or three words in boldface caps to start the caption. In stand-alones a small head or *catchline* might be placed over the caption.

*(Continued)*

## *Writing the Caption (continued)*

- **Give the location.** If the dateline is knocked out in the caption, make sure that the location is mentioned. Example: "GUARDING GOATS —Joe Fair, a 70-year-old pensioner, looks over his goats Rosebud and Tagalong, the subject of much furor in this northeastern Missouri community, boyhood home of Mark Twain." The Missouri community was Hannibal, but the caption didn't say so.
- **Rewrite wire service cutlines.** The same pictures from news agencies and syndicates appear in smaller dailies as well as in metropolitan dailies. Some papers merely reset the caption supplied with the pictures. Most, if not all, such captions should be rewritten to add to the story told in the picture and to indicate some originality on the paper's part.
- **Watch the mood.** The mood of the caption should match the mood of the picture. The caption for a feature photo may stress light writing. Restraint is observed for pictures showing tragedy or dealing with a serious subject.

Although no one should try to write a caption without first looking at the picture, pictures frequently have to move to the production department quickly. The editor removes the captions (from wire service and syndicated pictures) and jots down the slug and size of the pictures and any revealing elements in the pictures that might be added to the cutlines. A photocopy of the picture may be made. Time permitting, a proofsheet showing pictures and their captions should be given to the photo editor.

When the caption has been composed, the writer should compare the message with the picture. The number of people in the picture should be checked against the number of names in the caption. Everyone appearing prominently in the picture should be identified. If a person is so obscured in the crowd that the person is not easily identifiable, that fact need not be brought to the reader's attention.

## INFORMATION GRAPHICS

### The Role of Graphics

Photographs help the reader gain a visual appreciation of reality; information graphics help the reader understand the massive, the intangible or the hidden. A map of part of the city is usually a better locater than an aerial photo of the same area. A chart often helps the reader track trends over time. An artist's cutaway can show the undersea levels of a sinking ship.

Information graphics, as these devices are known, have appeared with increasing regularity in newspapers and magazines, large and small, in recent years. Personal computers are ideal for creating such graphics quickly and inexpensively, and newspapers rushed to embrace them in the 1980s. Suddenly, publications that once had only photographers and artists added graphic designers to create charts, graphs and maps. Contests added categories for graphic design, and the Society of Newspaper Design was formed. Clearly, the 1980s was a decade of unprecedented emphasis on graphics and design.

Many believe that transformation was sparked by the arrival of *USA Today* in 1979, and there is no doubt that it set new standards for the use of color and graphics. Colorful charts, graphs and maps played a key role in the concept of *USA Today,* which the Gannett Co. designed for the busy reader. Charts, graphs and maps are a good way of communicating lots of information in a hurry, and *USA Today*'s editors were quick to embrace them. The newspaper's colorful weather map was emulated by papers worldwide (see Figure 11–9).

But *USA Today* was far from alone in adopting information graphics. *The New York Times* and other prestigious newspapers began publishing full-page graphics on section fronts, *Time* magazine produced startling graphics under the genius of Nigel Holmes, and syndicated services of graphics material proliferated. The Associated Press greatly increased the quality and quantity of the graphics it provided. The AP and Knight Ridder Newspapers even created separate delivery services just for information graphics.

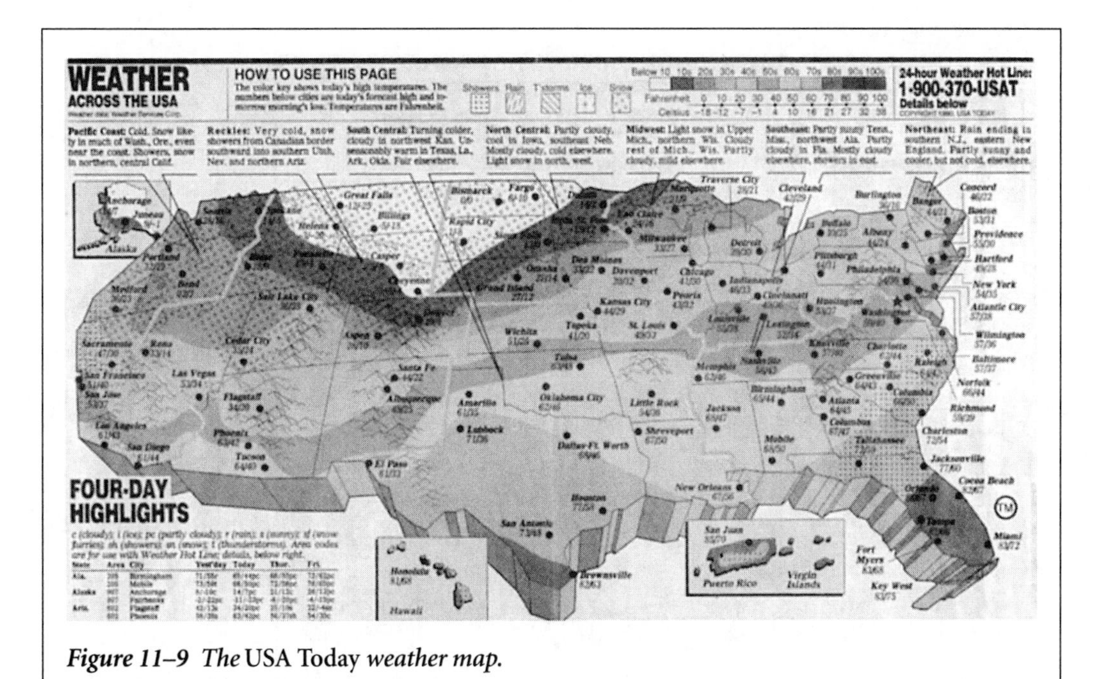


***Figure 11–9** The* USA Today *weather map.*

Courtesy of *USA Today.*

Like photographs, however, information graphics can be good or bad. Confusing charts, graphs and maps hinder readers rather than help them. Graphics that are too busy may obfuscate rather than enlighten.

## Types of Information Graphics

*Illustrations* help the reader understand complex things or concepts, including information that may be extremely difficult or impossible to describe in words. They take many forms, from the simple illustration to the complex diagram.

*Maps,* of course, help the reader locate things. Local maps help the reader locate places within the city. Maps of counties, states, regions or nations help readers locate sites with which they may not be familiar (see Figure 11–10). Good graphics departments maintain computerized base maps of various locales. When an event occurs, it is a simple matter to add specific locater information to the base map to create a helpful aid for the reader. Secondary windows within maps sometimes are used to pin down specific areas within geographic regions or to show the location of the specific region within a larger area.

The *table* is used to display numerical information graphically (see Figure 11–11). Tables are useful when lots of numbers are involved, as in precinct-by-precinct election results.

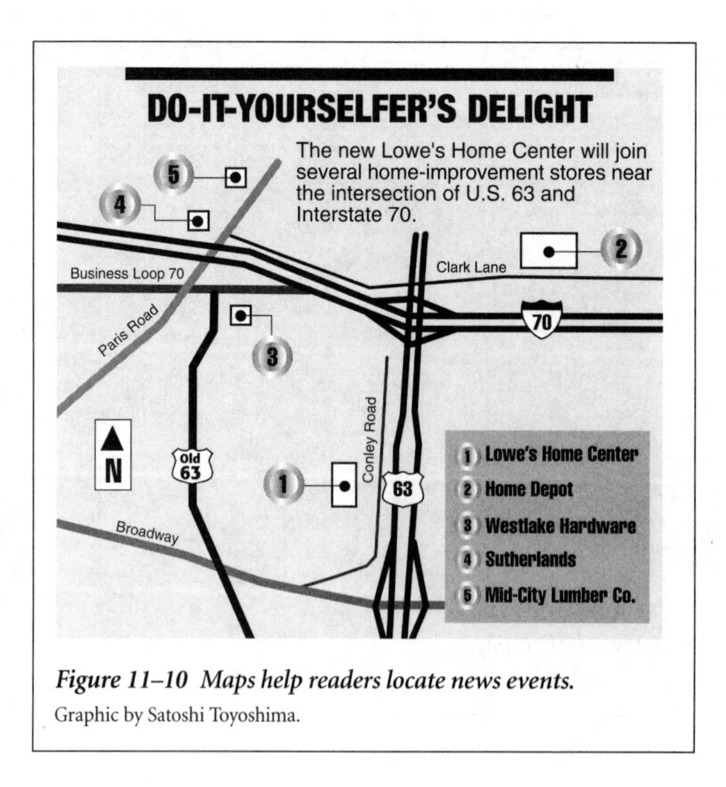


***Figure 11–10 Maps help readers locate news events.***
Graphic by Satoshi Toyoshima.

**Hot, Hot, Hot! The discomfort index**

| Relative Humidity | Air Temperature (Degrees Fahrenheit) | | | | | | | | | | |
|---|---|---|---|---|---|---|---|---|---|---|---|
| | 70 | 75 | 80 | 85 | 90 | 98 | 100 | 105 | 110 | 115 | 120 |
| | Heat Index | | | | | | | | | | |
| 0% | 64 | 69 | 73 | 78 | 83 | 87 | 91 | 95 | 99 | 103 | 107 |
| 10% | 65 | 70 | 75 | 80 | 85 | 90 | 95 | 100 | 105 | 111 | 116 |
| 20% | 66 | 72 | 77 | 82 | 87 | 93 | 99 | 105 | 112 | 120 | 130 |
| 30% | 67 | 73 | 78 | 84 | 90 | 96 | 104 | 113 | 123 | 135 | 148 |
| 40% | 68 | 74 | 79 | 86 | 93 | 101 | 110 | 123 | 137 | | |
| 50% | 69 | 75 | 81 | 88 | 96 | 107 | 120 | 135 | | | |
| 60% | 70 | 76 | 82 | 90 | 100 | 114 | 132 | | | | |
| 70% | 70 | 77 | 85 | 93 | 106 | 124 | | | | | |
| 80% | 71 | 78 | 86 | 97 | 113 | 136 | | | | | |
| 90% | 71 | 79 | 88 | 102 | 122 | | | | | | |
| 100% | 72 | 80 | 91 | 108 | | | | | | | |

Source: The World Almanac — NEA GRAPHICS

As temperatures rise, humidity increasingly becomes a factor in reducing the body's ability to cool itself. The discomfort index measures what various temperatures and humidities "feel like" to the average person.

***Figure 11–11** Tables are a form of information graphics used to make numerical information easier to read.*

Courtesy of Newspaper Enterprise Association.

*Bar charts* help the reader visualize quantities (see Figure 11–12), and *fever charts* show quantities over time (see Figure 11–13). *Pie charts* are used to show the division of the whole into components (see Figure 11–14).

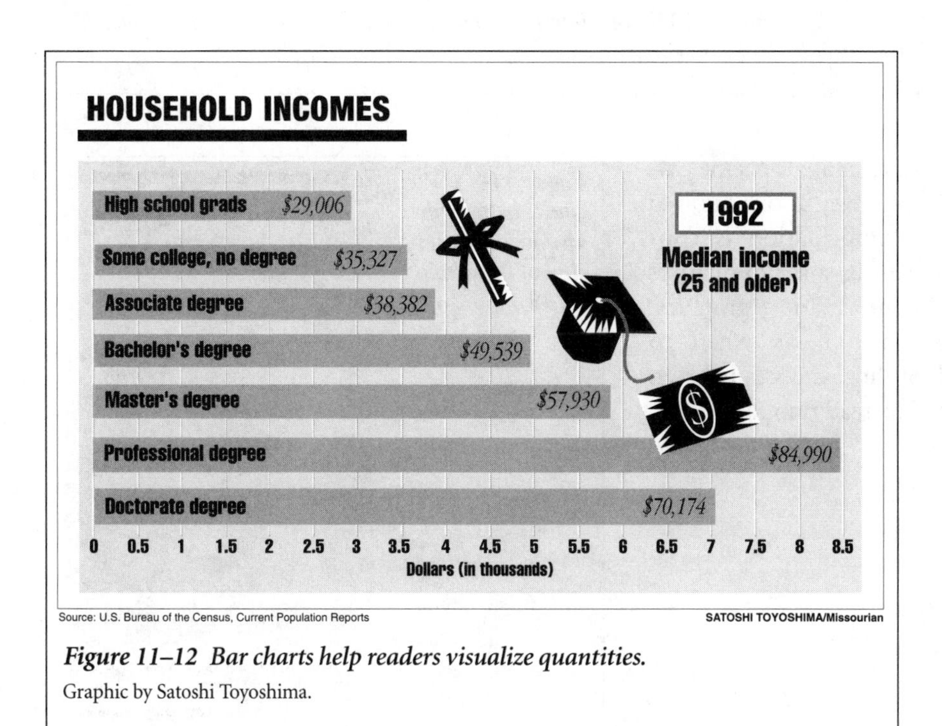


***Figure 11–12** Bar charts help readers visualize quantities.*

Graphic by Satoshi Toyoshima.

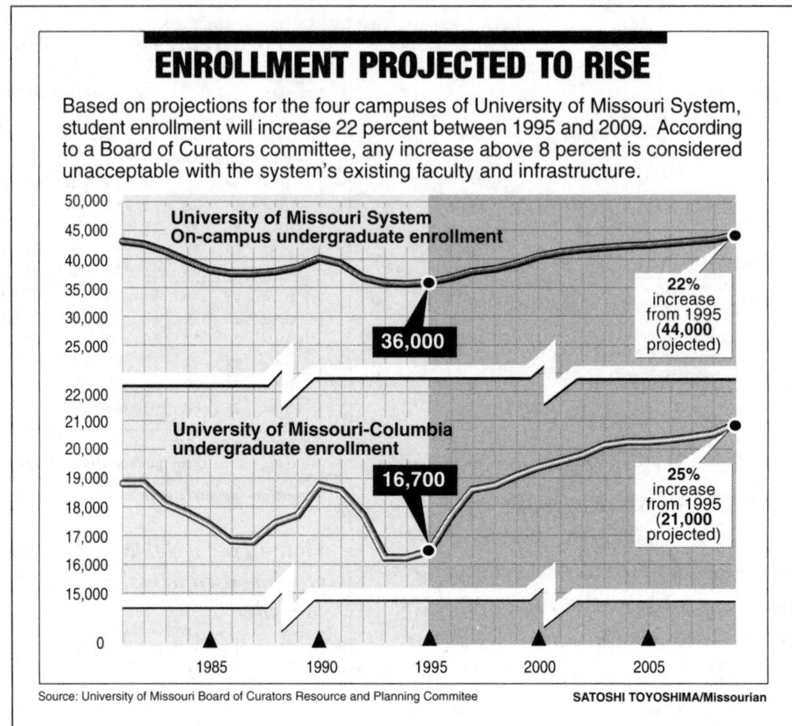


*Figure 11–13 Fever charts help readers visualize quantities over time.*
Graphic by Satoshi Toyoshima.

All these devices are tools the editor uses to help convey information to readers. Some information is conveyed best in words, some in pictures and some in information graphics. The editor who knows when to choose each of those devices helps to create an easy-to-read publication.

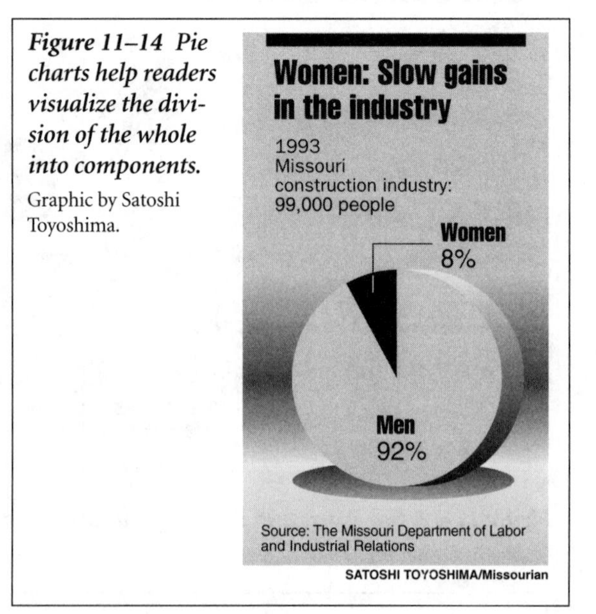


*Figure 11–14 Pie charts help readers visualize the division of the whole into components.*
Graphic by Satoshi Toyoshima.

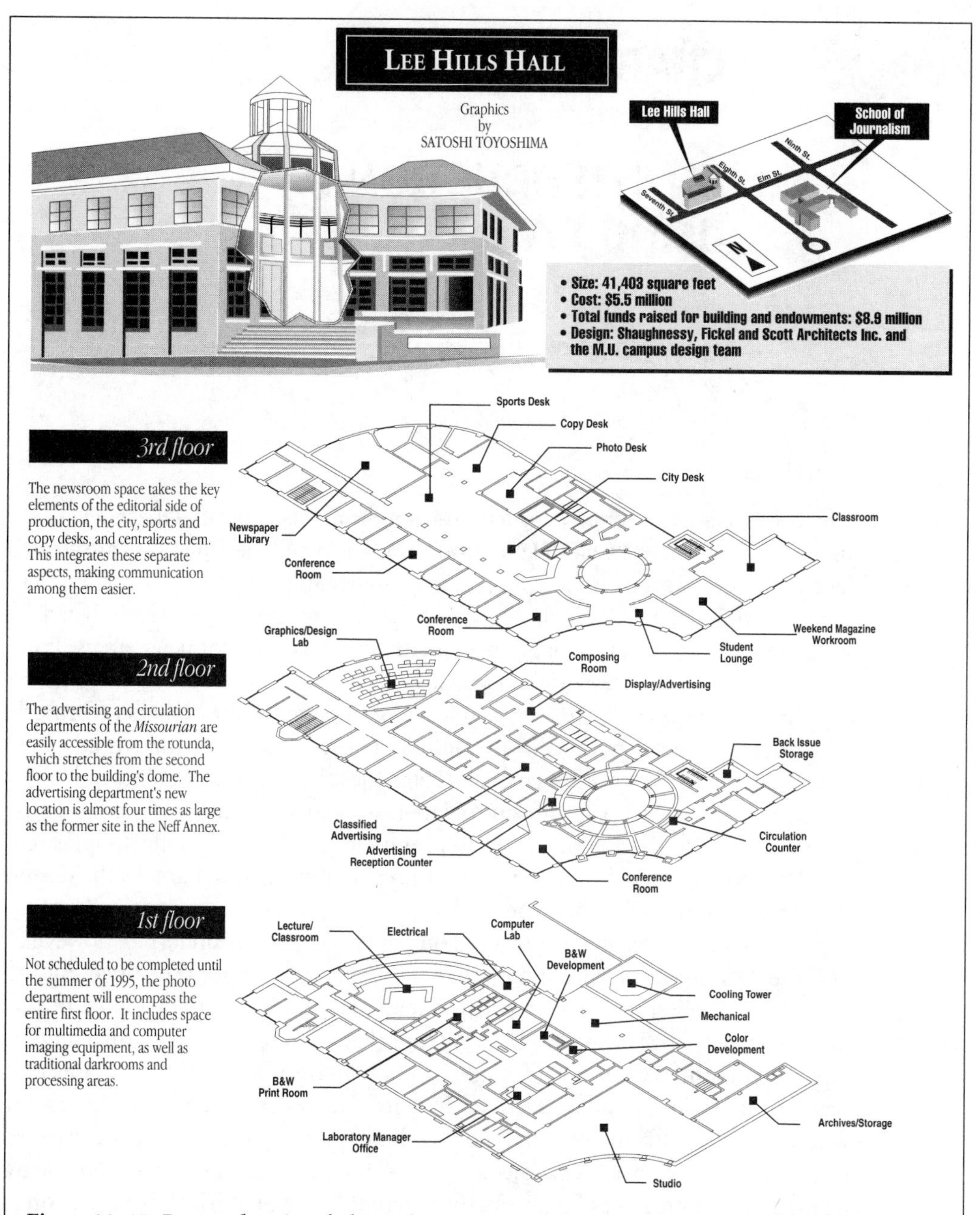


***Figure 11–15 Process drawings help readers visualize progression of action or detailed plans.***
Graphic by Satoshi Toyoshima.

# CHAPTER 12

# AN INTRODUCTION TO USING TYPE

Type, or typefaces, as editors, printers and graphic artists know it, is the letter shapes used to distinguish one style of letter from another. Type style differences are created by designers to meet the needs of users who prepare type for hundreds of different kinds of printed materials, including newspapers. Some typefaces were designed for wedding invitations, others for advertisements and still others for newspapers. Typeface varieties therefore are a design tool for editors (see Figure 12–1).

*Type* or *typeface?* The term *type* refers to each letter of the alphabet (formerly cast from hot metal poured from molds). Those individual letter-molds were assembled first, by hand, into sentences. Later, these letter-molds were assembled by a machine, from which an entire line of letters were cast into lines of type (Linotype slugs). Metal type (both individual letters and Linotype) has virtually disappeared.

The term *typeface* has the same meaning it formerly did, namely, the design of letters. Even though most type is now set electronically, the terms typeface and type still refer to what you see printed on paper. Typeface is preferred, however.

Typefaces with unique styling give page editors an option of breaking the monotony of newspaper pages, which all tend to look alike. Most headlines of stories on a newspaper page have been *set* (another term for *composed)* using typefaces from families of the same design. (A type *family* is a group of typefaces that have strong style resemblances to each other, much like human families do.) If you notice radically different styles of headline typefaces on the same newspaper page, it is probably because someone wanted to make the page more readable by using different typefaces for distinctly different kinds of stories. Headlines offer such opportunities more often than body type. This is but one example of how type can be used in editing, even though most newspapers already have a headline schedule that they ordinarily follow closely.

e e e e

**Typefaces:** Bookman Helvetica Tiffany Times

*Figure 12–1 Typefaces that show different styling. These letters are all set on a computer in 24-point type, yet each has a different basic design. The names of the typefaces are underneath.*

The purpose of this chapter is to introduce editors to some basic ways of using type, so that, ultimately, they can help readers read easier and faster. This need is more critical now than ever before, as other media such as television, radio and the information superhighway are all competing for newspaper readers' time and attention. The winner of the competition may get more attention but will not necessarily provide more understanding of the news and its implications for society. Type plays a major role in translating and communicating ideas from one person to another. The new media are more convenient but not a better interpreter for readers. There is a need to make the news understandable.

## MAKING TYPE EASIER AND FASTER TO READ

Most journalism students are computer-literate and tend to think that all typefaces are equally readable, and therefore it doesn't matter which typeface is used. Typefaces in common use are about equally readable, but there are many typefaces that are not in common use, and all typefaces are *not* equally easy to read.

For example, the line of type in Figure 12–2 is very difficult to read because the letter designs call too much attention to themselves. The letters are examples of contemporary typefaces and are used to harmonize with an unusual story.

These unusual typefaces may have a place in journalism but not usually in the news section or headlines of daily newspapers. These typefaces tend to create feelings and usually stop a typical reader who wants a closer inspection of the special design. Such typefaces would normally be used for advertising and promotional work; some newspapers may use this kind of type in feature stories.

Most often the editor seeks to help a reader read faster, easier and, as a result, read more of what has been written. Older readers often take unlimited time to read a newspaper, sometimes reading it from cover to cover. But younger people tend not to take as much time to read newspapers, if at all. Efforts are under way to encourage younger readers to give more time to the daily paper by appealing to their special interests.

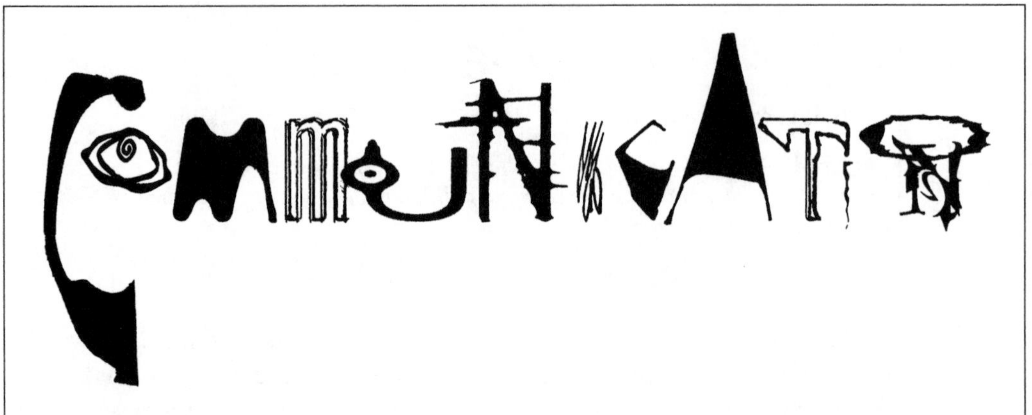

*Figure 12–2 Sample of contemporary typefaces. These letters are not designed to be shown as they are here (10 different styles all combined into one word), but each letter has an individual and distinct style, and each are used with common type forms.*

From Hugh Hart, "It takes All types," in Chicago Tribune, June 2, 1995, Tempo Section.

## Avoiding Type and Arrangements of Type That Impede Reading

It is essential that the kinds of type, or arrangements chosen for news stories, be set in ways that do not impede reading. In newspapers, the letters on any page do not necessarily communicate simply because they are there. Type is supposed to facilitate reading. When readers are paying attention to type design, rather than the meaning of words, they may not get the message. Typographers who work with advertising have often found this to be the case.

Type can also increase or impede reading speed. For example, one technique thought to make reading faster does the opposite. Setting type in all capital letters slows reading, even though capital letters appear larger than lowercase letters. Research on reading all-cap headlines shows that they are read about 20 percent slower than headlines set in caps and lowercase.

There is no question that type set in some unusual style can be charming and, perhaps, provide more meaning to a sentence than if it were set in an ordinary typeface. With extra white space around an unusual typeface, a typical reader can read a word or sentence without spending unnecessary time studying the novelty of the typeface. But a reader who spends too much time looking at novel typefaces will be wasting time that was probably allocated for reading purposes. Therefore, an editor should have an understanding of the effects type may have on readers. Most necessary is an understanding of what typeface readability means.

## Readability or Legibility?

In the United States, most persons involved with the use of type tend to use the term *readability* to refer to typefaces and their arrangement on a page as that which can be read easily and quickly. These people do not necessarily have a quantitative measure to define speed of reading and therefore use the term loosely. But they mean that a readable typeface is one that can be read faster than a nonreadable typeface in certain settings.

When two psychologists, Donald G. Paterson and Miles A. Tinker, both professors of psychology at the University of Minnesota, reported on their experiments on reading speed of a small sample of typefaces, this casual term suddenly had some quantitative data to define it. However, the two psychologists also raised a question about whether the more proper term should be *legibility.* Professor Tinker defined optimal legibility of print as:

> Typographical arrangement in which shape of letters and other symbols, characteristic word forms, and all other typographical factors such as type size, line width, leading, etc., are coordinated to produce comfortable vision and easy and rapid reading with comprehension.[1]

Nevertheless, both Paterson and Tinker used the term readability, not legibility, in the title of their early book.

Much of what readability has come to mean in newspapers came from practitioners such as printers, editors and journalists, who needed a term to define it, even if it was somewhat vague. However, despite the research on readability and legibility, these two terms are often used as if they were the same thing.

In Professor Tinker's later book, he noted that one of the problems of using the term readability was that it could cause confusion in the field of psychology because that term tended to conflict with what are known as readability formulas.[2] These formulas are devised to measure the level of mental difficulty of reading material, and they may result in an entirely different meaning.

In retrospect then, two major disciplines have been using the same term, readability, for their own purposes. The publishing industry has used readability for years, with a general agreement in mind about its relationship with legibility. As a result it is recommended that journalists, and people in related fields, *continue* to use the term readability for most uses in publishing, as they probably will anyway, despite Paterson or Tinker. But if they should be preparing material for the psychology field, legibility is obviously the proper term. There is much agreement that readability and legibility mean *about* the same thing.

[1]Tinker, Miles A. *Legibility of Print,* Iowa State University Press, Ames, Iowa, 1963, Page 5.
[2]Tinker, Page 4.

## WHAT EDITORS SHOULD KNOW ABOUT TYPEFACES

Maintaining high standards of readability and keeping a publication up-to-date require that editors understand the basics of type. The following is a checklist of typographical concepts that an editor should know:

- The meaning and use of *typeface.*
- How to measure type for printing.
- What to look for in differentiating typefaces.
- Details of construction of letter forms.
- General style of type classification, such as roman, italic.
- Type sizes (body types and headlines) and their importance relative to the age of readers.
- Typeface weight.
- X-heights.
- Descender and ascender dimensions.
- Family styling.
- Line widths.
  General concepts.
  Common sense on extra-wide or extra-short widths.
  Rules of thumb on widths.
- Spacing (line or word).
- How many extra points leading are needed.
- Kinds of type that require more or less spacing.
- Word and letter spacing.

## GENERAL TYPE MEASUREMENTS

Most type measurements are carryovers from the old days of hot type and are identical to those used by traditional printers. Although desktop computers allow measurement in inches, many people retain the printers' practice of measuring in picas and points. Following are the basic traditional measurements:

- *Points*—72 points equal an inch (a point is 1/72 of an inch). All type is measured in points. An editor can specify 12-point or six-point type or any point size. Type may even be quite large, such as 120-point type. It also is correct to specify such large type by its height in inches, such as six-inch or eight-inch type.

- *Picas*—Six picas equal one inch (a pica is $\frac{1}{6}$ of an inch). Twelve points equal one pica. The width of a line of type is usually expressed in picas, for example, *14 picas* or *35 picas wide.*
- *Page measurements*—These are usually measured in inches. For example, a page may measure 7 by 10 inches (or 7″ x 10″). (Width is usually expressed first and length second.)

## DIFFERENTIATING TYPEFACES

An editor specifies a particular typeface when preparing copy to be set in at least five ways: by point size, by family, by weight, by letter width and by style, as shown in Figure 12–3.

### Differences in Letter Design

One way typefaces differ is in their use of *serifs.* The body type in most newspapers have serifs. A serif is a relatively fine line at the bottom and top of letters such as *d, h, i, l, m, n, p, r, u, x* and *z.* Serifs also appear on capital letters.

Type letters also differ in their use of thick and thin elements, which can be placed in slightly different positions on each letter. Figure 12–4 illustrates the differences among three typefaces. The type user, however, should not pay too much attention to slight variations of single letters but should concentrate on their appearance in mass form, as in a typical newspaper paragraph. A paragraph (of about 50 words) can show the editor what the mass effect looks like.

**Five Ways Typefaces Differ and Sample Specifications**

| How these could be specified: | (1) By Point Size | (2) By Type Family | (3) By Weight | (4) By Width | (5) By Style |
|---|---|---|---|---|---|
| Example 1 | 24 pt. | Bodoni | Lightface | Regular | Italic |
| Example 2 | 60 pt. | Century | Boldface | Condensed | Roman |

*Figure 12–3 Basic information that should be on type specification sheets.*

## Differences Between Typefaces

Shown below are three typefaces often used on desktop computers. Study the differences and read the comments.

Here is an example of 24 pt. Bookman

Here is an example of 24 pt. Times

Here is an example of 24 pt. Palatino

**Comments:**

a. The letters in one typeface are designed to be wider than others.
b. Some letters in one face are taller than others.
c. Compare the design of an "e" or "a" in each typeface. Note the differences.
d. Each typeface has its own set of peculiar characteristics.
e. Note the mass effect of each typeface below.

**How Sample Typefaces (above) Look in Paragraphs**

10 pt. Bookman

Typefaces have been designed to have unique characteristics. In selecting typefaces for a story, think of the connotation (or feeling) of the typeface and its relationship to the story. Is one typeface better than another? Some typefaces are warmer, more legible, more powerful or more delicate. Some are more feminine or masculine.

10 pt. Times

Typefaces have been designed to have unique characteristics. In selecting typefaces for a story, think of the connotation (or feeling) of the typeface and its relationship to the story. Is one typeface better than another? Some typefaces are warmer, more legible, more powerful or more delicate. Some are more feminine or masculine.

10 pt. Palatino

Typefaces have been designed to have unique characteristics. In selecting typefaces for a story, think of the connotation (or feeling) of the typeface and its relationship to the story. Is one typeface better than another? Some typefaces are warmer, more legible, more powerful or more delicate. Some are more feminine or masculine.

***Figure 12–4** Some of the more important ways typefaces differ.*

Futura Light

Futura Regular

**Futura Bold**

**Futura Extrabold**

*Futura Regular Italic*

***Futura Bold Italic***

Futura Light Condensed

Futura Regular Condensed

**Futura Bold Condensed**

**Futura Extrabold Condensed**

Futura Bold Outline

*Figure 12–5 Some members of the Futura type family. Although each typeface is different, each also has family characteristics. When mixing different typefaces on the same page, it is a better choice to use different members of the same type family.*

## Differences in Type Family

Just as members of the same human family tend to have similar facial characteristics, so do members of a type family. A type family includes all variations of a given type with common characteristics. Some type families have many variations; others have few. Figure 12–5 shows one of the large families of typefaces.

LITHOS EXTRA LIGHT

LITHOS LIGHT

LITHOS REGULAR

**LITHOS BOLD**

**LITHOS BLACK**

*Figure 12–6 Various weights of typefaces within the Lithos family.*

## Differences in Type Weight

Type may be differentiated by the weight of the letter. Most typefaces are manufactured in lightface and boldface. Some faces are manufactured in medium, demibold heavy and ultrabold as well. The terminology tends to be confusing. One manufacturer calls its medium-weight type *demibold,* whereas another calls a corresponding type *medium.* The terms *heavy, bold* or *black* also may mean the same thing. Figure 12–6 shows common examples of type weights.

## Differences in Letter Width

Most typefaces are manufactured in normal (or regular) widths. Regular widths are used in most reading matter, but wide and narrow widths also are available. Type manufacturers have created extra-condensed, condensed, expanded and extended typefaces in addition to regular. These additional widths, however, are not manufactured in all type sizes or families. Therefore, it is necessary to check your newspaper's headline schedule to see if a desired width is available (see Figure 12–7).

## Differences in Type Style

There are a number of ways to differentiate type by style. Each of them helps the editor find some unique quality in most typefaces. Here are some common style differentiations and classifications:

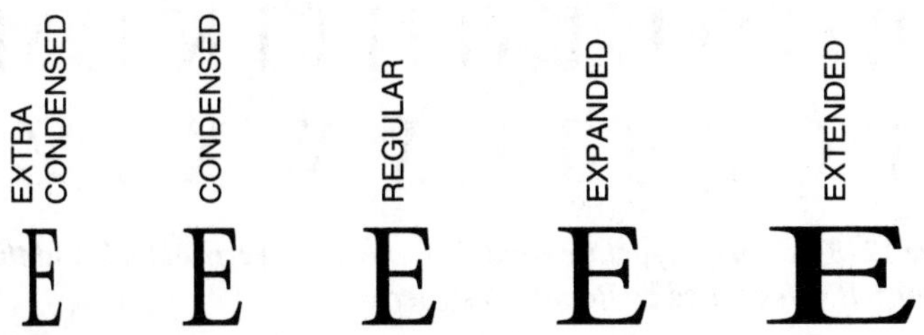

EXTRA CONDENSED E  CONDENSED E  REGULAR E  EXPANDED E  EXTENDED E

***Figure 12–7** Variations in widths of type. Each variation is chosen for a specific purpose.*

Seven general styles: Roman
Italic
Sans serif
Text
Script
Cursive
Square serif

Two very broad styles: Oldstyle
Modern (sometimes called transitional)

Two more broad styles: Traditional
Contemporary

Company of production: Each manufacturing company may make the same type family

The most common means of classification divides type into broad categories termed *roman, italic, text, sans serif, script* and *cursive,* sometimes called the *races* of type. It is best to think of these classifications as style characteristics that help in differentiating and identifying typefaces. Square serif is sometimes added to make a seventh style.

**ROMAN TYPE**

Roman type (see Figure 12–8) has a vertical shape; it has serifs; it usually has combinations of thick and thin elements in each letter (called *stem* and *hairline,* respectively). Some type experts consider all vertical letters to be roman, even those without serifs or with no variations in the width of letter elements (stem and hairline). This form of classification, therefore, may be confusing to the beginner because the roman designation will have two purposes: one to distinguish it from sans serif and

A B C D E F G H I J K L M N
O P Q R S T U V W X Y Z

*Figure 12–8 A roman typeface (Bodoni) showing a vertical look and thick and thin elements. It was created by Roman designers.*

*ABCDEFGHIJKLMNOP qrstuvwxyz*

*Figure 12–9 Italic versions of the Bodoni family.*

the other to distinguish it from italic. For simplicity's sake, it is best to use the classification as first described.

### ITALIC TYPE

Italic types are characterized by their slanted letter shapes. Although italic types were originally designed to print many letters in relatively little space, their use today is limited to citations or words that must be emphasized. They are also used in headlines and body types. Today, italic types are designed to accompany roman types, to provide consistency in the family of design. Figure 12–9 shows an italic type of the same family.

### SANS SERIF (OR GOTHIC) TYPE

In the United States, printers use two terms to identify typefaces with no serifs. One is *sans serif;* the other is *Gothic.* The French word *sans* means "without," thus *without* serifs. The other term, Gothic, is a misnomer. Originally, Gothic type meant the churchy-looking types Americans often called "Old English." But today printers use Gothic also to refer to serifless type (see Figure 12–10).

### TEXT TYPE

Text type is often incorrectly called Gothic because it looks like the Gothic architecture of the Middle Ages. But printers call it *text* because it appears to have a texture, like cloth, when printed in large masses. These letters were originally drawn with a broad-nibbed pen and were created to show a minimum of curves. Today the type is used for church printing or conservative headlines. Students should

24 pt. Lithos

ABCDEFGHIJKLMNOPQR

24 pt. Frutiger Bold

**ABCDEFGHIJKLMNOPQR**

***Figure 12–10*** *Different styles of sans serif typefaces. There are many variations, all called sans serif.*

HARD TO READ

Easier to read

***Figure 12–11*** *Engraver's Old English text typeface set in all caps and lowercase. "Text" should never be set in all caps. (Note: Printers' use of the word text is different from artists' usage. Printers call a churchy-looking face "text" because when seen en masse, it has a texture.)*

never have text set in all capital letters for two reasons: (1) It was never written that way originally, having always appeared in capital and lowercase letters; (2) it is difficult to read when set in all capital letters (see Figure 12–11).

### SCRIPT TYPE

Script letters resemble handwriting. Although the type designers have tried to make it appear as if all the letters are joined, small spaces can be seen between them. Some script letters appear to have been written with a brush, others with a calligraphic pen. Like text type, script type should never be prepared in all capital letters because it is hard to read in that form (see Figure 12–12).

### CURSIVE TYPE

Although cursive type styles look much like script typefaces, they are easily differentiated by their ornateness. Cursives are used mostly in advertising but occasionally for compartmentalized headlines on the Lifestyle page (see Figure 12–13).

HARD TO READ

Easier to read

*Figure 12–12 Brush script also should never be set in all caps. Notice how hard it is to read in all caps.*

ABCFGHIJKLMNOPQ

abcdefghijklmnopqrstuvwxyz 12345

*Figure 12–13 A cursive typeface.*

#### OLDSTYLE V. MODERN TYPEFACES

Editors may note that some typefaces are labeled "oldstyle" or "modern." The term *oldstyle* may be interpreted literally because it is a style designation. But the term *modern* be misleading. Modern type does not refer to contemporary times, but to style, and both styles are being designed today.

## HOW TO MEASURE TYPE FROM A PRINTED PAGE

There is no problem in measuring a piece of metal type, but because metal type is a thing of the past, editors must be able to measure type from a printed page, as in a type specimen book or another newspaper. Specimen books usually indicate the size of typefaces, but occasionally they do not.

### Type Shoulders

When measuring printed type, the space underneath the lowercase letters must be accounted for (see Figure 12–14). This space is called a *shoulder*, and it is there to allow for descenders of letters such as *g*, *p*, *q* and *y*. But shoulders vary from typeface to typeface. Some have large shoulders, others have small ones. How can the shoulder space be determined?

PROFITS

***Figure 12–14** It is difficult to measure this line of type because it does not show descenders.*

***Figure 12–15** Drawing of a letter showing the imaginary lines that define ascenders, descenders and x-height.*

equipped

Measure this distance

***Figure 12–16** How to measure type: Measure the distance between the top ascender and lowest descender.*

To dramatize the problem Figure 12–15 shows imaginary lines by which letters are created on a piece of metal type. These lines are called (1) *base line,* on which all letters other than *g, p, q,* and *y* rest; (2) *cap line,* to which most capital letters and tall letters rise; (3) *lowercase line,* where small letters align (called the *x-height* of letters); and (4) *descender line.* Each line helps in aligning letters.

For example, the ascender of the letter *h* rises above the lowercase line, but there is no descender below the base line. Thus, the shoulder underneath the letter must be included in the measurement. To accurately determine ascender space, simply look for a capital letter or one with a high ascender. If it is necessary to measure the point size of a line of capital letters, take the space normally used for descenders into consideration (see Figures 12–16 and 12–17).

E E E E E E E

*Figure 12–17 Why printed type is so difficult to measure. Each of these E's is 48 points high, but the shoulder space (or descender space) of each varies.*

## AN INTRODUCTION TO LEADING

The space between lines of type is called *leading* (pronounced *ledding*). Remember that leading for body type usually has been determined when a newspaper creates its basic design. Therefore, computerized typesetting systems may be preset to control the amount of line spacing that the designer recommended. The machine automatically provides the leading desired.

However, do not assume that because the leading has been preset, it never can or should vary. In most operations, provisions have been made for changing the leading occasionally to meet certain needs, most often in feature stories or other special kinds of stories. When editors want to feature a momentous story, they should use additional leading. Leading usually makes lines of type easier to read because it gives readers more white space between lines, which helps them focus their eyes on the type. Also, when lines of type are 20 picas or longer and the shoulders of typefaces are very short, some leading should be used.

What then are the leading options? If a newspaper uses no leading at all, that practice is called setting type "solid." (Actually there is a tiny bit of space between lines of type when it is set solid, caused by the descender space at the bottom, or shoulder, of each letter.)

Options range from one-half-point to six-point leading for body type, and the range is extended from three to about 12 points of leading for headlines. The only way to know how type will look after it is leaded is to set a sample paragraph or two and study its readability. See Figure 12–18 for samples of alternative leading. Also see the discussion of spacing in newspaper typography that follows.[3]

### Spacing in Newspaper Typography

Good typography depends so much on the way words and lines are spaced that a separate discussion is in order here. There are a number of different places in newspaper layout where type spacing is important.

[3]Herold, Don, *ATA Handbook*, Advertising Typographers Associates of America Inc., 1975, Page 24.

**LINE SPACING**

The principle of making each line of type an easy eyeful can be aided by the generous (but not too generous) use of space between the lines. This provides a "right-of-way" for the eye along the top and bottom of each line. Types with short ascenders and descenders and large lowercase letters need more space between lines than faces with long ascenders and descenders and small lowercase letters. A fairly safe rule is to let the spacing between the lines approximately equal the space between words.

The above paragraph is set with generous spacing (3-point leads) between lines, while *this* paragraph is set with no leads and is consequently tougher ploughing for the eye. The type is the same size but looks smaller. Educated instinct will in time tell you the difference between jamming and scattering type lines.

***Figure 12–18** The top paragraph has been leaded three points. The bottom has been set solid, meaning that there is no leading between lines. Leaded lines usually are easier to read than solid.*

Watch for spacing:

- Between the lines of a headline.
- Between a headline and the story below.
- Between a headline and the story above.
- Between body type and illustrations or ads.
- Between paragraphs.
- Between subheads and body type above and below.
- Between words.
- Between letters.

### HEADLINE

Research on line spacing is inconclusive. Yet there is a feeling that lines with generous space between them are easier to read than those tightly spaced. On the other hand, too much space between the lines of a headline is undesirable. The editor must judge whether all lines can be read as a single entity (desirable). When there is too much space, each line receives too much emphasis (see Figures 12–19 and 12–20).

The responsibility for maintaining good headline spacing lies not only with the printer or typesetter but with the editor as well.

### BETWEEN A HEADLINE AND THE STORY BELOW

One of the most unattractive ways of placing headlines on top of stories is to have little or no space underneath the head. The objective is to position the headline so

**Heat Wave
Hits Cities
Near Coast**

*Figure 12–19 Headlines usually are easier to read when leaded. These lines have acceptable spacing.*

**Heat Wave
Hits Cities
Near Coast**

*Figure 12–20 There is too much leading here.*

**Bob Mann stages rally
to take Pensacola lead**

PENSACOLA, FLA. (AP) Bob Mann fired seven birdies and one eagle for a 9-under par 63 Saturday to take a one-stroke lead Mann, share the National Team Play title with Wayne Levi, but he has never won an individual title on three years on the

*Figure 12–21 Headline is too close to body type.*

that it gets attention and does not crowd the first line of the body copy underneath. The amount of space underneath a head depends on the depth of shoulder that accompanies the headline typeface. If the shoulder is large, then little extra space is necessary. If it is small, additional space is required. Essentially, there ought to be about 12 points of space between the base line of the bottom line of a headline and the top of the first line underneath for typefaces from 14 points to about 36. Typefaces larger than 36 points may need an additional two to four points of space. Smaller headlines may need only eight to 10 points space between the base line and the top of the line underneath (see Figure 12–21).

This same principle of spacing applies to standing heads. Such heads also need adequate space above and below them.

### BETWEEN A HEADLINE AND THE STORY ABOVE

A general rule of thumb about the spacing of headlines is that there should be at least $1\frac{1}{2}$ times as much space above as below. This is to make it clear that a story above is not part of the headline below. In bygone days, stories ended with a "30"

The size, timing and details of the tax cut envisioned by many Senate Democrats would be set by the Finance Committee.

## Smith's view

WASHINGTON (AP)–Senator Thomas Smith said today he sees no possibility that a tax cut will be enacted in the near future.

Criticizing Republican senators for trying to push through a tax cut in the Senate Monday, Smith said it was "very unstable."

***Figure 12–22** There should be more space above and less space below a headline. In this example, some space underneath the head should be removed.*

dash. Today, white space is substituted for the dash. But the reader must never be confused by spacing. Therefore, a generous amount of space should be placed above headlines. For larger-sized headlines, from 36-point type and larger, this space may be as much as two picas. For smaller sizes (14 to 36-point type), the space above the head may be anywhere from 18 points to two picas (see Figure 12–22).

### BETWEEN BODY TYPE AND ILLUSTRATIONS OR ADS

An area of typography that is overlooked many times in the contemporary newspaper is the spacing of type on top of illustrations and ads (and sometimes below illustrations). The problem usually is that the type is positioned too close to these elements (see Figure 12–23). As a result, the reader's attention is diverted from news copy to either an illustration or an ad. Again, as a rule of thumb, there ought to be no less than one pica of space separating elements, and more if possible. The one pica of space allows both the story and the illustration below to get the attention each deserves. Furthermore, space of more than one pica between body type and illustrations simply looks much more attractive than less than one pica of space.

### BETWEEN PARAGRAPHS

It has been the practice in setting type not to place any space between paragraphs of a story. However, when a particular story is a bit short for the column space allotted (in pasting the news on a page), a common practice is to place extra space between each paragraph to lengthen the story.

SOUTHERN ILLINOISAN, FRIDAY, NOVEMBER 1

West Germany to cash in on America's coal mine modernization

WAL-MART

Store Hours: Mon-Sat 9-9 (Both Stores) Sun.-C'dale 12-6 M'boro 1-6
Location: C'dale: 1702 W. Main M'boro: Jackson Square Shopping Center

Mens & Boys Outerwear sale

Super Savings on these plus many more styles now thru Saturday

*Figure 12–23 The body type has been positioned too close to an advertisement (partly shown underneath). The body copy in the story above may be hard to read as a consequence. At least one pica of space or more should separate the two.*

It is recommended, however, that one or two points of space be deliberately placed between paragraphs in order to make the story easier to read, especially if it is long. Paragraph spacing tends to bring more light into a story and give the reader short pauses while reading paragraphs. There is no research to prove the value of this extra space between paragraphs, but a long story is usually more attractive with extra paragraph spacing than without. It is a matter of artistic judgment for the most part.

## ABOVE AND BELOW SUBHEADS

When subheads are used between paragraphs of a long story, there should be more space above the subhead than below. Generally, about four points are placed above and two points below. The purpose of this spacing is to allow the subhead to be seen and perform its function.

Subheads are used to break up large masses of type in a long story. Without subheads, this type has a gray, boring appearance that tends to discourage the reader. When the gray area is broken up, the type looks more interesting. Furthermore, a subhead allows the reader a half-second breather before continuing. If a story is interesting, subheads may not be needed. But not all stories can be equally

**(A)**

Police said they presented a search and seizure warrant at 6:30 p.m. Monday at her second floor apartment. They said they found a package wrapped in foil allegedly containing hashish.
**WAITED FOR SECOND**
The officers said they spent several hours searching the basement for further evidence.

**(B)**

Police said they presented a search and seizure warrant at 6:30 p.m. Monday at her second floor apartment. They said they found a package wrapped in foil allegedly containing hashish.

**WAITED FOR SECOND**

The officers said they spent several hours searching the basement for further evidence.

*Figure 12–24 All headlines should have a bit more space above than below. Make it clear that a new idea is being discussed. In paragraphs (A), the subhead is set solid, meaning no space above or below. The space underneath that head is caused by its shoulder, not a true space. (B) has about three points of space above the subhead and is easier to read.*

interesting to all readers, and there is logic in adding subheads. The space above and below a subhead should help rather than hinder reading. Generally, a subhead is added to a story about every fourth or fifth paragraph (see Figure 12–24).

## WORD SPACING

Spacing between words should be narrow rather than wide. Narrow word spacing helps the reader see more words at one sighting and therefore speeds up the reading process. However, there ought to be enough space between words for the reader to recognize where one word ends and another begins. Phototypesetting sometimes increases the space between words too much (see Figure 12–25). These gaps of white space slow down reading because the words must be read individually instead of in groups. Of course, the other extreme of long words in narrow line widths is also undesirable. Generally, the space between words should be about one-third the size of type used. Nine-point type would have three-point word spacing.

## LETTER SPACING

Letter spacing usually is done for headlines that are too short. Added spacing between letters makes the line appear longer. Letter spacing usually is not programmed for body type on desktop computers, but it is possible. It was easily done when type was set on a Linotype. If letter spacing needs to be done, no more than one point should be added between the letters in headlines under 14 points. Two

While in Urbana-Champaign, the students will learn how to convert the computer language of their experiments into the language of the supercomputer and compete for a year's worth of on-line time to complete their experiments.

*Figure 12–25 Too much space between some words.*

points of space may be added between type that is more than 14 points high, and even more for large type. Generally, letter spacing should be avoided, if possible, because it tends to call attention to the words that receive the extra space. The reader may stop and notice that a word has been spaced out. Any time the reader is made to stop the flow of reading, typography is relatively poor.

## TYPE SELECTION PRINCIPLES

To use type effectively, editors must be able to differentiate type. That isn't all that is required, but it certainly is the first step along the way. What follows are some basic principles on making intelligent type selections and then arranging the type in the most readable manner.

### Choose Legible Typefaces

Any typefaces selected for printing must be legible. That is, every letter must be easily deciphered. When a typeface is not legible, readers may read a lowercase *c* for an *e*, or perhaps be confused by any letters that look like other letters. Illegible type may require readers to spend extra time trying to figure out the meaning of words. If the news is fascinating, readers will read a story regardless of legibility, but it is the editor's job to make reading as easy as possible.

Newspaper editors tend to use the most legible type for their main news. The place where less legible type could appear would be feature story headlines and/or body types where an editor is looking for some unique typeface that would call the reader's attention to the story or perhaps fit its unique nature.

When the more legible types are selected and reading becomes easier, readers may be able to read more of what has been written in the amount of time normally

Education

**EDUCATION**

*Figure 12–26 Legible versus less legible type. The bottom typeface may be used occasionally for a special purpose when surrounded with enough white space. It has some use when a few words are needed to attract attention.*

allocated for reading newspapers. Whether they *will* read more depends on factors other than typography.

Almost all typefaces in common printing use today are *somewhat* legible. It isn't necessary to consult research to learn which types are most legible, because legibility isn't an either-or evaluation. It is a matter of choosing the best available alternate. Figure 12–26 shows two typefaces: Minion and Koloss. Readers might well say both are legible, but there is no question that Minion is much more legible than Koloss. Perhaps a short headline of Koloss may cause no reading difficulties, but Minion could be used in short or long headlines and still be readable.

## Choose Attractive Typefaces

Of the thousands of typefaces available for printing, some are much more attractive than others. In order to decide which typeface is most attractive, editors should have a sense of artistic appreciation and good taste. Editors will sometimes select a typeface that is cleverly designed but not the most attractive choice available. When artists are asked which typefaces are most attractive, they tend to select simple, plain typefaces rather than clever or ornate ones.

## Choose the Best Type Size and Weight

It is important to consider point size and weight while selecting a typeface. To determine the size and weight needed for type, consider the column width in which type is to be set. A 36-point headline set in a one-column width rarely looks readable. The problem is that only one short word could be set in a one-column space. That's hardly a headline worth having, unless it is a *hammer head* (a one-word headline used in conjunction with other, longer headlines underneath to tell a striking fact about a piece of news).

The editor should therefore decide how many words of any size type are needed within a given column width. This is easy to determine with a computer. When no computer is available, the editor has to count the number of characters of a given typeface that can be set within a desired column width. How many words of a headline should appear on one line is a matter of editorial judgment.

A second consideration is the appropriate weight for the type size. Headlines may be too small and insignificant in lightface type but look very nice in boldface of the same size. Therefore, the editor has to balance type size and weight at the same time in order to make an intelligent choice. Sometimes an editor can achieve the effect of more weight by using larger typefaces even though they are set in lightface letters. More often, however, a larger and bolder typeface is needed to get the reader's attention. Readers tend to skim headlines as they look for interesting stories. Editors can help readers in this task.

## Consider the Age of the Population

As the population of the United States continues to get older, readers may have difficulty reading newspapers because the type is so small. For that reason, the editor should pay special attention to both body type and accompanying headline type sizes. What is the optimum?

There is no valid research on the subject, so correct type sizes are a matter of opinion. Generally, body-type sizes run from about nine points to about 11 points. Recommended body-type sizes for older readers are from 10 to 14 points with extra leading between the lines.

In attempting to decide on the proper size of body type, the editor should ask, how large is the body of the type size now being used? If the type has a large body and very little shoulder, then larger typefaces may not be needed. But if the type now being used has a small body and large shoulder, larger sizes may be required.

Headline point sizes depend on how many column widths the head covers. Headline sizes may be as follows:

| | HEADLINES: PRESENT SIZES | HEADLINES: NEWER SIZES |
|---|---|---|
| 1 column width | 14 to 18 points | 16 to 20 points |
| 2 column widths | 18 to 24 points | 20 to 30 points* |
| 3 column widths | 24 to 36 points | 30 to 42 points* |
| 4 column widths | 36 to 48 points | 42 to 60 points* |

Many, though not all, condensed typefaces are poor choices for headlines and hardly ever appropriate for body type because they are difficult to read. Some type designers have chosen condensed typefaces for headlines as a means of making the

*A slightly condensed type may be required here, but not extra-condensed.

type sizes larger and still retaining the same number of words as regular type. This is supposed to make the headlines more readable for older people (see Figure 12–27).

That idea can easily be challenged if the typeface is too condensed. The reason is that few adult readers have an opportunity to read much condensed type in print anywhere: not in books, magazines, newspapers or advertisements. Therefore, they are not used to reading it. On the other hand, if the choice of condensed type is not extreme, older readers' eyes may become acclimated to the narrowed letters with little difficulty.

A much better alternative, however, is a typeface that is not condensed but has somewhat narrowed lowercase letters. The best of these is Times Roman, but another possibility is Century Schoolbook.

**Colorful, detailed Indian fashions find Western fans**

Her designs are a kaleidoscope of color. Sequins, mirrors, brocade or intricate embroidery accent each piece differently and—if the crowds can be a gauge—successfully.

"I'd like to communicate all the moods and colors of Asia in each piece

**Going to greater lengths**

**Those unpredictable British designers take the lead in feminine, wearable fashions.**

By Pat Morgan

*Writing from London*

We have seen fashion's future, and it is long.

After an extended infatuation with skirts that could seemingly go no shorter, but always did, several new looks have come our way. This could be the season to change all that, however.

With the most fanciful, and virtually unwearable, clothing presented during the recent runway shows in Italy—the European country that traditionally offers the most classic, tailored styles—

*Figure 12–27 Some condensed type is difficult to read quickly.*

It is important to note that even when condensed type is mixed into a primarily roman headline schedule, the combination can be difficult to read.

## Avoid Unusual Typefaces

Any typeface that calls attention to itself is usually undesirable. The only exception would be one- or two-word headlines that sometimes appear on feature stories and are surrounded by generous amounts of white space. They can be suitable. The extra white space seems to allow readers a little more time to decipher the words than is needed for ordinary letters. Be wary of selecting letters for their novelty rather than their suitability. The result may appear amateurish rather than professional.

## Avoid All-Capital Headlines

Type set in all capitals is usually more difficult to read than type set in caps and lowercase, or *downstyle* (only the first letter is in caps). Most readers read by recognizing word shapes of caps and lowercase letters with their ascenders and descenders. Letters set in all capitals have none of these differences, and readers must put the letters together to form words: a slower process. Figure 12–28 illustrates the problem. Also see Figure 12–29.

## Avoid Type That Causes Fatigue

Although an editor may select a legible type for a story, that type may cause readers' eyes to tire. One kind of type that is considered to be particularly fatiguing is sans serif. Sans serif typefaces in headlines seem to cause no fatigue problems because letters are fairly large and there is a generous amount of space around headlines. But sans serif typefaces, especially in the lightweight versions, tend to be so monotonous in color that they are boring.

Another kind of type that falls into this category is condensed body type. Letters that look "squeezed" tend to be harder to read in mass and tend to be fatiguing.

The solution to reader fatigue, of course, is to avoid those typefaces. However, for certain purposes, they look attractive, and some editors might want to use them anyway. If sans serif or condensed typefaces must be used in large quantities, they can be made more readable by adding space between the lines. This will let the reader see the words in a more contrasting environment.

## Select the Best Type Family

Which type family is best for a headline on a particular kind of story? There usually isn't much choice in most newspaper offices. But even with a limited number of choices, there may be good or bad options from which an editor may choose. So the

**COMPLETE**

**complete**

***Figure 12–28** Why caps are read slower: The reader must place each letter together to form a word. In lowercase, the shape of the word is recognized and helps identify it.*

**CAPS VS. LOWERCASE**

There is little doubt that lowercase letters are more easily read than caps. Some typographers like to avoid capitals, even in headlines.

In fact, the recognizability of words seems to lie chiefly in the upper half of the lowercase letters as illustrated below.

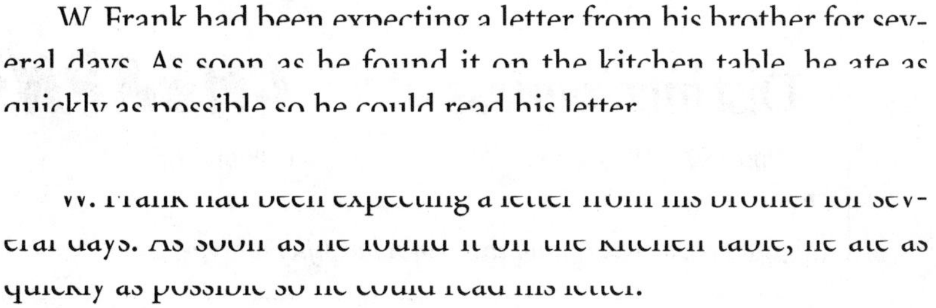

W. Frank had been expecting a letter from his brother for several days. As soon as he found it on the kitchen table, he ate as quickly as possible so he could read his letter.

***Figure 12–29** The legibility of type seems to lie in the upper half of lowercase letters.*

answer is that the type selected ought to match the nature of the story, if possible.

Matching a typeface to the nature of a story is based on the connotation of each type family. Type families differ just as humans do. Therefore, it is necessary to know the connotations of each typeface that a newspaper owns. It is not a difficult task to evaluate most of these connotations (see Figure 12–30).

*Connotation* answers the question of how readers will discern meaning. Typefaces can be warm, cold, strong, weak, feminine or masculine, to name only a few connotations. The connotation of type, matched with the nature of a particular story, enables type to add to or distract from the news. Some typefaces have very

| | |
|---|---|
| Elegance | MODERNISM |
| DucDeBerry | Bauhaus |
| DIGNITY | UNUSUALNESS |
| Castellar | Bees Knees |
| Antiquity | Nature |
| Post Antiqua | Amadeus Open |
| Sincerity | Strength |
| Announcement | Koloss |
| Distinctiveness | CHEAPNESS |
| Delphian 2 | Balloon Extra Bold |

*Figure 12–30 The connotations of some different typefaces.*

little connotation and neither add nor distract. It is also possible that two or more persons can look at the same typeface and disagree about its connotation. But most often there is little disagreement about connotation.

The person making a type selection may sometimes have to spend some time searching for an appropriate face for a story. At other times, it is easy and quick to find what one is looking for.

## TYPOGRAPHY ON A DESKTOP COMPUTER

### The High-Resolution Type Problem

Many persons who use desktop computers are unaware of the problems related to good typography. They usually think that the desktop computer produces rather nice-looking printed type that is equivalent or nearly equivalent to type from a phototypesetting machine. Many desktop computer users simply have not developed a sharp enough perception of the difference between typical desktop printing and fine typography set on a high-resolution machine like the Linotronic phototypesetter. There's a big difference.

Most desktop computer printers produce type at 300 dots per inch (dpi). At that rate, type looks superficially like phototypesetting, but on closer examination it is a poor reproduction because most phototypesetters print somewhere around 1,200 dpi. In typefaces smaller than 18 points, the dots that make up a single letter are invisible. It will take a strong magnifying glass to see them. But in larger typefaces, they tend to be visible. Many companies now have a Linotronic machine that can produce type up to 2,540 dpi. Therefore publishers can buy additional equipment to smooth out these irregularities (called *jaggies*), but they often find it too expensive and continue using 300 dpi. While 300-dpi type is better than dot-matrix printing, it is not in the same league as the higher dpi printing.

High-quality typography is not possible if the type resolution is only 300 dpi. Some smaller newspapers use 300 dpi type with virtually no complaints because their readers do not know or care about the difference. But some publishers produce type on their desktop machines and then have it reproduced at a higher dpi level by companies that turn out camera-ready copy at the high-resolution levels for a fee.

Shall the typesetter use the highest level of type resolution? For a newspaper on which quality printing is a standard to be upheld, the answer is yes. Some newspapers cannot afford such a high standard. Many readers don't care as long as the type is large and clear enough without more resolution.

## Desktop Type Printed Over Tint Screens

One of the newspaper typographic techniques that became popular even before desktop publishing was that of printing type over a tint screen. A tint is a colored mass of dots over which type is printed. It is often used in boxes. The objective is to make a certain story more noticeable. In the desktop publishing age, the technique has become even more popular. The problem with this technique is that unless the tint is very light (gray or a light color), type can be difficult to read. Unreadable tint boxes are often seen in newspapers. An editor who wants to use a screened background tint should always place it over a very light screen, using a light color such as yellow or orange or a very light gray. If this admonition is ignored, all the best efforts at increasing readability will have been defeated (see Figure 12–31).

## Appropriate Use of Contrast

Attractive typography often brings contrast to a page. Contrast is achieved by selecting typefaces in sizes and weights that make some type print a bit heavier and some print lighter or by using italic with roman headlines or serif with sans serif

**Public Aid in area counties**

The percentage of residents in east central Illinois counties who receive aid (Aid to Families with Dependent Children, food stamps, Medicaid, etc.) from the state Department of Public Aid.

| | 1988 | 1989 |
|---|---|---|
| Champaign | 4.7% | 5.0% |
| Coles | 7.0% | 7.4% |
| Douglas | 5.2% | 5.4% |
| DeWitt | 6.1% | 5.8% |
| Edgar | 8.5% | 9.4% |
| Ford | 5.8% | 5.8% |
| Iroquois | 5.2% | 6.0% |

Source: Illinois Department of Public Aid

**Poverty threshold**

The following are the maximum incomes for households to be considered in poverty by federal guidelines.

| Household size | Annual income |
|---|---|
| 1 | $6,620 |
| 2 | $8,880 |
| 3 | $11,140 |
| 4 | $13,400 |
| 5 | $15,660 |
| 6 | $17,920 |
| 7 | $20,180 |
| 8 | $22,440 |

Source: U.S. Department of Health and Human Services

***Figure 12–31** Dark gray or color tints underneath type may make it difficult to read. Choose light grays or light colors for tints.*

heads. The editor should strive for pleasant, attractive contrast rather than strong, overbearing contrast. Too much contrast is as bad as no contrast.

### Offbeat Typefaces

Many people who are setting type on a desktop computer for the first time are excited at having the opportunity to choose from so many different type options. Some will inevitably be attracted to offbeat typefaces. (An offbeat typeface is not commonly used and tends to be garish.) Some people think that these typefaces will add a dash of charm to a feature article, and indeed they may. Most often, however, offbeat typefaces call too much attention to themselves and not necessarily to the ideas. Therefore, the editor who chooses typefaces for headlines should not make it a practice to select offbeat typefaces.

## DEVELOPING ARTISTIC JUDGMENT

The best way to understand type is to be able to differentiate the many kinds of existing typefaces. Unless the slight variations in letter shapes are known, one typeface may appear as good as another. Once the differences in typefaces have been learned, however, the editor can develop an aesthetic sense for what looks attractive in print, based on some elementary artistic principles.

Developing a sense of artistic judgment in type is best done by studying type in print, no matter where it appears. For example, beautifully set type often appears in a magazine, a book or a financial report. An editor who works at developing a sensitivity to what looks good in print will be alert to such printing. Then it is advisable to make a mental note of the way type was set or why it looks so good. An editor may adjust these techniques to the newspaper environment, if at all possible. In this way he or she uses type to aid communication.

The editor should also know and keep up with typographic research. Almost every year, new studies add to knowledge of how to use type. Although research cannot answer whether one typeface is more appropriate than another for a headline, there are many helpful things to be learned from research. In fact, as more studies are conducted, some of the older ideas on using type will be rejected or replaced.

# CHAPTER 13

# AN INTRODUCTION TO NEWSPAPER LAYOUT AND DESIGN

Newspaper layout and design are about the arrangement and appearance of almost every page in a newspaper, so that each is attractive and readable. They use basic design principles and techniques created just for that purpose. Design is about the theoretical planning of each page, and layout represents a written design plan for each page.

## MAKEUP, LAYOUT OR DESIGN?

Three widely used terms in newspaper production refer to somewhat the same activities discussed in this chapter: makeup, layout and design. Their differences reflect the evolution of the newspaper production process.

*Makeup* is associated with an older era of newspaper production that means assembling the newspaper's content, page by page. In bygone days, news was set on Linotype machines and, after having been proofread and corrected, was brought to a special printer, whose job it was to assemble the columns of type into pages. In early years, makeup men did not necessarily work from a written plan (now called a *dummy* or *layout*). However, in later years, the plans were incorporated into a layout for the placement of key stories, at least. Makeup men were then responsible for filling the remainder of a page with unused news type. In some cases, editors simply told the printer where the key stories should be placed, and no written plan was used at all. That was makeup. Eventually, almost all pages of a newspaper required a dummy, showing where news should be placed. The term *makeup* is still acceptable in most plants, but *layout* is a more professional word.

*Layout* refers to a document or written version of a plan that covers selecting, shaping and arranging the contents of each newspaper issue. Contents include news, advertising, features, news directories and many other items.

*Design* is a relatively new term that incorporates many of the activities of the other terms but with more artistic and creative judgments about planning and arranging the contents of a newspaper. What has happened, however, is that all three terms tend to be used interchangeably. In this book we use the term *layout* to mean the plan for each page's arrangement of news and features. We use *design* to include layout arrangement and all artistic decisions related to the appearance of individual pages.

## HOW IMPORTANT IS GOOD NEWSPAPER DESIGN?

The most important part of selling a newspaper is its content, first, and then its design, but the two elements are not divided fifty-fifty. People buy newspapers that carry stories they want to read. There is considerable variance in what each person wants to read. Children and some adults may want to read the comics, but with different priorities. Children may want only comics, but adults usually want something more, including news, sports scores, sports stories, stock market reports and so on.

How much does layout and design matter to readers? The answer is that while most readers want an attractive and easy-to-read paper, those things are not usually their highest priority. More often they have other priorities, such as newspapers that are honest or unbiased, papers whose administrators are known and liked, papers that do a good job covering local and/or national news or papers that best represent their personal views and special interests.

Many of the leading U.S. newspapers are well-designed, but they are not purchased *primarily* because of their design. Yet editors and publishers of these newspapers will argue that a good design is a basic necessity that affects readership to a varying extent.

Well-designed newspapers can be appealing to readers and help encourage them to spend more time reading. While good design may help satisfy readers, it cannot replace more important factors that also affect readability and salability. (See Chapter 12 for a discussion of readability and legibility.) For example, good design can't substitute for the kind of news readers want. If people want to read financial news and can't find it, they will choose some other paper or seek out some other news medium.

Good design can't replace the need for speed of delivery. In the 1995 Simpson trial, for example, many people were glued to their television sets because they could have their news almost instantaneously, and along with it came instant interpretation. Better-designed newspapers could not have reversed that need. Many families want their news fast in order to have more time to do other things with their lives. So design does not play a competitive role in speed of news.

In the competition for younger newspaper readers' time, outstanding design seems not to have been able to attract younger readers. However, newspapers carry important news and ideas, much of which the public should read to be better-informed citizens. Once readers have chosen a newspaper, good layout and design are facilitators of reading.

## HOW TO RECOGNIZE A WELL-DESIGNED NEWSPAPER

Although graphic designers may disagree about the precise criteria of good design, there are enough agreements among them to build a body of knowledge that can help people recognize good design for themselves. The general characteristics of a well-designed newspaper are given here, and the following sections discuss ways to achieve them.

**Good Organization.** Good design organizes the news to help readers easily find whatever they are looking for. Similar kinds of news ought to be in proximity of each other, if possible. Readers dislike reading stories in one section of the paper and then having to look for additional stories of the same kind in some distant part of the paper. Organization also covers the smooth transition of news from one column to the next. Modern newspapers have been categorizing news for years, and that helps the organizational problem somewhat.

**An Attractive Newspaper.** There usually is a sense of agreement about what is attractive and what is not. Generally, an attractive newspaper has an adequate amount of space between lines of type and between stories, or between columns, on almost every page. Avoid what is known as a *tight page design,* in which there is so little space between stories that readers have difficulty concentrating their attention on any one. Good design tends to have generous amounts of white space carefully distributed on a page.

**Attractive Display of Illustrations.** Photographs with a fine screen (above 100 lines to the square inch) are usually the ideal kind of illustration. This makes it easy to see the details. However, the quality of newsprint often determines screen sizes, and better-quality paper makes fine screens and reading somewhat easier. With the cost of newsprint rising, the better-quality paper may be too expensive.

Photographs must be given adequate space, and enough of them should be used to illustrate at least the highlights of the news. They enhance the total page design if they are the right size and number. There are no rules that require a precise number of photographs be used each day or how large they should be. Artistic judgment is usually required for the good use of illustrations.

A great deal of artistic judgment is necessary to place illustrations on a page. Some designers like at least one dominant illustration on every page, accompanied by one or more smaller illustrations. The worst-designed pages tend to use illustrations that are all about the same size.

**The News Is Easy to Follow.** Pages should be arranged in a way that makes it easy to follow a story from column to column. It is not a simple matter to wrap stories from one column to the next right-hand column for an easy-to-read style. The art of wrapping stories requires good judgment as to *where* in the next right-hand column a story should be continued.

**Pages Have Contrast.** Of all the basic principles of design, contrast is the most important for attractiveness. Pages should be designed to have some, but *not overwhelming,* contrast. Contrast generally provides attractive pages by offering readers a change of pace. The contrast may be large vs. small photographs, dark vs. light sections of the page, regular- or irregular-shaped stories or illustrations, vertically vs. horizontally shaped photographs.

**Unity.** Attractive pages usually look unified, or, as if everything on the page were carefully placed just where it is.

**Different Typefaces Are Kept to a Minimum.** Computer users may be tempted to use too many different typefaces on the same page. They are used to having many different type choices. The professional knows how many alternative faces are just right. Again, it takes artistic experience to decide.

**Pages Have Balance.** The best pages usually are not top- or bottom-heavy. Although readers usually cannot discern balance in particular, they get the feeling that a page is or is not overbalanced, one way or another.

**Generous Line Spacing.** Good typography usually requires adequate to generous amounts of leading between lines. But there may be a point of diminishing returns on leading. In other words, leading can be too large. Many art directors for advertisements tend to use too much leading when they already have an adequate amount of space on a page. Newspapers are not trying to sell each story separately, but it takes a keen sensibility for spacing to know what leading is ideal. Wider column widths usually require more leading than shorter widths.

**Placement of Stories Is Well Thought Out.** The top editors and writers review the content of each issue (before it is laid out) from the point of view of story importance. They decide where in the newspaper leading stories should be positioned.

Smaller or less important stories usually don't require long discussions about positioning and may be placed after all important stories have been positioned.

## VISUALIZING TOTAL PAGE STRUCTURE

Page layout begins with some idea of general structure. Will pages be horizontally or vertically shaped? If the designer doesn't make up his or her mind about this early, pages may have somewhat of a circus design, with little order on them. Modular designs, which are so widely used throughout the country today, tend to be distinctly vertical or horizontal.

In the early designs of U.S. newspapers, pages were all vertical. If there were eight columns to a page, the stories all started at the top, and readers read down and then up to the top of the next right-hand column. Reading down and up describes vertical designs.

Today, although there are some totally horizontally styled pages, most pages are designed to be a mixture of vertical and horizontal, but not in equal proportions. Generally, pages are divided into 25 percent-75 percent or 67 percent-33 percent, either vertically or horizontally. Too much of either one is not necessarily attractive. (See Figures 13–1 and 13–2.)

If the design of a page isn't obvious to you, draw lines around each story, as shown in Figure 13–1. If you made this kind of drawing, you would discover that the page is strongly vertical. A vertically designed page is usually quite orderly and easy to read. Notice that different column widths (left-to-right arrangement) are not 1-2-3, but 1-3-2. (There is a one-column at the left, then a three-column to the right of column one and, finally, a two-column width.) The reasoning for this is that some readers dislike the obvious, and 1-3-2 column arrangement is not an obvious division of space.

Figure 13–2 shows another example of strong vertical design, using boxes and photographs. The top box is covered by a 2-point rule, while the bottom box is covered by a 1-point rule, suggesting that the top story is more important than the lower story. The three action photographs are interesting and artistically placed. This page is also readable.

The horizontally designed pages shown in Figures 13–3 and 13–4 are used as a change of pace from vertically styled pages. Neither of these basic designs is better than the other; they are just different. Notice that the editorial cartoon in Figure 13–4 has not been placed in the traditional upper-right position because readers will search for the cartoon wherever it is. Therefore, the upper part of the page is now open for a story or column of presumably more significance. The page has a cleaner, orderly look about it and is easy to read. The lead editorial uses subheads to clarify the story.

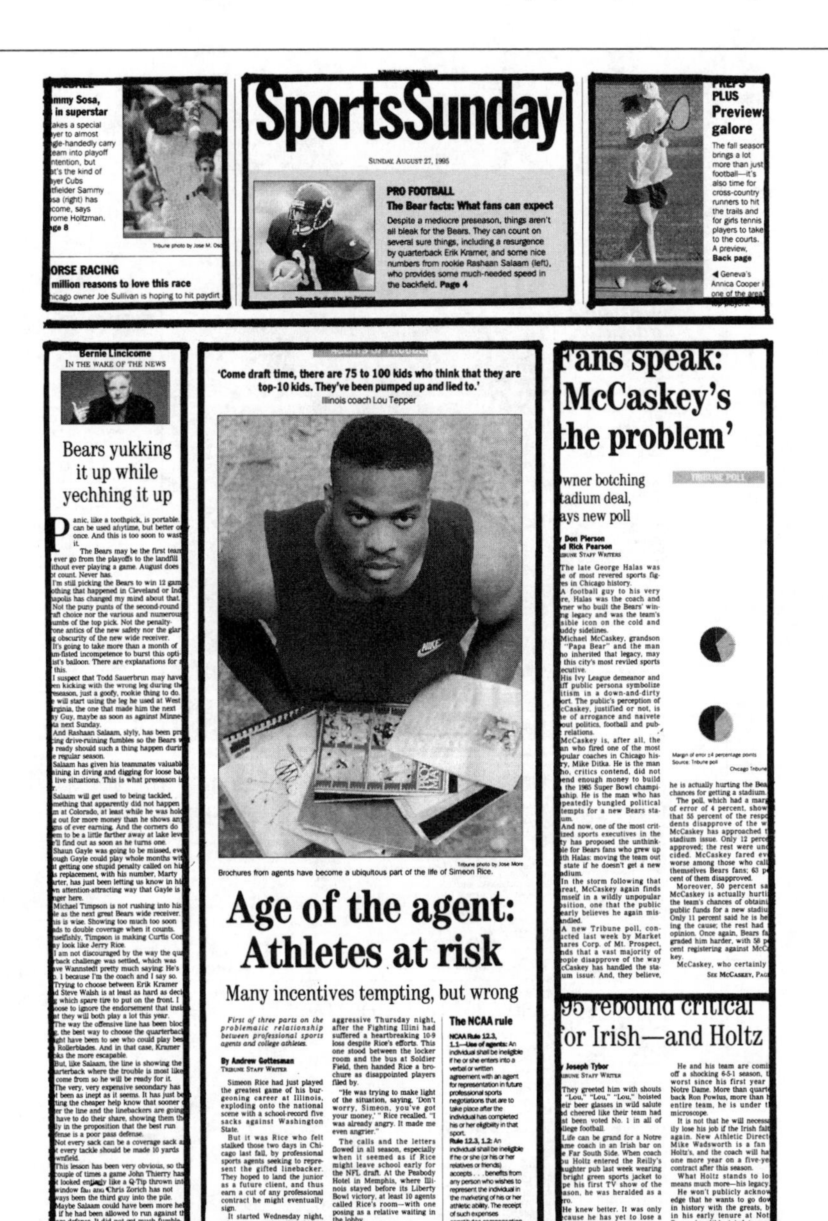

# SportsSunday

SUNDAY, AUGUST 27, 1995

**PRO FOOTBALL**

**The Bear facts: What fans can expect**

Despite a mediocre preseason, things aren't all bleak for the Bears. They can count on several sure things, including a resurgence by quarterback Erik Kramer, and some nice numbers from rookie Rashaan Salaam (left), who provides some much-needed speed in the backfield. **Page 4**

**HORSE RACING**

**PLUS Previews galore**

Bernie Lincicome

IN THE WAKE OF THE NEWS

## Bears yukking it up while yechhing it up

**'Come draft time, there are 75 to 100 kids who think that they are top-10 kids. They've been pumped up and lied to.'**

Illinois coach Lou Tepper

Brochures from agents have become a ubiquitous part of the life of Simeon Rice.

# Age of the agent: Athletes at risk

## Many incentives tempting, but wrong

*First of three parts on the problematic relationship between professional sports agents and college athletes.*

**By Andrew Gottesman**
TRIBUNE STAFF WRITER

Simeon Rice had just played the greatest game of his burgeoning career at Illinois, exploding onto the national scene with a school-record five sacks against Washington State.

But it was Rice who felt stalked those two days in Chicago last fall, by professional sports agents seeking to represent the gifted linebacker. They hoped to land the junior as a future client, and thus earn a cut of any professional contract he might eventually sign.

It started Wednesday night, when an agent called Rice's room at the Chicago Hilton and Towers. That one proposed, unsuccessfully, a meeting in the hotel lobby.

Another agent was more aggressive Thursday night, after the Fighting Illini had suffered a heartbreaking 10-9 loss despite Rice's efforts. This one stood between the locker room and the bus at Soldier Field, then handed Rice a brochure as disappointed players filed by.

"He was trying to make light of the situation, saying, 'Don't worry, Simeon, you've got your money,' " Rice recalled. "I was already angry. It made me even angrier."

The calls and the letters flowed in all season, especially when it seemed as if Rice might leave school early for the NFL draft. At the Peabody Hotel in Memphis, where Illinois stayed before its Liberty Bowl victory, at least 10 agents called Rice's room—with one posing as a relative waiting in the lobby.

It sometimes went further: One agent offered Rice $100 per sack, which he declined, while another joked about get-

SEE AGENTS, PAGE 5

**The NCAA rule**

**NCAA Rule 12.3, 1.1—Use of agents:** An individual shall be ineligible if he or she enters into a verbal or written agreement with an agent for representation in future professional sports negotiations that are to take place after the individual has completed his or her eligibility in that sport.

**Rule 12.3, 1.2:** An individual shall be ineligible if he or she (or his or her relatives or friends) accepts . . . benefits from any person who wishes to represent the individual in the marketing of his or her athletic ability. The receipt of such expenses constitutes compensation based on athletics skill and is an extra benefit not available to the student body in general.

Source: NCAA

## 'Fans speak: McCaskey's the problem'

Owner botching stadium deal, says new poll

TRIBUNE POLL

## '95 rebound critical for Irish—and Holtz

***Figure 13–1** This page is clean and readable because it is well-organized. The full column spaces that run from the bottom to the top (three horizontal panels at the top) also help make it orderly. The sum of the three top panels give the page a bit of horizontal feeling.*

FEBRUARY 1, 1995

CALIFORNIAN

# Wednesday SPORTS

SECTION B

## A note to Al Davis: Just make a decision, baby

BOB KEISSER / Long Beach Press-Telegram

Gasping for breath in the toxic air of despair and confusion that hangs over Raiderville:

Before we get around to examining trace levels of disease, let's update the basic Raider medical report.

Owner Al Davis is paralyzed. He can't decide

COMMENTARY

whether to fire head coach Art Shell, where to reassign Shell in the organization if he does fire him, and how to personally avoid the recriminations if he fires him.

Shell is heartsick that his steadfast loyalty may be repaid with a dismissal, especially since he's thrown himself on so many grenades the last three years aimed at the commander-in-white.

Mike White is dizzy, sitting between his owner and his coach while wondering if this is the last chance he will ever get at a head coaching job.

Joe Bugel is disoriented, wondering how he fits into this silver-and-black stew. Is he a coach-in-waiting, a well-paid assistant, a personnel guy or a thankless liaison between the whims of a strong-minded owner and an edgy, scared coaching staff?

The players are hamstrung, aware that they had Super Bowl talent but Freedom Bowl organization and that as long as Davis retains his vice-like grip on the team, they will have more than just an opponent to beat every Sunday. . . .

The easy thing to do is wear a gleeful smirk about the Raiders' new commitment to anguish. An organization that prides itself on fear and gloating doesn't encourage much sympathy.

But there is real pain and heartache here and no small amount of paralysis as the Raiders deal not only with a disappointing season but the realization that the franchise is in serious decline.

Start with the owner. Davis is agonizing over his team's nine-year malaise and mortified that opponents he once dominated, as he might say, are now snickering behind his back and claiming the glory he has always considered his.

The San Diego Chargers in the Super Bowl. The Kansas City Chiefs dominating them with a player he persecuted. The San Francisco 49ers' offensive coordinator, a man he hired and fired, being called a genius and now headed to the hated Broncos.

Worse yet, Davis is torn over what to do with his head coach. Shell has been a loyal corporal for 27 years, as a player and coach. It was Davis who inducted Shell into the Pro Football Hall of Fame. For someone who has propagated the image of Raider loyalty, this is a very bitter decision to consider. . . .

No less so for Shell.

Please see RAIDERS, B2 ➤

THE ASSIST MAN

## No passing fancy

STEVE WILSON / Associated Press

Utah guard John Stockton passes the ball past Minnesota's Winston Garland on Monday for an assist.

## Utah's Stockton on brink of record

BOB MIMS / Associated Press

SALT LAKE CITY — John Stockton likely will break Magic Johnson's career assist record at home tonight against Denver — and the Utah Jazz guard can't wait to be finished with the hype.

It began with the start of the season, his 11th with the Jazz. Stockton had 9,383 assists, 538 short of Johnson's mark. What had been a trickle of requests for interviews has now become a flood.

After 14 assists in Monday's 115-80 victory over Minnesota, Stockton was 11 shy of the record by the Los Angeles Lakers great.

Stockton dodged reporters after the game, but dutifully made the rounds with broadcasters and sports writers following Tuesday's practice. The Jazz public relations department, he said, made it clear he must deal with this part of his job.

"I'll be glad when it's over with and we can go back to normal," Stockton smiled through gritted teeth. "Normal is that we have a pleasant but distant relationship with you guys."

Stockton, 32, acknowledged he has begun to feel the pressure.

"My wife (Nada) really wants to get it over with because I'm far from friendly around the house right now," he said. "It's just more than we're used to."

It has taken the 6-foot-1 product of Gonzaga University in Spokane, Wash., 859 games over 10½ seasons — the past seven as the NBA's assist leader — to reach Johnson's record, set in 874 games from 1979 to 1991.

Karl Malone, a favorite target of Stockton's passing artistry, said the assault on the record

Please see STOCKTON, B2 ➤

## Zanin lifts Murrieta over Elsinore

■ **GIRLS BASKETBALL:** *The forward scores 17 points and pulls down 18 rebounds in league victory*

RICH SMITH / The Californian

MURRIETA — When the Murrieta Valley High girls basketball team needed it the most, Maggie Zanin stepped up her game.

Zanin helped propel the Nighthawks to a 44-41 victory Tuesday night over Mountain View League rival Elsinore with 17 points and 18 rebounds. The victory keeps Murrieta's playoff hopes alive as it improves to 6-12 overall and 3-4 in league.

"She really stepped it up tonight," said Murrieta coach Ted Berry. "Just look at the stats, 18 boards, 17 points, that says it all."

As far as the playoff hunt goes, Murrieta is just one game behind Centennial for third place.

"We still have a shot," Berry said. "I'm not ready to give up

Please see MURRIETA, B2 ➤

CYNDY SULLIVAN / For The Californian

Murrieta's Maggie Zanin (right) fights for a rebound with Elsinore's Colleen Sullivan (left) during the Nighthawks win over the Tigers Tuesday.

## A bad ending to a good day for Ceballos

■ **PRO BASKETBALL:** *Lakers forward makes the All-Star team but can't lift Los Angeles past Chicago*

JIM McCURDIE / Long Beach Press-Telegram

LOS ANGELES — The phone rang at 6 a.m., and Cedric Ceballos pried open his eyes, and wondered who was responsible for the unexpected wake-up call.

"I thought, 'Man, this better be an emergency,'" he said.

As it turned out, there was good news on the other end of the line. Ceballos' uncle was calling to tell him he'd just heard on the radio that his nephew was an NBA All-Star. Ceballos' initial reaction? "I told him to call me back at 8 o'clock."

The rude awakening came hours later, when Ceballos' day of celebration ended in the Lakers' frustrating, 119-115 loss to the Chicago Bulls in front of a sellout Forum crowd of 17,505.

Scottie Pippen had 34 points and 13 rebounds and B.J. Armstrong had 25 points as the Bulls snapped the Lakers' three-game winning streak, and provided the not-so-happy ending to Ceballos' otherwise happy day.

After scoring 17 points in the first half, Ceballos was held to just four in the second, and didn't his first field goal of the half until there were 7.7 seconds left in the game. Meanwhile, the Laker defense collapsed in the third quarter, and quickly saw an eight-point halftime lead disappear.

Pippen and Armstrong combined to make 9 of 11 shots to help the Bulls outscore the Lakers by 14 points in the third quarter. The Bulls made the most of some easy transition opportunities, shooting 77.8 percent (14 of 18) in the third period.

The Lakers came back to tie it at 97-97 on Nick Van Exel's three-point play with 7:55 to play, but the Bulls responded with an 11-4 run.

"We really didn't get the job done defensively," Laker coach Del Harris said. "That's been my biggest concern . . . that we just aren't as dedicated to playing defense as we need to be to be a consistently good team."

Ceballos, victimized several times by Pippen's offensive quickness, expressed disappointment at the fact that he played only 15 minutes in the second half.

LEON ALGEE / Associated Press

Anaheim goalie Guy Hebert knocks St. Louis' Craig Janney to the ice. Janney would end up being instrumental in the Blues' 7-2 victory.

## Out of the doghouse, Janney bites Ducks

■ **HOCKEY:** *St. Louis center returns with two assists in a 7-2 rout of Anaheim*

R.B. FALLSTROM / Associated Press

ST. LOUIS — Now it's time for Mike Keenan to pick on somebody else.

Craig Janney emerged from two games in coach Mike Keenan's doghouse and set up two goals, and goalie Geoff Sarjeant had an easy game in his first NHL start as the St. Louis Blues beat the Anaheim Mighty Ducks 7-2 Tuesday.

"I was just trying as hard as I could," Janney said. "That's all he expects of me, to play as hard as I can and to play smart."

Janney didn't assume he was back on Keenan's good side, however.

"I think you earn it back day to day," Janney said.

Keenan pulled Janney, a deliberate skater with precision passing skills, in the third period of the Blues' 6-4 loss at Calgary on Jan. 24. Janney, who had 68 assists last year and scored 106 points in 1992-93, then spent the next two games on the bench.

"He played with more tempo," Keenan said. "That's what I was looking for."

Janney also said his lack of attention on defense could have landed him on the bench.

"Sometimes my offensive instincts get in the way," Janney said. "That's my problem. Always has been."

Janney got an ovation after he set up the Blues' second goal. Janney held the puck at the side of the net and drew two defenders his way before centering it to Hull for an easy goal at 13:27 of the first period. He also set up Tikkanen's power-play goal, a quick wrist shot from the top of the right circle, that made it 5-1 at 14:53 of the second.

Esa Tikkanen led the Blues with two goals and an assist despite coming up empty on two breakaways. Brett Hull and Ian Laperriere each had a goal and an assist as St. Louis outshot Anaheim 47-19.

"Twenty guys were on the rope tonight pulling, and it's fun to see," Janney said. "Hopefully we can do that often."

It was one of the most lopsided losses in the short history of the Mighty Ducks, who debuted

Please see DUCKS, B2 ➤

*Figure 13–2 Another example of a strong vertical design.*

C4 SUNDAY NEWS JOURNAL ••• OCT. 15, 1995

## Comment

John Paul II makes many of us feel a little uncomfortable.

# Pope posed moral challenges to just about everyone

By E.J. DIONNE JR.

Whenever Pope John Paul II visits the United States, one piece of analysis echoes from one end of the country to the other: that Americans "love the pope but do not listen to his teachings." There always follows a lot of perfectly accurate polling data on the pope's approval ratings (high) and on American Catholic views on questions of sex, birth control and the role of women in the church (American Catholics are, indeed, more "liberal" than the pope).

The next step in logic varies. The most sophisticated version of the theory holds that American Catholic enthusiasm for the pope grows out of admiration for his personal strength (in standing up to the Soviet Union, for example) and his support for human rights. This is combined with a reverence for the leader of an institution about which they care deeply. The man is, after all, known as "the successor to St. Peter."

The less sophisticated, one might even say insulting, version uses a lot of mumbo-jumbo about charisma, "superstardom" and "blind faith" to suggest — often with condescending sympathy of the "isn't that cute" variety — that all the folks who cheer the pope are in the grip of powerful emotions that no one outside their cult could possibly fathom. It's as if the faithful are akin to 3-year-olds who love Barney or teen-agers infatuated with a rock virtuoso.

The sophisticated version at least contains some truth. But after this particular papal visit, the whole of the accepted theory about the pope and American Catholics ought to be junked. It's entirely true, of course, that there are differences between many American Catholics and the pope on matters related to sexuality and gender — and differences on the nature of papal authority itself.

But the pope's sermons underscored that there is a lot more to what he has to say than sex and gender, and that a lot of Americans (not just Catholics) are happy to have some of their own assumptions challenged. Put another way, even within their particular disagreements with the pope (on birth control or women's roles, for example), American Catholics find

a lot of agreement with him (e.g., on the responsibilities of family life).

The best thing about this trip is that those who write off John Paul as nothing but a "conservative" got a big surprise. He stood up repeatedly for two groups decidedly unpopular these days, the poor and the immigrants. A lot of conservatives on Capitol Hill must have been uncomfortable on hearing things such as this: "Is present-day America becoming less sensitive, less caring toward the poor, the weak, the strange, the needy? It must not!"

"America's sometimes extravagant affluence," he said at another point, "often conceals much hardship and poverty." He asked New Yorkers: "Have the people living in this huge metropolis lost sight of the blessings which belong to the poor in spirit?" As for immigrants, he suggested that the United States "persevere in its own best tradition of openness."

Now it's important to note that this is not all the pope said about the poor. His view is rooted in the idea that a lack of money is not all that matters.

"The poor have needs which are not only material or economic," he said, "but also involve liberating their potential to work out their own destiny and to provide for the well-being of their families and communities." In this telling, the poor are not simply victims but also actors with responsibilities to themselves and others — which, of course, is how most poor people view the matter.

And he was as uncompromising as ever in his talk about family life and abortion. "The well-being of individuals and communities depends on the healthy state of the family," he said. "Society must strongly reaffirm the right of the child to grow up in a family in which, as far as possible, both parents are present." There was no ambiguity on abortion, either: "Love life, cherish life, defend life, from conception to natural death."

There will now be a lot of jostling between more liberal and more conservative Catholics over the real import of what the pope said. Already, conservatives are reacting a bit defensively to all the pope's talk about the poor, arguing that helping the needy is not the same as endorsing the Department of Housing and Urban Development's approach to their problems. Well, sure. But even less was it an endorsement of the prevailing mood toward the poor, especially in Congress. If conservatives didn't feel challenged, they weren't listening.

Liberals jumping on the Republicans' anti-welfare bandwagon should feel challenged, too. But liberals of all stripes could usefully note how the pope linked compassion with responsibility, uplift with strong families.

What the pope underscored is the extent to which what he and the Catholic Church have to say does not fit the prefabricated boxes into which we stuff most of the ideas we run across. John Paul was saying that it's not possible to build a moral society without also pursuing economic justice. But he insisted that economic justice itself is impossible unless it is rooted in a moral sense. "Democracy," he said, "cannot be sustained without a shared commitment to certain moral truths about the human person and the human community."

If you want to know why Catholics admire the pope even if they disagree with him on some or even many things, my guess is that they welcome the fact that he poses a moral challenge to just about everybody. He's not peddling cheap grace. He makes all of us feel a little uncomfortable. That's the task of all serious moral teachers.

**Write to E.J. Dionne at The Washington Post, 1150 15th St., N.W., Washington, DC 20071.**

# The Earned Income Tax Credit program is out of hand

By JAMES K. GLASSMAN

The road to a $5 trillion national debt is paved with good intentions.

Look at the Earned Income Tax Credit (EITC). Launched by Gerald Ford, lauded by Ronald Reagan, expanded by George Bush and Bill Clinton, it's based on welfare principles that even a Republican (or a professed New Democrat) can love. The only problem is that, like many other good ideas in Washington, it's gotten completely out of hand.

Currently, the EITC is the fastest-growing program in the federal budget. It will cost the Treasury $24 billion this year, up from less than $2 billion 10 years ago.

In their giant reconciliation bill — the final budget measure of the year — Republicans are trying to restrain this growth. Under the Senate version, EITC costs will rise to $32 billion in 2002. In the budget language of Washington, that's a cut. In any other language it's an increase -- although not so large as projected under the current law, which has costs rising to $36 billion by 2002.

The EITC is a sort of negative income tax. If you fall into a certain earnings bracket, you don't pay the government; the government pays you.

The idea of the EITC is to put more money in the pockets of low-income working families. If you don't work, you don't qualify. Since the benefits are paid in cash and the rules are simple, the Internal Revenue Service can administer the EITC easily and cheaply.

Believers in the free market like the notion that the EITC doesn't force recipients to use funds for a particular purpose like other federal programs (housing, food stamps). Instead, it gives them money and lets them make their own choices.

The EITC is not only the fastest-growing entitlement program, it's the broadest. In 1986 some 7 million families were covered by the EITC, and the average outlay by the government was $281. This year 18 million families are covered at an average of $1,265. In 1986 the maximum credit a tax

payer could receive was $550; today, it's $3,111.

In Mississippi, a whopping 39 percent of families receive the EITC; in Texas, 26 percent; California, 22 percent. With this kind of penetration, the EITC follows a welfare tradition invented by Franklin Roosevelt: To keep a program alive, make sure money flows not just to the poor but to the middle class. That's been the key to success for Social Security, Medicare, student loans and farm subsidies.

The EITC was begun as a modest program to help offset the burden of payroll taxes on the poor and, through its unique structure, to encourage them to work more. But the philosophy soon became: "Hey, if a little bit is good, then more is better," says Bruce Bartlett, an economist who served in the Bush Treasury Department.

Today, the EITC is enjoyed by families making as much as $26,672 a year, and that doesn't include outside income. Under the tax law that President Clinton promoted and signed two years ago, by 2002 families making

$34,612 will qualify for EITC benefits. The Senate wants to scale that figure back to $30,200 — which seems pretty sensible for a government that already owes its creditors $4.9 trillion.

At its core, the EITC is a massive income-transfer scheme. New IRS figures show that in 1993 the top 5 percent of American earners paid 47 percent of the federal income taxes, up from 37 percent in 1981. Meanwhile, the bottom 50 percent of earners — thanks in large measure to the EITC — paid 5 percent of the taxes.

The EITC, in other words, has created a veritable tax holiday for about half the families in America.

Many would say that's fair. But there's another question raised by the EITC: Does it really encourage work? There's doubt.

For 1996, families with two or more children will earn credits of 40 percent of their income until they reach earnings of $8,910 annually. Then, they max out at a credit (in nearly all cases, a cash payment) of $3,564. So far, so good. Clearly, there's a big incen

tive to work, since a dollar paid on the job becomes $1.40 in the pocket (minus modest payroll taxes).

If you earn between $8,910 and $11,630, you still receive the maximum credit. Then the disincentive begins — you start losing 21 cents of credits for every additional dollar you earn. When your income reaches $28,533, your credits hit zero.

Again, this sounds fair. But the problem is that the EITC forces lower-income Americans to face marginal tax rates that are higher than those faced by the richest Americans.

As Bartlett wrote recently in a brief for the National Center for Policy Analysis: "Families with incomes between $11,000 and $26,000 are being taxed at the rate of 60 percent on each additional dollar earned. ... This total tax rate includes federal, state and local taxes plus the reduction in the EITC."

And these high marginal taxes definitely discourage work. Economist Edgar Browning of Texas A&M reported in the National

Tax Journal that nearly half of all families receiving the EITC had less income than they would have had without the tax credit-because the credit enticed them to work less. And a University of Wisconsin study found that "on balance the EITC reduces the total hours worked."

Is there a solution to the EITC conundrum? One answer is to remove the phase-out of benefits: Simply give all taxpayers an extra 40 percent credit for the first $10,000 or so of income. But that would be hugely expensive. Another answer is to kill the EITC entirely. But that would be politically impossible.

The third course is to try to restrain a program gone bonkers. That's what the Republicans are doing. At the same time, however, they should admit that the EITC isn't quite so glorious as they once thought. Maybe luring people out of poverty is something that government just can't do.

**▶ Write to James K. Glassman at the Washington Post, 1150 15th St., N.W., Washington, DC 20071.**

# From Capitol Hill to home, the generations divide

By CHARLOTTE GRIMES

In a recent joint House and Senate session of poignant reminiscences and stirring patriotism, Democrats and Republicans alike paid tribute to the military veterans and home front supporters who won World War II.

Not far away in another House office building, police hauled away a rowdy few of that generation from a congressional hearing room. They'd come to protest a Republican overhaul of Medicare.

Even as I work to sort out what the debate over Medicare means for the nation — I keep thinking of someone else.

My father is coming for a visit. .At 72, he is . deeply rooted to his small farm. So I am dividing my thoughts between keeping up with Congress and getting ready for my father's arrival.

Time to call the airline. Confirm his flight. Be sure he's got a guide through unfamiliar, chaotic Atlanta airport to change planes.

"I'm going to fly one more time in my life." As an 18-year old, my father flew a lot: 72 bombing missions over Europe in World War II in a B-26 Marauder. He has not had a taste for flying since.

Medicare statistics are mind-boggling and are ammunition for both parties and every faction.

Medicare's costs are now about 11 percent of the federal budget. Its expenses are rising at about 10 percent a year. Without the additional money, its hospital trust fund will go broke in seven years. By then Medicare would account for 15 percent of the budget.

To control that growth — and remake the program the Republicans have proposed saving $270 billion from Medicare's expected spending over the next seven years. The savings will also help balance the budget. Democrats charge, and Republicans deny, that it will also help pay for a $245 billion tax cut they've proposed for wealthier Americans.

The argument has touched something basic in the country's conscience: What is our responsibility to different generations? Today's Medicare beneficiaries will get the equivalent (for an average working couple) of $117,000 more in benefits than they paid in taxes. Today's children will pay just about that much each as their share of the national debt.

To pay for us baby boomers, more and more analysts agree that Medicare will have to change fundamentally. But as we try to sort out what's going on in Congress, it might be worth remembering: Sometimes a generation pays its share in more than dollars.

At the Smithsonian Air & Space Museum, there's the fuselage of a B-26, "Flak Bait." I've visited it often, trying to imagine myself in my father's place, edging up into his little glass turret to face the enemy fighters.

In a strange way, this become the heart of my father's visit, our reaching out to each other through the shell of a World War II plane. His memories. My search to understand them. It wasn't an easy moment for either of us. But it was a priceless one.

From Capitol Hill to home, the generations divide

**▶ Write to Charlotte Grimes at the St. Louis Post-Dispatch, 900 N. Tucker Blvd.,St. Louis MO 63101.**

## Rights: Not enough progress made

FROM PAGE C1

right thing. It was invented as a means of implementing what was, for a very brief moment, something approximating a national consensus that opening all of this country's manifold opportunities to all of its citizens was the proper business of the nation. What no one understood is that social change takes a long time and is disruptive not merely for those who may have to sacrifice to achieve it but also for those whom it is meant to benefit. We didn't realize that we were in for the long haul; as soon as we figured that out, as soon as we saw just how hard it was going to be, we pulled the emergency brake and jumped off the train.

That we had so little appetite or patience for the task is not, as some surely would have us believe, a peculiarly American trait. It is true that we undergo more interracial tension than any other country, but that is because no country has opened its borders so widely as we have to people of all races and origins. The bad temper many of us display when confronted with the demands of people who aren't exactly like us is a human, not an American, trait.

What a pity it is, though, to see our baser instincts once again triumphing over our better ones. A desegregated society, from which we seek to retreat not merely in our schools but in other parts of our common lives, was a noble goal, and the commitment with which we briefly pursued it was admirable, bordering on amazing. That era, we now can see, was the last spasm of national optimism we are likely to see for some time. As Applebome writes, "... one thing that critics and proponents (of busing) agree on is that the activism of the '60s and '70s to foster desegregation, whether from the courts or the federal government, was (an) aberration, with few parallels in the nation's experience, that is unlikely to resurface any time soon."

The end of the civil rights era brings with it, I think, a general if unarticulated sense of disappointment and failure. This is felt not merely within the black middle class, which has painful experience of how far short of expectations its great "success" has fallen, and the black underclass, the disillusionment of which may well be beyond remedy. It is also felt among white Americans, and by no means only among those most committed to integration. Many Americans were reluctant converts to civil rights, but supported it out of a belief that it was right to do so and that the nation would be better for it. This support has waned if not vanished altogether, and with it has faded a common purpose that will not easily, if ever in the foreseeable future, be retrieved.

It was common, especially among the tough-guy illuminati, to mock Jimmy Carter when he talked about "malaise," and to make sport now of Bill Clinton as he mentions — and then, characteristically, beats a cowardly retreat from — a national "funk." In fact both analyses, if they can claim to be that, are accurate. The American people have little apparent commitment to the national good and, for that matter, little apparent conviction that any good is within reach. Our failure in Vietnam has a lot to do with this, as do the massive hangover we're suffering from the most ill-advised excesses of the 1960s and the decline of community spirit in the age of electronic home entertainment and communications. But so too, surely, does the end of the civil rights era. Much progress was made as a result of it, but not enough. We turned out to be nowhere as good as, for a fleeting moment, we thought we were. Not even Dr. Pangloss would regard that as cause for cheer.

## Prisons: Better treatment needed

FROM PAGE C1

weaknesses of will and criminal activity to be met with increasingly punitive and costly (and often ineffective and counterproductive) law enforcement efforts.

The current and coming advances in this kind of medical research contain tremendous implications for correctional policy. As medical treatments are developed to more sophisticated and effective levels, and as they become available for other popular illegal drugs such as cocaine and crack, the state and national focus on criminalization of drug use, interdiction, law enforcement, and incarceration of drug offenders needs complete reassessment and redirection. These medical advances call for a new mindset founded on the promise of a real solution to one of the most intractable problems of our times. Based on this promise and the commitment we have to finding cures for all diseases, we need to hasten these medical remedies by diverting some of the millions of tax dollars we now spend on prisons to treatment, and the development of the medications that enhance the effectiveness of treatment. In the past, we have sought and found ways to prevent and cure smallpox, measles, tuberculosis, polio and many other diseases. We are now seeking them for cancer, heart disease and AIDS. The diseases of alcoholism and addiction call for no less.

With redirection to this profoundly promising perspective, a cure for alcoholism and addictions can, and will be, found in our lifetimes. We can make Delawareans secure and safe in their homes and on the streets without making it the first state in incarceration rates.

## Giving highway funds to Amtrak is unfair

**Rebuttal**
William D. Fay

How long, oh Lord, how long will it take for the beneficiaries of government subsidies to under stand that the federal budget will never be balanced until they stop holding their hand out. In his Sept. 17 column, Amtrak rider John H. Taylor Jr. asserts, "Congress should get on track with Amtrak funding."

He apparently believes that the massive subsidies Amtrak needs should come from Delaware highway users, whose roads are already strained and congested due to underinvestment. The following Delaware examples are among the 240,000 miles of American highways in poor or mediocre condition, which the Highway Trust Fund is supposed to repair:

Concord Pike; U.S. 13 Relief Route from Dover to north of the C&D Canal; Delaware 1; Delaware 9; U.S. 113; U.S. 13; and all Delaware roads with potholes, which are hazards.

The Highway Trust was created to collect money from highway users to improve the safety of highways and bridges. It's a simple concept: Those who use the roads, pay for them; those who don't, don't. It's a "user pays, user benefits" concept that would be violated by giving money to Amtrak. Highway users deserve better. They need every penny of the Highway Trust Fund to fix highways and bridges.

If that makes the Highway Users Federation "the NRA of the truck and automobile industries," that pejorative term should be applied to any group that stands for principle and tries to keep its members safe.

Amtrak — with its bloated labor and maintenance costs — may be the $700 toilet seat of the transportation sector.

If Amtrak enjoys so much support in Delaware, why did a Delaware Department of Transportation study conducted by the University of Delaware show that Delawareans will continue to drive until mass transit can get them from doorstep to office more quickly? Quoting the study: "Most respondents are probably saying (mass transit) is a good idea in principle, but let someone else ride!"

If riding Amtrak is a "relaxing, cost-effective" ride for Mr. Taylor, perhaps we should ask him to stop looking for government handouts (more than $50 per Amtrak ride) and to pay the cost of the ride himself.

**▶ William D. Fay is president of the Highway Users Federation in Washington,D.C.**

*Figure 13–3 All stories are horizontally shaped on this page. Note that the top two stories are a full six-column width, while bottom stories are each three columns wide. Bottom stories do not use boldface type.*

8A — The Macon Telegraph — Monday, Nov. 6, 1995

# Opinion

The Macon Telegraph

Macon was chartered in 1823
*The Macon Telegraph*
was established in 1826

KNIGHT RIDDER

**Carol Hudler**
President & Publisher

**Richard D. Thomas**
Editor & Vice President

**Ron Woodgeard**
Managing Editor/News

**Ed Corson**
Editorial Page Editor

**Barbara Stinson**
Managing Editor/Features

**R.L. Day**
Associate Editor

## Our views

Unsigned editorials in larger type below, are generally a consensus of The Macon Telegraph editorial board, although the publisher and the editor have the final say. The opinions expressed in syndicated and local columns, cartoons and letters to the editor are expressly those of their creators.

## Rabin murder must increase peace resolve

Israeli Prime Minister Yitzhak Rabin, 73, martyr for peace, was buried in Jerusalem shortly after 7 a.m., our time. His legacy is the Middle East peace process for which this tough warrior-statesman risked and gave his life. It must be preserved and enriched by Israeli, Arab and all people of good will.

More than two score heads of state and governments attended; so did a former president and two congressional leaders from Georgia. Even the rulers of Egypt, Jordan and other Arab countries came — once his foes, now his mourners who know his work must be carried on.

He was the first prime minister to be assassinated in the tumultuous and threatened 47 years of Israel's existence. Some Israelis compare the shock of his murder to what our nation experienced with its first presidential assassination. President Abraham Lincoln also was killed by a countryman for ideological reasons as he tried to lead a divided nation through traumatic change.

Yitzhak Rabin won fame as the military chief of staff who conducted the brilliant Six-Day War in 1967. In it, Israel seized the West Bank and other territories. And in the 1980s, as defense minister, he fought the Arab uprisings there with an iron fist.

### Toughness made peace moves palatable

Only he, with such hard-line security credentials, could have persuaded his countryman to accept a deal with their nation's violent foe, Yasser Arafat. The secret negotiations leading to the 1993 peace agreement between Israel and the Palestine Liberation Organization won Rabin a share in the 1994 Nobel Peace Prize. (PLO leader Arafat and Israel's Foreign Minister Shimon Peres, who succeeds him as prime minister, shared the prize.)

It's deeply ironic that what has brought extremist Jewish residents and their sympathizers to fever pitch was the follow-up agreement Sept. 28 setting up the mechanism for creating Palestinian self-rule in the same West Bank Rabin had won 27 years before. Breathing that miasma of hatred, a 25-year-old Israeli law student could murder Rabin without remorse at a peace rally, "on orders from God."

We can learn from how Rabin changed over the years: In 1967 he was sure that seizing the disputed territories would provide the tiny nation surrounded by Arab foes with a security buffer. By the 1990s he realized that land had not brought security after all.

We can learn from how Rabin changed toward Arafat: His body language at first revealed shock, horror and disgust at the idea of shaking Arafat's hand when the PLO leader extended it to him (with President Bill Clinton's encouragement) in the Sept. 13, 1994, agreement ceremony. But two years of negotiations with Arafat led to a sort of bonding in mutual respect and understanding. Arafat appeared genuinely shaken and saddened by the death of his one-time blood foe.

### 'Talking' can bring political miracles too

When foes talk instead of fight, human miracles can happen. (Remember the Reagan-Gorbachev talks when the Soviet Union was still the Evil Empire?)

Some say it will take a political miracle to maintain the progress of peace with the Palestinians and get the talks with Syria moving. Shimon Peres is equally committed to the peace process, but is thought too ready to make concessions; he doesn't inspire such trust as his predecessor.

But there is good reason to hope that revulsion against this murder will extend to the violent rhetoric and intransigence of Rabin's opponents and cause a backlash for peace.

Israel is a democracy and a proud people. The majority, who want peace, will not allow the anger and violence of a few compatriots to scare them from pursuing peace any more than they'd buckle under terrorism from outside.

As President Clinton said, "Peace must be, and peace will be, Prime Minister Rabin's lasting legacy."

Pray that costly legacy will not be squandered.



## Why beat up on television talk show 'sleaze'?

**Carl Rowan**

WASHINGTON — What's wrong with me? I can't seem to get into a spasm of cultural and moral superiority that would allow me to join in the attacks on the talk shows of Sally and Ricki and Montel and the like.

Former Education Secretary Bill Bennett, Sen. Joseph Lieberman, D-Conn., and others are trying to shame TV hosts and stations into abandoning what they call "sleazy" shows that feature and exploit "freaks."

I admit to being amazed that so many Americans are willing to appear on these "talk" shows to spill their guts about intimate, abnormal, often painful, aspects of their lives. But my dictionary says a "freak" is some kind of monstrosity, a word I would not apply to a young woman who complains that her mother stole her boyfriend and gets both Mom and the boyfriend to talk about it on TV.

Am I supposed to look down on a battered woman who tells of the horrors of her marriage — or "live-in" relationship — on a television talk show? The Justice Department says there are 572,000 cases of battered females reported every year and probably a larger number of unreported cases. It is not sleazy to air real cases of spousal abuse on television just because some "high-class" women are made uncomfortable when "low-class" women tell the ugly, self-humiliating truth about a very serious problem.

Most talk show producers find amazingly titillating and provocative ways to deal with issues such as incest, marital infidelity, runaway teen-agers, teen-age pregnancy, venereal diseases, homosexuality, alcohol and drug addictions, interracial sex and, of course, fornication in all its permutations. But these are all compelling problems and widely felt issues in America. I cannot say honestly that the talk shows deal with them in worse ways than do the afternoon soap operas or the prime-time TV and Hollywood movie fare.

The talk show people merely TALK trash; the soaps simulate lurid, dirty behavior in living color; and HBO, Showtime, many other networks and Hollywood give us graphic, steamy versions of sexual intercourse as early as 8 p.m., complete with gratuitous breasts-and-buttocks nudity.

So Bennett and Lieberman declare the mere trash talkers to be the great villains? Please, repossess my idiot box! Now!!

Bennett and Co., like all the morality cops of our time, swear that they don't advocate censorship — just embarrassment. But many of their conservative disciples are screaming for boycotts and other pressures on advertisers who sponsor Donahue, Jenny, Leeza, Geraldo and others. Such boycotts are a form of censorship.

I just can't reconcile myself with the arrogance of saying to 10 million Americans, "You slobs who regularly watch these talk shows are a cultural and moral curse on the rest of America. Now shape up!!!"

There's a measure of self-preservation in my refusal to join Bennett and Lieberman. I've been on a political talk show, "Inside Washington," for more than 30 years. During such times as the Watergate scandal, the Iranian hostage crisis, the S&L thievery, and this very day, some offended pols have regarded OUR talk show as blasphemous of presidents and congressional leaders and thus inimical to "a stable government."

I figure that to protect my right to free speech I'd better also protect those of Maury Povich, and even the rights of Jerry Springer.

*Carl Rowan is a syndicated columnist and a former head of the U.S. Information Agency.*

## Unless precautions are taken, Bosnia could happen elsewhere

**Jeane Kirkpatrick**

Have there ever been as many defenseless civilians slaughtered in the full view of as many well-armed bystanders as in Bosnia?

Has there ever been as much indifference, incompetence and complicity on the part of as many governments?

Has there ever been as much money spent and as many forces mobilized, with as little effect, as for peacekeeping in Bosnia?

It is nearly unbearable to read again descriptions of the murder of up to 10,000 Bosnian men and boys following the fall of Srebrenica to a Serbian attack of last July 9-11. But how interesting that three-and-a-half months after these grisly murders, leading American newspapers are again providing eyewitness accounts of U.N. peacekeepers, Bosnian government officials, survivors and others of the Serb attack on the defenseless U.N. "safe haven" of Srebrenica and its swollen population of refugees.

Perhaps the reason is that this time the accounts are documented by aerial photographs of mass murder just released by the U.S. government.

It is not clear why the government decided to make these available and why leading American newspapers decided to feature this documentation of Serb war crimes now, on the eve of the "peace talks" in Dayton. But they did. The new accounts contain some new details of the massacres, including information on the direct participation of military forces of the government of Serbia, some new details on when and how the U.S. government learned that a massacre was under way, and what the U.S. officials did and did not do about these events.

But the essential facts had been reported in Europe and the United States as the massacres occurred last July.

It may be that the U.S. government desired to release the story before the Dutch government issued its report on the behavior of Dutch peacekeeping forces in Srebrenica. Perhaps the Clinton administration preferred to lay out its version of events before the debate began in Congress on providing 20,000 U.S. troops for peacekeeping in Bosnia.

In any case, the new accounts of Serb massacres and international inaction that have been spread across the front pages of the U.S. press recall other great crimes of this violent century and raise a similar question: Why didn't the American government, the United Nations or the Allies do something to stop the murders?

For example, we learn from the new accounts that U.S. Ambassador to the United Nations Madeleine Albright was informed by Bosnia's Ambassador Mohammed Sacirbey on July 11 that Bosnian men and boys who had been forcibly removed from Srebrenica were being slaughtered in cold blood, and also that one day later, July 12, U.S. reconnaissance planes photographed these atrocities.

But the United States did not bring the matter before the U.N. Security Council for one month, until Aug. 10. Why did the U.S. government wait so long to seek action on this most urgent matter?

Can we learn anything that will be useful in avoiding future such catastrophes from the study of unrealistic rules of engagement, cumbersome organization and the perverse priorities that kept NATO planes circling above Srebrenica while civilians were slaughtered below?

Perhaps.

We can face the fact that this "peacekeeping" operation most assuredly did not achieve its goals. It did not "mitigate and reduce the impact of the violence on innocent civilians," which Bill Perry described as a chief concern. It did not prevent a humanitarian catastrophe.

U.N. peacekeeping and the Clinton administration failed in their goals. I believe they failed because they valued their multilateral tools more than the human lives the tools should have protected.

Peacekeeping has value only as it is able to keep peace. NATO has value only as it is able to preserve freedom and peace and repel aggression. Both these require recognizing when aggression has occurred. Nothing useful can be accomplished in Bosnia by 20,000 U.S. troops or 100,000 NATO forces until the governments are able to make the essential moral and political distinctions between the perpetrators and victims of violence and define the task as protecting victims from aggressors.

*Jeane Kirkpatrick is a former U.S. ambassador to the United Nations.*

## Women's rights remain elusive

**Larry Fennelly**

When one looks at the enormous progress that women have made during the last 20 years toward equal opportunity in the workplace, it's easy to conclude that the opponents of the Equal Rights Amendment have been vindicated by subsequent history. Or so I was thinking recently when, searching for my autumn-weight exercise clothes, I encountered a faded "200 Years Is Long Enough" T-shirt, a veteran of the rally at the state Capitol in 1976 by supporters of the notion that both male and female ought to enjoy the protection of the law in equal measure.

Passed by Congress in 1972, the ERA was ratified by only 35 of the required 38 states. Opponents whispered that it would destroy the family, that separate rest rooms would be declared unlawful. Publicly, they argued that women's rights were already fully protected.

How many people still believe that? My old T-shirts are stored in a spare drawer in my daughter's room, next to the drawer that houses the shirt I bought her for Christmas last year. With the caption "Future President," it seemed like a reasonable gift for someone intelligent, talented and altruistic. Since then has appeared the news report of the Wal-Mart that withdrew a T-shirt proclaiming, "Some Day a Woman Will Be President." The corporate spokesperson, a woman no less, was quoted as saying, "The shirt goes against Wal-Mart's family values."

While I read that the store experienced a change of heart, the incident is a vivid reminder that it was a relatively recent development that women were allowed to vote. When the 19th Amendment was ratified in 1920, 14 states had already taken action on their own, but virtually all of the Southern states — including Georgia — were "men only." Part of the problem that Southern states had with women's suffragists was that many of them, led by Quakers such as Susan B. Anthony, had been opposed to slavery.

Does this seem long ago? By 1920 the airplane and the radio were on the scene. Yet it wasn't until the 1960s that public colleges began to welcome female students, although they were subject to stringent regulations — for example, rigid curfew, no pants, no shorts. The strange name for the new arrivals still endures — coeds.

While in 1995 few debate women's rights to vote and to have full access to higher education, recent developments have put renewed focus on the status of women and their so-called gains. Seventy-five years after they won the right to vote, women account for barely 20 percent of state legislators. That's the national average. Women in Washington state hold 39 percent of the seats, women in Alabama less than 4 percent. Georgia's 18.2 percent is near the top among Southern states.

Indeed, those states traditionally viewed as the Bible Belt are conspicuous by their low numbers of females in the state house. It may not be coincidence that these are the states where the recent outcry for "family values" has found the warmest reception. While many parents have embraced family values in the belief that they are opposing sex and violence in movies, television and popular music, instead they have found that the "family" referred to is the old female-submissive variety prevalent at the outset of the women's suffrage movement — when women were seen as chattel and the "age of consent" was 10.

The current landscape presents a chilling view to those who believe that God created male and female to be equal partners — not merely in the workplace but in life. It's not just family values; it's not just the misogynistic lyrics of contemporary music or the elevation to national hero status of a rapist or, more recently, a habitual wife-beater. Nor is it the rise of all-male movements such as Promise Keepers or the Million Man March, although viewed in the aggregate, all of these give rise to grave concern.

What turns these shadows into full-fledged horrors is the budget legislation currently on its way to passage in Washington. Our male-dominated Congress — with its long tradition of philanderers — seems to have a Contract on Women, such is its focus on female behaviors. What about all of the men who don't know who or where their children are?

When is Congress going to ask men to demonstrate some family values? How long before women awake to the realization that, while they don't have the ERA, they do have the 19th Amendment and the power to send someone to ask that question in Washington, if necessary — Wal-Mart not withstanding — all the way to the White House?

*Larry Fennelly is chair of the department of developmental studies at Macon College and a reviewer for The Macon Telegraph.*

***Figure 13–4 Another horizontally designed page, and one that is very readable. The page represents a change of pace from other vertical or mixed story pages.***

The Macon Telegraph.

# LAYOUT AND DESIGN GOALS FOR AN EDITOR

## The Objectives of Newspaper Layout and Design

Five major objectives should guide an editor in working effectively with newspaper layout and design:

- Layout should organize the news.
- Layout should create an attractive page.
- Layout connotations should be consistent with the nature of stories.
- Layout should be interesting and dramatic, if possible.
- Layout should reflect the current scene.

### LAYOUT SHOULD REFLECT THE ORGANIZATION OF NEWS

Good layout may be defined as a plan for bringing order to an entire newspaper or a specific page. Every layout should show how the news has been organized for the day so that readers can know which stories are most and least important. Those who spend a great deal of time reading will have to spend less time searching for stories of interest if the organization of the paper is clear and simple.

One element of organization is story placement throughout the newspaper. The layout staff should place stories adjacent to other, similar stories in certain sections of the paper.

Organization also covers each individual page, where stories often are graded from the top down in importance. This does not mean that important stories cannot be placed at the bottom, because occasionally, for variety's sake, an important story is placed there. But layout personnel have many ways of telling readers which stories are most important: by placing them at or near the top of a page, by placing them in boxes and by setting their headlines in large type.

Another dimension of this same objective is the need to help readers follow a story when it is continued from column to column and page to page. Under intense pressure to get the paper out, the layout personnel may place the last six inches of a story in an obscure column or position that does not help the reader. Good layout avoids this.

### LAYOUT SHOULD MAKE A PAGE APPEAR ATTRACTIVE

Because attractiveness is a major cultural value, good layout should be attractive to most readers. We all like things that look nice, and we use this principle in selecting beautiful clothes, furniture and cosmetics that enhance personal appearance.

An attractive layout invites readers to come into a page and read the news. An unattractive layout could do the opposite.

Readers often skim a page before reading it. In other words, they may be looking for an interesting story, and a skim may save time in finding it. But pages that are *too dark,* or *too light,* may not be read. Of course, scintillating news stories of great importance may be strong enough to motivate a reader to read a page regardless of the impediments. Readers devoted to poorly designed newspapers often defend their paper's design with great fervor because they have become acclimated to it. They don't recognize poor design, and they perceive only their favorite paper. Even in this situation, readers will often approve of a new and easy-to-read layout if they can be persuaded that it is indeed better than the old one.

### LAYOUT CONNOTATIONS CAN COMPLEMENT STORIES

Occasionally, a layout can match the nature of a story. For example, various artistic devices can be used to surround stories for Thanksgiving Day, Christmas or dramatic political or sporting events.

### STYLING DEVICES CAN CREATE INTERESTING OR DRAMATIC LAYOUTS

To avoid creating a monotonous page, a layout editor can vary story shapes or lay stories out into strongly vertical or horizontal shapes. Stories can be boxed, run over colored tints or run with multiple photographs. All these are styling devices to avoid boredom. After many years of reading a newspaper, readers may simply wish for some variety, even though they may not write to the editor. Slight layout or dramatic variations in design may be just what are needed to avoid dull-looking pages.

### LAYOUT SHOULD REFLECT THE CONTEMPORARY SCENE

Some of the great graphic designers of the 20th century have noted that design should reflect contemporary culture. For newspaper design, this is important because news is the essence of the contemporary scene. It seems illogical to place contemporary news in an old-fashioned format.

One may then ask, "Are there various degrees of contemporary design, and if so, which is the best?" The answer is not easy to determine, but most designers would agree that some degree of contemporary design belongs in a newspaper layout. Nevertheless, there are always some readers, perhaps a small number, who dislike modernism in almost everything and possibly in design as well.

## The Search for the "Best" Design

No single design can be called the "best" in the United States. A study of a large sample of U.S. newspapers would show that many different styles are in use, with some common characteristics among all. *USA Today* is widely imitated but not necessarily because it is best. It simply calls attention to itself, a quality that many editors want.

Because there is no "best" design, there is room for innovation. The main value of knowing that there is no best design is that leaders in the business ought to have a certain amount of tolerance in judging the work of layout editors. In the old days of makeup, strict rules had to be followed slavishly. Any deviation from those rules was judged to be bad design. More tolerance would allow more experimentation in design. However, discipline is needed—not everything goes.

One consequence of more tolerance in design is a pressing need for design research. Artists typically dislike the regimen of research imposed on an activity like art because they assume that it will restrict their freedom. But when design is intended for a specific purpose, such as improving the readability of the news, research is necessary to overcome the habit of making quick judgments that "this layout is good and the next one is bad." Research can provide more definitive answers to the question, What is good design? At present, many design judgments are totally subjective.

## Graphic Responsibilities of Layout Editors

Although the objectives of layout and design are quite clear, they do not provide enough details about what editors face when laying out a page. A layout editor has the following responsibilities:

- Telling readers which stories are most important.
- Helping readers follow stories from column to column.
- Helping readers know on which page a story is continued and where the continuation appears.
- Helping readers know which other stories are related to the one they are reading.
- Helping readers know which pictures are related to which stories.
- Helping readers know where stories end and new ones begin.
- Keeping readers from losing interest in long stories.
- Setting type large enough, in the best typeface, in an optimum line width and with optimum line spacing to maximize readability.
- Making the entire page interesting—not just the top.
- Helping readers with limited time find the stories they want.

## PRINCIPLES OF ARTISTIC DESIGN APPLIED TO NEWSPAPERS

The application of artistic design principles can help layout editors achieve their objectives and carry out their responsibilities. The newspaper is a graphic art form, using words, pictures, color, lines and masses subject to the same principles of artistic design as other graphic art forms. The principles most applicable to newspapers are known as balance, contrast and unity.

### Balance: A Means of Making the Page Appear Restful

Balance means equilibrium. It means that a page should not be overwhelmingly heavy in one section or extremely light in another. An unbalanced page may give readers a vague feeling of uneasiness because of the concentration of weight in only one or two sections of the page. As mentioned earlier, most readers do not know whether a page is balanced or unbalanced. They are not artists and do not care about the principles of artistic design. Yet they often know that a certain page "feels" better to read than do other pages. Top-heaviness is the most common form of imbalance in newspaper design, caused by placing large and bold headlines at the top while using almost insignificantly light headlines at the bottom. Another cause of imbalance is the practice of placing a large, dark picture at the top without having one of similar size or weight at the bottom. As a result, readers' eyes tend to gravitate toward the bolder sections of the page and away from the lighter portions. Assuming that every element on a page has value, an unbalanced page is more difficult to read than a balanced page.

Balance in newspaper design is achieved by visually weighing one element on a page with another on the opposite side of the page, using the optical center as a fulcrum. The optical center is a point where most persons think the true mathematical center is located. It is a little above and to the left of the actual mathematical center (see Figure 13–5). This practice does not lead to precise balancing, but there is no need for that degree of precision. All that is required is a feeling of equilibrium on a page, not precise mathematical weighing.

Which elements need balancing? Any element on a page that has visual weight should be balanced. To determine which elements have visual weight, squint at a page and notice that much of the printed material disappears. What remains are pictures, headlines and black type rules of any kind. Although body type does have some weight, it isn't significant enough for consideration in visual weighing. The goal is to distribute prominently weighted objects pleasantly on the page.

Balance is most often done between top and bottom rather than side to side. The process is similar to balancing a heavy person with a light person on a seesaw.

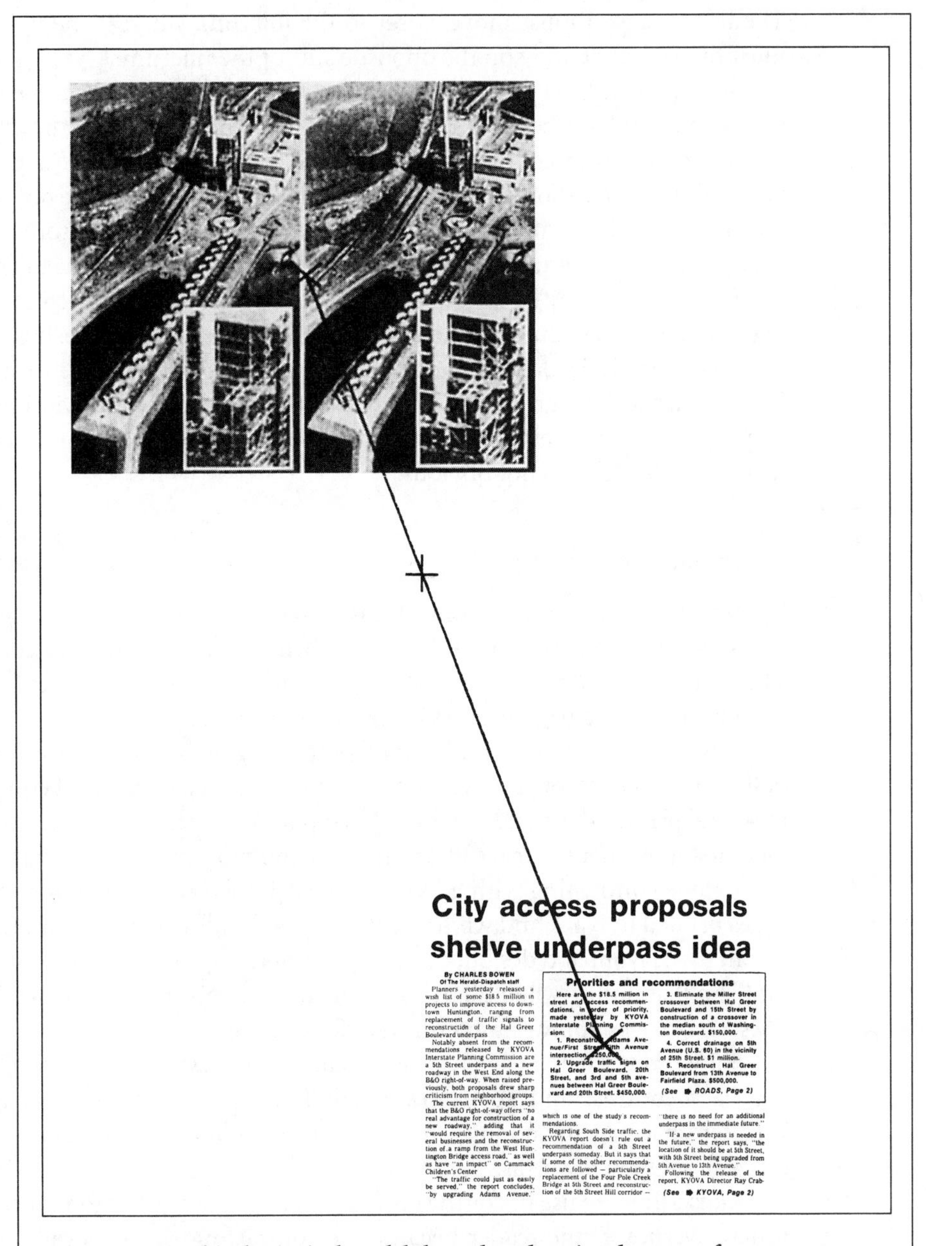

# City access proposals shelve underpass idea

By CHARLES BOWEN
Of The Herald-Dispatch staff

Planners yesterday released a wish list of some $18.5 million in projects to improve access to downtown Huntington, ranging from replacement of traffic signals to reconstruction of the Hal Greer Boulevard underpass.

Notably absent from the recommendations released by KYOVA Interstate Planning Commission are a 5th Street underpass and a new roadway in the West End along the B&O right-of-way. When raised previously, both proposals drew sharp criticism from neighborhood groups.

The current KYOVA report says that the B&O right-of-way offers "no real advantage for construction of a new roadway," adding that it "would require the removal of several businesses and the reconstruction of a ramp from the West Huntington Bridge access road," as well as have "an impact" on Cammack Children's Center.

"The traffic could just as easily be served," the report concludes, "by upgrading Adams Avenue," which is one of the study's recommendations.

Regarding South Side traffic, the KYOVA report doesn't rule out a recommendation of a 5th Street underpass someday. But it says that if some of the other recommendations are followed — particularly a replacement of the Four Pole Creek Bridge at 5th Street and reconstruction of the 5th Street Hill corridor — "there is no need for an additional underpass in the immediate future."

"If a new underpass is needed in the future," the report says, "the location of it should be at 5th Street, with 5th Street being upgraded from 5th Avenue to 13th Avenue."

Following the release of the report, KYOVA Director Ray Crab-

*(See ➧ KYOVA, Page 2)*

**Priorities and recommendations**

**Here are the $18.5 million in street and access recommendations, in order of priority, made yesterday by KYOVA Interstate Planning Commission:**

**1. Reconstruct Adams Avenue/First Street/Fifth Avenue intersection, $250,000.**

**2. Upgrade traffic signs on Hal Greer Boulevard, 20th Street, and 3rd and 5th avenues between Hal Greer Boulevard and 20th Street. $450,000.**

**3. Eliminate the Miller Street crossover between Hal Greer Boulevard and 15th Street by construction of a crossover in the median south of Washington Boulevard. $150,000.**

**4. Correct drainage on 5th Avenue (U.S. 60) in the vicinity of 25th Street. $1 million.**

**5. Reconstruct Hal Greer Boulevard from 13th Avenue to Fairfield Plaza. $500,000.**

*(See ➧ ROADS, Page 2)*

*Figure 13–5 This theoretical model shows how heavier elements of a page are balanced against lighter elements. The fulcrum is the optical center, a point where readers think the true center is located. Heavier elements are placed closer to the optical center, and lighter ones are more distant.*

The heavy person must move closer to the fulcrum, whereas the lighter person must move farther away on the opposite side of the fulcrum.

To implement the principle of balance, weigh the most outstanding elements, such as bold or large headlines, at the top of a page against similar headlines at the bottom. If the bottom of the page has no bold or large headline, the page is likely to be top-heavy. Plans should be made to include such headlines at the bottom. Follow the same procedure with pictures. A headline or picture at the bottom need not be as large or bold as one at the top because it is farther away from the fulcrum.

Page balance may be formal or informal. Formal balance is achieved by placing headlines and pictures of the same size on either side of a page. It is sometimes called symmetrical balance because one side of the page tends to mirror the other. But symmetrical design may be unbalanced from top to bottom. Most newspapers employ an informal balance from top to bottom. The feeling of equilibrium is there even though it is not obvious.

## Contrast

Contrast is the principle of using at least two or more elements on a page, each of which is dramatically different from the other. A light headline may contrast with a bold headline, a small picture with a larger one. Because one element is different from the other, the page appears lively and interesting.

Contrast, therefore, is a means of preventing artistic pieces from becoming dull. Almost all art forms have some contrast—especially musical compositions, theatrical plays and printed material. A symphony, for example, contrasts a fast and loud first movement with a soft and slow second movement. A play has a relatively quiet scene contrasting with a lively scene. A book or magazine may have most pages printed in black and white contrasting with full-color illustrations.

In page layout and design, contrast prevents a page from appearing too gray, a problem that occurs when there is too much body copy and too many light headlines. Gray pages appear uninviting and forbidding.

Sometimes when a page has been deliberately designed to feature balance, it may lack contrast and appear rather dull. The editor therefore may have to brighten that page by adding another picture and/or large, bolder headlines to bring about better contrast.

Indiscriminate use of contrast is undesirable. If a page has too much contrast, it may overpower the reader because the contrasting elements call attention to themselves and not to the page as a whole. The goal is to provide pleasant, not overpowering, contrast. To achieve this goal the layout person will have to develop a sense of good taste.

Contrast may be achieved in four general ways: by shape, size, weight and direction.

- Shape contrast may consist of a story set flush on both sides in opposition to another story set flush left, ragged right. Or an outline picture may be used with a rectangular picture.
- Size contrast may be achieved by using a large illustration on the same page as a smaller one, or large type with smaller type.
- Weight contrast may employ a picture that appears very black with a lighter picture, or a story set in boldface type contrasted with one set in lighter typefaces.
- Direction contrast could show vertically shaped stories contrasted with horizontally shaped stories.

These contrast alternatives are but a few of many that are possible on any given page. An objective of designing a page, however, is to achieve pleasant contrast.

### Unity

The principle of unity creates a single impression in a page design. Stories on a unified page each appear to contribute a significant share to the total page design. A page that does not have unity appears as a collection of stories, each fighting for the reader's attention to the detriment of a unified page appearance.

Lack of unity often results when stories are laid out from the top of the page downward. The layout editor is building a page piece by piece and cannot be sure how each story will contribute to the total page design until the layout is complete. At that point, however, the layout person may find that there is not enough time to shift stories around to achieve unity. The result is that readers may find it difficult to concentrate on any one part of a page because of too many centers of interest. A unified page, on the other hand, appears as if everything is in its correct position, and the page is therefore interesting.

The editor plans for a unified page by keeping the design of the entire page in mind at all times while working on any part of it. Each story, therefore, must be visually weighed against all other stories in terms of the probable appearance of the entire page. In page layouts, the editor may have to shift some stories around on the dummy until a satisfactory arrangement has been found. As with the other principles of artistic design, layout editors must develop an appreciation of unity through sensitivity to good design.

## THE MECHANICS OF PAGE LAYOUT

### Vertical and Horizontal Designs

A major consideration in layouts is the appearance of story shapes. In past years, story shapes were not a major consideration of layout editors. But as editors have sought ways of making pages more attractive, story shapes have become important. The selection of the most appropriate shapes involves a number of considerations. The first has to do with preventing a page from becoming one-directional. Too many vertically shaped stories, all leading the reader's eyes downward, make the page look old-fashioned and unattractive. Newspapers circa 1850 were all vertical, and vertical makeup is distinctly old-fashioned.

A simple design is preferable. If it isn't simple, it may compete with the message. Such a layout is distinguished by horizontally shaped stories that are continued into three or more adjacent columns. Although a story may continue into the adjacent column, this does not necessarily produce horizontal layout if the shape of the story is vertical. Another distinguishing feature of horizontal layout is that stories are squared off at the bottom. This means that the depth of each column where the story continues is the same. However, a page using horizontally shaped stories exclusively may be as monotonous as one with all vertically shaped stories. The best-looking pages have a mixture of shapes. (See Figure 13–6.)

## Avoiding Oddly Shaped Stories

In traditional layout many stories are not squared off and take odd shapes. Too many uneven wraps that look like inverted L's, such as those shown in Figure 13–7, will make the page appear to lack unity. One then would have a story wrapped underneath for three columns, presumably, and squared off.

Another kind of odd shape occurs when a story is continued to adjacent columns but each column depth containing the story is of a different length (Figure 13–8). Such shapes also tend to make the page look unattractive.

## Wraps

A major consideration in layout is the problem of what to do with stories that must be continued into adjacent columns because they will not fit in the space dummied. Should they be jumped or wrapped into the next right-hand column? Jumps are undesirable for reasons that will be discussed shortly. The best procedure is to wrap, or turn, a story underneath a headline or a two-column lead paragraph. If the wrap is under a headline, as shown in Figure 13–9, it should be very clear to the reader where the second column has been continued.

Some editors prefer to start a story with a two-column lead paragraph and wrap the second column under the lead. In such situations, two variations are commonly used: (1) wrapping the second column with a cutoff rule to separate the wrap from the lead or (2) wrapping the second column without a cutoff rule, as shown in Figure 13–10. The first technique is preferred. It is unfortunate that many

WN
TURNING POINT

Chicago Tribune, Sunday, August 20, 1995 Section 6 3 WOMANEWS

Susan Smith and her daughter: "I wanted to make a better life for Alicia and myself."

# Building a new life

## Prison term opens doors as mother helps inmates

By Candace O'Connor
Special to the Tribune

Night is always a lonely time in the women's prison at Chillicothe, Mo., but the bleakest of Susan Smith's 22-month stay for theft came on Christmas Eve 1989. That evening she called home to talk with her 8-year-old daughter, Alicia, a severe diabetic with a host of medical conditions.

"Alicia was in the hospital, and no one had told me," says Smith, a 45-year-old St. Louis resident who still sobs at the memory. "She had tried to commit suicide. She thought that if she died, they would let me come back."

Distraught at her mother's absence, Alicia had stopped taking insulin and other medications. Until she agreed to begin again, doctors barred all contact with Smith, who was in despair at the enforced silence.

The shock of this experience jolted Smith into some anguished soul-searching—and a two-part decision about her future.

A high school graduate who had never dreamed of college, Smith realized that her only chance for a better life was to get more education.

Looking at the misery around her, she also formed a new philosophy: to reach out to other people in trouble, especially prisoners.

"I was feeling like that's what I'm on this Earth for, to be a benefit to others," she says.

Four years after her release, she is only a few credits away from a bachelor's degree, and she directs a transportation service that helps St. Louis families visit their relatives who are inmates in Missouri's 12 prisons.

Smith has managed to save enough money to buy a bungalow, and Alicia is 14 and doing well in school. She has been declared a disabled child and eligible for Medicaid, which pays her hospital expenses. With threats of budget cuts in Washington, Smith worries that Alicia may again lose coverage.

After all, it was Alicia's medical bills that got Smith in trouble in the first place. When Alicia was born in 1980, she had a bronchial blockage, and a series of other medical problems soon developed.

"She's an acute asthmatic and allergic to everything," says Smith. "I mean when the allergist did tests, he said, 'Ms. Smith, did you ever think about moving to the North Pole?' "

Smith's job, as office manager for a St. Louis sewer-cleaning firm, did not include health insurance. Alicia's father, who had been with Smith for 13 years, left the family when Alicia was 5.

As hospital bills mounted, Smith says, "I started taking money from work, a little at a time, to pay the medical bills. And I didn't hide it—I always thought, 'She's going to get better, and I'm going to pay it back.' "

In 1988, Smith missed two weeks of work when Alicia was in the hospital. While she was gone, her company uncovered the missing funds—by now in the thousands of dollars—and pressed charges.

In August 1989, Smith was sent to Chillicothe for a maximum seven-year sentence, while her daughter stayed with Smith's mother and stepfather.

After the Christmastime incident with Alicia, Smith immediately began to turn her life around. In January 1990 she started taking vocational classes. With the help of Pell grants, which are no longer available for prisoners, she also took college-level courses, becoming the first member of her family to do so.

Her self-confidence slowly began to grow. Because of her medical experiences with Alicia, she was chosen as tutor for the prison's certified nursing assistant class.

To her astonishment, she became a leader among the inmates. Even total strangers came to her for advice.

"Other people said, 'Susan, why are you talking to them?' But I would say, 'Why shouldn't I? They're not doing anything to me, and I'm no better than they are. I'm in here too.' "

Today, her commitment to those prisoners remains strong as she arranges bus trips for families to visit inmates. When small children want to visit their mothers but don't have an adult to go with them, Smith volunteers her time to make the trip herself.

"Having been there, I know. The children need to see that their mom's all right, and Mom needs to see that her babies are all right."

Recently, Smith was offered a job as bookkeeper for a St. Louis company, a position that would have nearly doubled her salary.

"I did pray over that," she says. "The money would have come in handy. But I wouldn't have been happy knowing how much they need me here. This program has to go on."

Many prisoners, she says, never muster the strength to turn their lives around.

"The main reason why some do and some don't is the desire inside of you," she says. "You must have that desire and want to change."

In Smith's case, that incentive was Alicia, an only child with big eyes, dimples and an irresistible grin.

"She is the best thing that ever happened to me," Smith says. "I wanted to make a better life for Alicia and myself."

The encouragement she received in prison and afterwards from teachers, mentors and friends also stiffened her resolve to do the same for others.

"Maybe I can help someone else turn their life around," she says. "It's kind of like a repayment, you know?"

Ever since her early release in June 1991, Smith has proved her commitment to building a new life. The first weekend after her release, she searched the St. Louis want ads, clipping every conceivable job and arranging them geographically so she wouldn't waste gas.

One was a blind ad for a data entry job.

"I went there, and they were real busy," Smith says. "They asked if I could come back that evening. I'm just like 'Oh, yes,' " but I was scared to death because I was on house arrest, and they wanted me back after the time I had to be in."

She went home and called her parole officer, who gave her permission to stay out until 6 p.m.

"At 5:45 I'm sitting there saying, 'Please let this interview be over so I can just go home.' And finally they said, 'Well, you have the job. Can you start Wednesday morning?' "

So one week after her release, she was working, but she also went to a community college to see about more courses. There she met the head of the human services department, a woman whose death last year from cancer still fills Smith's eyes with tears.

"She became a mentor to me, and she knew about me being in prison and everything," says Smith. "But she said, 'Susan, you have too much to offer. Don't let anything pull you down.' "

In one course, Smith had to sketch four of her life-changing experiences.

"My first scene was about the prison door closing, just hearing that door slam shut," she says. "The second was the frustration and hurt of when Alicia was sick. The third was about going to school, and the fourth was coming out of that gate to my mom and daughter."

Smith's teacher was so moved that she kept the paper.

Smith's program at Florissant Valley Community College required two practicums. First, she volunteered in the children's hospital where Alicia had often been hospitalized. Second, she signed on to help Rev. Ted Schroeder, pastor of a small Lutheran church in a struggling neighborhood, run the rapidly growing Prisoner Family Services program, which provides bus trips for families visiting inmates.

At the end of the practicum in 1993, Smith had done so well that Schroeder asked her to stay on as director, and she has been there ever since.

After completing her associate's degree, Smith entered the University of Missouri at St. Louis, where she is only a few credits away from a bachelor's degree in social work, with an emphasis in criminal justice.

"At my mentor's funeral last year," Smith says, "I said, 'I guess I'll have to go on and get my master's for her too.' "

In her personal life, she credits her quick readjustment to Project COPE, a program that matches ex-convicts with congregations ready to help them return to society.

When Smith applied for the program, Linda Schroeder, Project COPE director and wife of the pastor, came to visit with a member of her prospective church team.

"It was like there was a connection there from the beginning," Smith says. "My team was a lot of help to me, emotionally and spiritually."

While Smith was on house arrest at her parents' home for four months after her release, she needed that support.

"I'm not making excuses," she says, "but house arrest is so constraining that I can really see why people go bad on it. At least when you're in prison, you can walk the yard, but you can't on house arrest because you have only so much footage that you can be away from your monitor."

One time she skimped on other groceries so she could buy expensive diabetic ice cream bars for Alicia. Then, minutes before supper, her brother's girlfriend walked into the kitchen, opened the freezer and casually handed one of the precious bars to her son.

"That was the last straw," says Smith. "I put food on the table and stormed out of the house, but I got outside and it was just like, 'I can't even go for a walk.' I sat down there and started crying. Then my mom came to the door and said I was wanted on the phone. It was Joan from my COPE team. I said I was fine, but she said, 'No, Susan, I can tell something's wrong. I had a feeling; that's why I called you.' COPE did so much to keep Alicia and me together."

Smith has learned from her prison time and her reintegration into society.

"For me, prison was an experience that I could have done without," she says. "But it was the turning point. It's nothing I'm proud of, and it's something I wish I hadn't gone through for the pain I caused my family and friends and myself.

"But it was also an experience that let me take advantage of some opportunities and better my life from them."

She hopes she can be an example to other prisoners.

"Remember that just because we've done something bad doesn't make us bad. We're still good inside, and don't let anyone take that goodness away from you because you are a good person."

**'Maybe I can help someone else turn their life around. It's kind of like a repayment, you know?'**

*Figure 13–6 The design of this feature page helps make it easy to read. The inordinate amount of white space contributes to its simplicity. The large horizontal photograph at the top is appealing and attractive to readers and the use of a large initial letter at the beginning of the first paragraph adds a bit of charm.*

***Negro Clergyman Defeated***

**Council of Churches Picks Woman Leader**

*From Our Wire Services*

DETROIT, Dec. 4. — Dr. Cynthia Wedel, an Episcopalian and ardent advocate of women's rights, won overwhelmingly over a Negro clergyman Thursday to become the first woman president of the National Council of Churches.

Mrs. Wedel, of Washington, defeated the Rev. Albert B. Cleage Jr. of Detroit 387-93 in secret balloting at the NCC's triennial general assembly.

When the vote was announced, Mr. Cleage, the first Negro candidate for the presi-

*Profile and Picture On Page 3*

dency, went to a microphone on the assembly floor and castigated what he called the "White racist establishment of the NCC."

"This organization is anti-Christ and until young people or oppressed people take over, you'll remain anti-Christ" Mr. Cleage declared. "Time is

**Korean Report Raps Operation of Center By Buck Foundation**

By EDWARD N. EISEN
*Of The Inquirer Staff*

The State Commission on Charitable Organizations made the Pearl S. Buck Foundation produce a letter Thursday from Korea's embassy highly critical of the foundation's work in that country.

The letter, by Sung Kwoo Kim, counsel and consul general at the embassy in Washington, said a 10-day inspection in August at the foundation's Sosa Opportunities Center, west of Seoul, showed unsanitary conditions, overcrowding, misuse of funds and other shortcomings.

**REASONS DIVULGED**

The letter was produced at a commission hearing on the foundation's appeal for a license to solicit funds in Pennsylvania. The foundation was to have presented new evidence, as demanded by Joseph J. Kelley Jr., Secretary of the Commonwealth.

*Figure 13–7 Many uneven wraps like these, used on the same page, will make the page appear to lack unity.*

*Figure 13–8 A page with columns of uneven length. This style also affects page orderliness. Stories that have too many different lengths may destroy the unity of a page.*

SAIGON (AP) — Gen. William C. Westmoreland, departing after four years in command of U.S. forces in Vietnam, said today American strength was greater than ever "but it is unrealistic to expect a quick and easy defeat of the Hanoi-led enemy."

"If he feels time is on his side, he can go on a long time," Westmoreland said of the enemy in a farewell news conference on the eve of his departure to become Army chief of staff in Washington.

He added that he could not predict what would happen at the peace talks in Paris, but from his view the enemy still appeared to be in search of major military victory.

**"Price Can Be Raised"**

Westmoreland said he felt that a classicial military victory was not possible in South Vietnam in view of U.S. policy decisions not to escalate the war or to enlarge its geographic boundaries.

He added: "But the enemy can be attrited, the price can be raised. It is being raised to the point that it could be intolerable for the enemy. It may reach the point of the question of the destruction of his country, and jeopardizing the future of his country, if he continues to pay the price he is now paying and destined to pay in the future."

Westmoreland said the enemy had lost 113,000 men since the first of the year and added "he doesn't have the manpower or resources to take these losses in stride."

**Sees Net Reduction**

Westmoreland said that although infiltration continued at a serious pace down the Ho Chi Minh trail through Laos, the Hanoi government was not able to make up its manpower losses in the South and that since midsummer last year enemy strength had shown a net reduction.

He said Hanoi's strategy appeared to be to continue pressure against Saigon and its political structure and to seek some major victory on the battlefield.

Westmoreland departs tomorrow and will turn over the Saigon command to his deputy for the past year, Gen. Creighton Abrams Jr.

"At this time our military posture is at its height since our commitment," he said. "We are now capable of bringing major military pressure on the enemy.

"This we are doing, and the enemy is beginning to show the effect. The Vietnamese armed forces are growing stronger in size and effectiveness."

**Headquarters Farewell**

Earlier today, Westmoreland said goodbye with a "good luck and bless you all" to the officers and men of his headquarters.

"Please accept my very best wishes for continued success," Westmoreland told his staff,

"and my fervent hope that peace and security for the long suffering and freedom loving people of Vietnam will soon reward your efforts."

In his swing north yesterday, Westmoreland visited the headquarters of the South Vietnamese Army's 1st Military Corps in Da Nang, took a helicopter to Provisional Corps headquarters at Phu Bai 35 miles away, and visited the headquarters of the 3rd U.S. Marine Amphibious Force.

He wound up his tour with a flight to the nuclear carrier Enterprise in the Gulf of Tonkin to bid farewell to the U.S. 7th Fleet.

# Bethlehem directors OK drydock work

The Bethlehem Steel Corp. board of directors officially plunged the company into the Pleasure Island drydock project by stamping its approval on the months of work by company officials.

Richard E. Blackinton, general manager of operations and facilities for Bethlehem Steel, said the operating agreement between Bethlehem and the Port of Port Arthur "still contains some details that have to be worked out, but things are proceeding on schedule."

He anticipates the drydock will be working by the end of March.

Plans call for transporting the drydock from Pearl Harbor, assembling it on Pleasure Island, and relocating and rebuilding about a mile of Texas 82.

Blackinton said the Army Corps of Engineers received four objections to the drydock project but that he didn't expect them to delay corps approval of the dredging and operating permits.

Blackinton said one response came from the Environmental Protection Agency — which had questions not objections — and that Bethlehem was preparing an answer.

Karen Brown, a public information assistant with the Environmental Protection Agency in Dallas, said the agency questioned a private firm using a designated corps spoils area for disposing of dredge materials. Other questions by EPA involve whether Bethlehem considered other ways to dispose the dredge material, if a sediment analysis of the dredge material was conducted and if the drydock would cause adverse impact on area water quality.

*Figure 13–9 When type is squared off (or set flush left and right) it is often called a module. A module is a self-contained unit with headline and body type that can be moved around the page easily. Modular designed papers are becoming popular.*

editors have abandoned both these techniques in the interest of saving production time. The benefits in graphic design outweigh the benefits in production time, and the practice of using two-column leads is a good one.

When a story is wrapped underneath a picture, there is little danger of confusing the reader, so a cutoff rule is usually unnecessary (see Figure 13–11). Notice the two-column cutline (or caption).

When a story is wrapped into an adjacent column at the top of a page without a covering headline it is called a *raw wrap* (Figure 13–12). In many instances raw wraps are undesirable; some newspapers forbid them entirely. The makeup editor faced with a raw wrap should ask that a headline be written to cover the wrap and make it clear that the wrap belongs to the headline above it. But occasionally it is permissible to use a raw wrap at the top of an advertisement where there is no doubt that the wrap belongs to the headline on the left.

When there is no time to reset a headline to a wider column measure, another layout procedure is to avoid raw wraps by filling the remaining space with shorter

*Figure 13–10 A story style in which the lead paragraph is two columns wide and the columns about half that width. This looks attractive, but it requires some computer adjustment to set type that way.*

# Lorelei McDonald bride of Jeffrey Earl Hammer

The wedding of Lorelei Shavonne McDonald, daughter of Mr. and Mrs. Lewis S. McDonald, 4313 Defiance Pike, Wayne, and Jeffrey Earl Hammer, son of Mr. and Mrs. Raymond Hammer, Gibsonburg, was solemnized Friday in St. Michael's Church. The Rev. Ray C. Przybyla officiated. The bride was given away by her father.

Altar decorations were two vases of yellow gladioli, ice blue carnations and white pompons with brown fall leaves.

With her lace-trimmed white wedding gown she wore a veil gathered to a lace bandeau and carried an old fashioned round bouquet of stephanotis, yellow silk rosebuds with baby's breath and ivy.

Her attendants were Mrs. Darlene Lentz of Bradner, matron of honor; and bridesmaids Mrs. Gary (Elaine) Lentz, Mrs. Alan (Patty) Adams and Ann Lentz, all of Bradner, and the bride's cousin, Mrs. Fred (Pam) Berno of Arcadia. Their colonial bouquets were light blue cosmos, white pompons and yellow silk rosebuds with baby's breath and ivy. The flower girl was the bride's niece, Michelle Cron of Portage.

The couple are at home at 110 County Road 26, Gibsonburg.

The bride, an Elmwood High School graduate, is employed at Fremont Memorial Hospital. The groom, who graduated from Gibsonburg High School, is employed by G.M. Sader Construction Co. of Bowling Green.

stories. Or perhaps a long story can be shortened to fit the space by cutting a story from the bottom.

In dummying a page, the layout editor should avoid the kind of wrap that requires the reader to jump from the bottom of a page to the very top above an ad. The size of the ad makes it appear as if the story has ended at the bottom. The makeup person should either cut the story and end it at the bottom (placing a new story above the ad) or find a shorter story to place at the left of the ad.

## Filling Remaining Space

When most of the page has been dummied, the layout procedure is complete. Small spaces may remain because not all stories fit precisely. The remaining space may be filled in two ways: If the space is large enough, fillers may be used. Editors assign someone the responsibility of providing a sufficient number of fillers each day. If the space is relatively small, leading is added between paragraphs until the column is filled. If there is time, a story may be filled by a new copy.

# Killer storm cuffs Rockies, Midwest

Associated Press

A winter storm already blamed for 11 deaths stretched from the southern Rockies to the upper Midwest today after burying parts of Arizona and Colorado under 20 inches of snow, unleashing tornadoes in Texas and downing ice-laden power lines in Kansas.

Six inches of new snow were already on the ground today in Kansas and Nebraska, and forecasters said more was on the way.

Raging thunderstorms spun off at least two tornadoes that damaged more than 100 homes and businesses in Texas yesterday and motorists in the western part of the state were warned that snow and freezing rain today would make driving hazardous.

In the Pacific Northwest, meanwhile, travelers' advisories were posted for the Cascade Mountains of Washington and Oregon as a new storm gathered strength in the Gulf of Alaska.

The mercury tumbled to 15 degrees below zero overnight in West Yellowstone, Mont., but in the East more than a dozen cities reported record high temperatures yesterday.

The storm that dumped up to 20 inches of snow over the southern Rockies yesterday, closing schools in parts of Colorado and Arizona and surprising residents as far south as Tucson, brought more snow but lesser accumulations to New Mexico and northeast Arizona.

"It's all going to slowly push east, but its hard to say how quickly," Steve Corfidi of the National Severe Storms Forecast Center in Kansas City, Mo., said today. "It's a more pronounced pattern than you normally

Associated Press

**A tornado yesterday in Mesquite, Texas, damaged more than a dozen houses and businesses. It first touched down in Ferris, a community 25 miles southeast of Dallas.**

expect . . . an interesting pattern. It keeps us busy."

Since Tuesday, the weather has been blamed for the deaths of seven motorists in Colorado, two in a 23-car pileup during a blinding dust storm yesterday about 70 miles southeast of Los Angeles, two in Oklahoma on slick roads and three in a fog bank in southeast Georgia.

Up to a foot of snow fell last night at Hawley Lake in Arizona's White Mountains, and 10 inches fell at Flagstaff in northern Arizona.

Thunderstorms raked north-central Texas yesterday, spawning tornadoes that damaged more than 100 homes Crow.

In Kansas, Wichita authorities said 15,000 homes and businesses temporarily lost electricity last night, as freezing rain snapped power lines. Service was restored within a few hours to about 12,000 residences, but Kansas Gas and Electric Co. officials said the rest were expected to go

*Figure 13–11 A wrap underneath a photograph. Cutlines are wrapped into two columns (used when photograph is wide).*

# *Diabetes and blood pressure*

The doctor says

by Lawrence E. Lamb, M.D.

About a year and a half ago my doctor said I had high blood pressure and put me on medicine. In a short time my tests showed I was low on potassium, and he put me on potassium medicine and gradually increased it because my level was so low.

Then he did a glucose test and said I'm a borderline diabetic. If I understand what I read, sometimes when there is no diabetes in the family and it shows up, it can be caused by high blood pressure medicine. Is this so?

If my blood pressure is causing the problem why can't they just give me other medicines? I know there are other medicines they can use.

Also, I'm 52 and going through the menopause. I'm somewhat overweight and trying to lose, as I need surgery for a bladder repair. I have a fibroid tumor, so the doctor won't give me hormones for my hot flashes. He says it will cause me to bleed badly. Is it true that fibroid tumors sometimes dry up after the change in life? I have been to two doctors, and one says surgery now, the other to wait until I get my weight down.

I see you are really having a time. First, please make every effort to lose weight as it may help relieve your blood pressure and high blood sugar problems. Why don't you try my weight losing diet? It has helped a lot of people to lose weight. Send 50 cents for The Health Letter number 4-7, Weight Losing Diet. Address your letter to me in care of Paddock Publications, Radio City Station, New York, N. Y. 10019.

You are right, some high blood pressure medicines will cause the blood sugar to be high. It is often stated, though, that they merely unmask an underlying diabetic, but you are beginning to get on theoretical ground there. The same types of medicine can and will cause the loss of potassium. And I would tend to agree that there are other medicines that could be used. The medicine used to eliminate salt and water that causes these problems, though, is very useful in combination with other medicines. The combination usually makes it possible to handle a patient's problem without so much risk of complications from the medicines.

Another problem with glucose tolerance tests is that they will give a result similar to that in diabetic if the person has not been eating any carbohydrates recently. Unless the patient is properly prepared for the test, it is of limited usefulness.

Estrogen hormones do enable fibroids to grow. If they are just under the lining of the uterus, hormones may cause you to bleed. And, some fibroids do shrink after the menopause. In general people do better during and after surgery if they have no weight problems. Nevertheless when it needs to be done surgery can be done in really quite heavy people. I suspect your surgeon thinks he will get a better result if he is able to operate after you have lost weight.

Meanwhile I would suggest making every effort you can to lose weight to try to get out of this combined mess you are confronted with.

(Newspaper Enterprise Assn.)

*Figure 13–12 A naked wrap. As long as the wrap is continued at the top of the page, it is usually not confusing.*

## Flexibility in Layout

In planning the layout of large newspapers, the editor should give some attention to flexibility of design to accommodate late-breaking news. There are two considerations in planning for a flexible design. The first is a mechanical consideration. Can one or two stories be replaced without too much effort? The task of remaking a page should be accomplished in the shortest amount of time to meet a press deadline. It may be necessary to rejustify as many as six columns of news in order to accommodate a late story. When the story to be replaced is oddly shaped, involving complex wraps, it will take more time to remake than it might if it were simply shaped. The new story may not be as long as the one it replaces, or it may be longer. Therefore, the design must be simple and flexible enough for any contingency.

A second consideration is the effect that a major story change will have on total page design. Although it is impossible to know how a late-breaking story will be shaped, it may be possible to anticipate how stories of various shapes will affect the design. If the original design is simple, chances are that any changes can be adapted easily to the old design without destroying the original appearance.

## Positioning of Nameplate

The nameplate of a newspaper (often incorrectly called the masthead) is the name usually appearing at the top of Page One. Editors sometimes want to move the nameplate to other positions on the page because they assume that readers know the name of the newspaper they are reading, and the space occupied by the nameplate might be better used for other purposes. Once a decision has been made to move the nameplate, a question arises: Where on the page would the nameplate be most appropriate? Editors sometimes want to move it indiscriminately because they consider the significance of news to be the most important criterion for positioning stories. Such editors feel that the nameplate is much less significant and therefore may be moved anywhere at almost any time. At other times, the nameplate seems to be moved around on Page One without any apparent reason.

On the one hand, it seems reasonable to move the nameplate from day to day. After all, most readers know the name of the newspaper without looking at the nameplate. On the other hand, there are a number of reasons for keeping it in the top position most of the time. In debating the reasonableness of moving it around on the front page, editors should consider all the purposes a nameplate serves.

Besides simply identifying the newspaper, a nameplate communicates the publisher's philosophical position. The typefaces chosen for nameplates usually look distinguished and have strong connotations. The best position for communicating these connotations is at the top of the page because the top position itself communicates a feeling of authority. Any object standing foremost among other objects is judged to be more significant. When an object is buried, its importance is diminished.

Another consideration has to do with the importance of top position in serving as a device that provides readers with a feeling of stability as they read the paper. The nameplate usually represents the starting point for examining the contents of Page One. If it is not in its traditional position, readers may have a sense of uneasiness and a slight loss of familiarity with the paper. But if the readers know that while the news may change the nameplate position will not, the page gains a sense of stability that tends to make reading comfortable.

Furthermore, postal regulations require newspapers to indicate that they are second-class material somewhere within the first five pages of the paper. In the past, this notice might be placed in the editorial page masthead, or it might be buried in a box on Page 2. But it is also placed in small type near the nameplate, perhaps within the dateline rules that appear underneath it. If the nameplate serves this purpose, it should remain at the top.

A final reason for keeping the nameplate at the top relates to its function in the total page design. A nameplate floated down in the page becomes a component of the page's design and thereby complicates the process of layout and design. It is usually easier to lay out a front page when the nameplate is at the top—with some exceptions, of course.

Yet there is some logic in moving the masthead if the move is not radical and if it is not done often. There are times when a six-column, horizontally arranged story might well be placed above the nameplate. This story may be of such significance that the editors want to be sure that everyone sees it. The very top position should provide such assurance. But if this practice becomes regular, it diminishes the importance of the nameplate.

There is also some logic to the idea that a nameplate may be moved from side to side, but must always remain at the top of the page. Among the most unsightly makeup devices of newspapers are the ears. *Ears* are pieces of news placed at either or both sides of a nameplate. The editors may eliminate them by moving the nameplate to either side of the page and moving a one- or two-column story to the top. Another reason for moving the nameplate to the side may be to make room for more news than would be possible otherwise. The editors most likely to experiment with moving the nameplate are those involved with publishing college newspapers. They are usually motivated by a desire to be creative. What is creative, however, may not be easy to read.

There are other uses for nameplates throughout the newspaper, but these do not involve moving the front-page nameplate. For example, some newspapers have identical nameplates on the first and third pages. The effect of this arrangement is to present a second front page to readers so that different kinds of news may be featured on each. National news might be used exclusively on Page One, and local news might be featured on Page 3.

Finally, modified nameplates may be used in various sections of the newspaper. Family, sports or financial pages also might have special nameplates. In reality, however, these are standing headlines rather than nameplates, and although they are designed to resemble the front-page nameplate, their function is to introduce special sections of the newspaper.

## Contemporary Headline Placement

The traditional-minded layout editor thinks of headlines in terms of large display typefaces placed at the top of stories. This treatment is logical for pages designed to be strongly vertical in appearance. But when makeup is conceived of as horizontal through the use of rectangularly shaped stories, headlines may be set in many different ways. These varied headlines not only summarize the news but make the page appear modern.

When horizontally shaped stories are used on a page, the headline may be placed in at least three places, and possibly a fourth: at the left or right side, in the center, or at the bottom of the story. The bottom position is not as desirable as the others (see Figure 13–13).

If the headline is placed at the side of a story, it looks best if it aligns with the top line of body type. But the headline may be set flush left or flush right, both being contemporary treatments. The best position is at the left side of the story because readers proceed from left to right. Occasionally, the headline may look attractive when placed at the right.

When a long story is given horizontal treatment, the headline may be embedded in the center with type on all sides. There is some danger that readers could be confused by this arrangement if they start reading directly underneath the headline instead of at the top of the left column. To avoid that, the layout person must not start a sentence directly under the headline. This arrangement also requires the typesetter to plan typesetting around the headline. Because such treatment takes extra time, it is used primarily for feature stories.

Finally, a headline works best at the bottom of a story that is clearly set off from the remainder of the page, as in a large box. When the headline is at the bottom of the box, it will be apparent that it belongs to the story above. If it is not set inside a box, the readers may assume it belongs to some other story.

## Treatment of Lead Paragraphs

The lead paragraphs of most stories are placed at the top of a story, set in type sizes that are larger than body type and sometimes leaded. The widths of most lead paragraphs are the same as body type, although occasionally they are set one or even two columns wider. Lead paragraphs are often made to stand out not only from the headlines above them but from the body type below.

In contemporary newspaper design, however, some lead paragraphs are often given more prominence than they would receive in traditional design. The feeling is that because lead paragraphs have replaced the old headline decks, they deserve more prominence. Decks were formerly used to summarize a story, each deck fea-

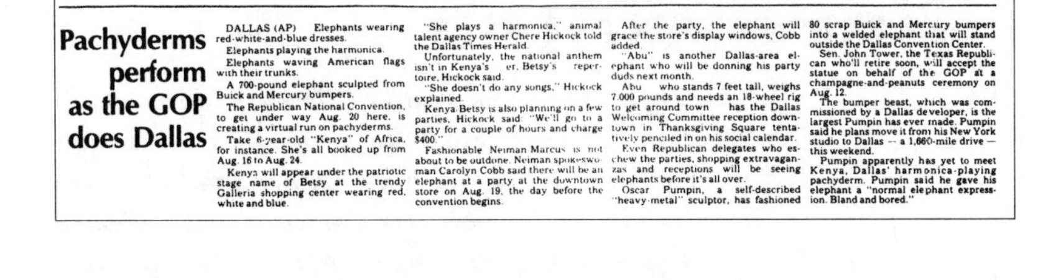

**Pachyderms perform as the GOP does Dallas**

DALLAS (AP) Elephants wearing red-white-and-blue dresses.

Elephants playing the harmonica.

Elephants waving American flags with their trunks.

A 700-pound elephant sculpted from Buick and Mercury bumpers.

The Republican National Convention, to get under way Aug. 20 here, is creating a virtual run on pachyderms.

Take 6-year-old "Kenya" of Africa, for instance. She's all booked up from Aug. 16 to Aug. 24.

Kenya will appear under the patriotic stage name of Betsy at the trendy Galleria shopping center wearing red, white and blue.

"She plays a harmonica," animal talent agency owner Chere Hickock told the Dallas Times Herald.

Unfortunately, the national anthem isn't in Kenya's er, Betsy's repertoire, Hickock said.

"She doesn't do any songs," Hickock explained.

Kenya-Betsy is also planning on a few parties, Hickock said: "We'll go to a party for a couple of hours and charge $400."

Fashionable Neiman-Marcus is not about to be outdone. Neiman spokeswoman Carolyn Cobb said there will be an elephant at a party at the downtown store on Aug. 19, the day before the convention begins.

After the party, the elephant will grace the store's display windows, Cobb added.

"Abu" is another Dallas-area elephant who will be donning his party duds next month.

Abu who stands 7 feet tall, weighs 7,000 pounds and needs an 18-wheel rig to get around town has the Dallas Welcoming Committee reception downtown in Thanksgiving Square tentatively penciled in on his social calendar.

Even Republican delegates who eschew the parties, shopping extravaganzas and receptions will be seeing elephants before it's all over.

Oscar Pumpin, a self-described "heavy-metal" sculptor, has fashioned 80 scrap Buick and Mercury bumpers into a welded elephant that will stand outside the Dallas Convention Center.

Sen. John Tower, the Texas Republican who'll retire soon, will accept the statue on behalf of the GOP at a champagne-and-peanuts ceremony on Aug. 12.

The bumper beast, which was commissioned by a Dallas developer, is the largest Pumpin has ever made. Pumpin said he plans move it from his New York studio to Dallas -- a 1,660-mile drive — this weekend.

Pumpin apparently has yet to meet Kenya, Dallas' harmonica-playing pachyderm. Pumpin said he gave his elephant a "normal elephant expression. Bland and bored."

**by dennis wheeler**

With Thanksgiving approaching, it is that time again to pause and consider the historical background for one of our most beloved holidays.

The Thanksgiving story, known to most of us from our earliest grade school days, is one of bravery in the face of giant adversity; one of achievement in the hour of supreme testing. It inspires reverence and hope in all of us. It is one of the true roots of our well-known American stick-to-itness.

You all remember, of course, how it happened back there in 1621. The Pilgrims had arrived at Plymouth Rock and had established their tiny foothold on the wild North American continent. Their lives beset by disease and tribulation, they managed to survive for some months. And then came winter, and near starvation. And the first Thanksgiving, at which the noble Pilgrims raised their eyes heavenward to thank the Almighty for providing them with enough food to make it through the long night of winter.

**THOSE, OF COURSE,** are the bare facts of the Thanksgiving tale. But people sometimes lose sight of how the Plymouth Rock adventure fit into the larger fabric of history. They overlook what was going on in the rest of the world; they forget against what historical backdrop the Pilgrim landscape was painted.

Well, I can help. A history buff of sorts, I can supply some of this additional information, the better to illuminate our heritage.

First of all, remember that the Pilgrims landed in North America at about the same time that old Will Shakespeare was writing his stuff back in England. In fact, some historians surmise that when Shakespeare wrote the famous line "Out, out, brief candle!" in MacBeth, he was referring to Candle Mather, the famous Bible-thumping leader of the Pilgrims, who was expelled from England by the King for complicity in burning Joan of Arc at the stake.

**IT IS ALSO** noteworthy that the landing of the Pilgrims took place shortly after the English defeated the Spanish in the famous sea battle of 1588. that's the one in which Sir Francis Tarkenton defeated the Spanish Armadillo mainly because the captains of the Armadillo's ships lost the toss and had to fight into the wind.

That's important because if the Armadillo had won, we might all be speaking Spanish today instead of English. But then on the other hand it is questionable whether the Spanish could have colonized North America as successfully as the English did. The Spanish are, after all, warm weather people, so they probably would have taken one look at cold Plymouth Rock and said, "No way, senors."

Besides, everybody knows you can't raise tacos in New England.

And don't forget the French.

**AT THIS TIME**, the French, led by Father University of Marquette and Louis XVI Joliet were selling booze to the Indians and canoeing down the Cal-Sag canal. One wonders what would have happened if communications in those days had been what they are today. If they were, the French would have heard on the 6 o'clock news that the Pilgrims had landed in Massachusetts, and (knowing them) they would have built the Erie canal across the Appalachian mountains and invaded Massachusetts from the west. Who knows? The battle of Bunker Hill could have happened 150 years early. I ask you: if that had occurred, where would Boston be today? In Montreal, that's where.

Not only that, we'd be speaking French today instead of English. Which means the entertainment center of New Orleans would be called be the German Quarter.

Why the German Quarter? Well, it's obvious. Because at about the same time the Pilgrims landed, the 73rd battle for Alsace-Lorraine was being waged between the French and the Germans in the Black Forest. The Germans won that one, and in the subsequent Peace of Westphalia, Adolph Hitler was able to wrest huge territorial concessions from the losing Frenchmen. And among these would have been a piece of New Orleans, which had just been explored by University of Marquette and Louis XIV Joliet and claimed for France.

**AND IT** would have stuck, because you can grow saurkraut in a bayou.

But let's not jump so lightly over the Indians in this analysis. Remember, the Indians were here first, before any of these other guys, unless you want to count the Vikings, who under the leadership of Eric the Weird invaded the continent in the 11th Century, tried unsuccessfully to turn New York harbor into a fiord, and left in disgust two days later.

Most historians agree the Indians could have easily held on to North America against all European intruders if they had done just three things:

Dumped the tea into Boston harbor themselves instead of letting revolutionaires dressed up as Indians do it. Every-

Everybody knows the economy of the colonies would have buckled under without tea.

— **LISTENED, WHEN** Sitting Bull solemnly warned, "You can't trust no white men" instead of retorting, "Sitting, that's a lot of Bull."

— Let Custer go instead of massacring him. Custer was such a lousy general he would have lost the West without any help from anybody.

But no. The Indians chose instead to attack trains and observe treaties. It was the latter act, of course, that spelled their downfall.

But what does all this have to do with Thanksgiving?

Well, it's simple. It all shows how differently the loose end of the sweater of history might have unraveled. It shows that almost any combination of events, had they occurred even a little differently, could have resulted in there being no Thanksgiving at all.

**AND THAT** would have been a terrible shame.

Because one thing is crystal clear: you can readily grow thankfulness for our many blessings anywhere in the North American continent.

***. . . Thanksgiving's Place in History?***

*Figure 13–13 (Top) Headline at the extreme left of a story (this is also a naked wrap). (Bottom) Headline at bottom.*

turing one outstanding aspect. The lead paragraph, by employing more words set in smaller type than decks, did a better job of summarizing the essential details. But as lead paragraphs came into common use, they were often accorded no better treatment than body type. In contemporary design, some have been made more dramatic in appearance and placed in more obvious positions relative to the remainder of the story.

When headlines are moved to the sides, center or bottom of a story, the lead paragraphs may be placed underneath the headline. However, in such cases, they should be set in a typeface and a size that clearly contrast with the remainder of the body type. Because the goal is to give them display treatment, lead paragraphs should appear markedly different from the body of the story (see Figure 13–14). The typeface should be one or two points larger than body type, set in sans serif, boldface or italic. Leading is necessary to make the lead paragraph stand out and yet be readable. One final treatment may be to set the lead flush left with ragged right, a distinctly contemporary appearance.

# Earthwatch

***Looking for something a little different to do on your vacation this year? Earthwatch can find you some work to do, helping scientists with menial — but worthwhile — tasks all over the world. How about counting ants? Or recording the habits of the spotted hyena?***

**By Daniel Q. Haney**

BELMONT, Mass. (AP) — For $1,000 or so, people buy the opportunity to lie on their bellies in 150-degree heat and count ants. Or, if it seems more attractive, they find out whether they can fool wild llamas by lugging their dung heaps back and forth.

his wife, Eleanor, consider themselves to be conservationists, so they picked a bird banding trip to Panama last year.

"We wanted something where we could actually participate actively ourselves," says Karl. "It was the most unpleasant one we could find, because it was in the wet season in the jungles of Panama with no housing or toilet facilities. But if we are really as ardent as we think we are, this would be the proof."

Before sunup each morning, Karl and his companions groped into the jungle ravines by flashlight to hang up 15-foot-long mesh nets. Then, once an hour, they ventured back out into the torrential rains to identify, weigh, band and then free the captured birds.

"**IT WAS** very hard, but

**(See Earthwatch, Page 2C)**

*Figure 13–14 A lead paragraph created in display format.*

## New Approaches to Cutlines, Overlines and Underlines

The traditional way of identifying photographs is to place cutlines underneath them. Overlines and underlines may be used alone or with cutlines. If the photograph is less than three columns wide, the cutlines are set full width, but if the photograph is wider, the cutlines may be wrapped in two or more columns underneath. When overlines and underlines are used, they are usually centered.

In contemporary layout, cutlines may be set flush left with ragged right, flush right with ragged left or justified (see Figure 13–15). When it is necessary to wrap cutlines into two adjacent columns, the flush-left or flush-right approach does not look pleasing. If, however, two pictures of the same size are placed next to each other (because they are related), it may be possible to set the left cutlines flush left and the right cutlines flush right.

A particularly modern style of cutline treatment, borrowed from magazine design, is to place them at the lower side of photographs. In such positions, they may be set in very narrow measures (from 6 to 9 picas in width), flush on both sides. But they also may be given the flush-left or flush-right treatment. When cutlines are placed on the right side, they should be flush left, and when placed on the left side, they should be flush right (see Figures 13–16 and 13–17) because the type nearest the photograph is aligned and looks more attractive.

Close call
Rescuers from Frenchtown Township near Monroe extend a ladder to help a man reach shore after he and four others got trapped on an ice floe on Lake Erie. The men were returning to Lake Sterling State Park after three hours of ice fishing when they realized they were on the chunk of floating ice. All were rescued and no one was injured.


*Figure 13–15 Cutlines justified on both sides.*

*Figure 13–16 Cutlines set flush left, ragged right. This fits certain kinds of layouts.*

**Former U.S. Sen. Paul Simon was on the Senate Judiciary, Foreign Relations and Labor and Human Resources Committees, and his contributors reflected those assignments.**

Overlines may be repositioned above cutlines placed at the sides of cuts. They, too, may receive the same flush-left or flush-right treatment as the cutlines. When they are positioned that way, there probably will be a large amount of white space above the overline (or above the cutlines when there is no overline). But this white space will enhance the appearance of the treatment and should not be considered wasted space.

*Figure 13–17*
*Cutlines set flush right, ragged left.*

**Oil spire plumbs ocean**

*Soaring 340 feet into the air like a spire, this crane is part of a new derrick for a pipe-laying vessel used to service offshore oil and gas fields. Owned by Standard Oil Co. (Ind.) and DeGroot Offshore Ltd., the vessel, Ocean Builder I, has the capacity to lift 2,000 tons.*

## Inserts and Page Design

*Inserts* are additions to the story sometimes placed within the story's body type. The ostensible purpose of breaking into a story is to provide information that helps the reader better understand the news. In traditional layout practices, editors often inserted freaks, refers or other material into the main body of a story. A *freak* is any material placed in a story that is set differently from the main story. A *refer* is something that refers to a related story. But no matter what the purpose is, the effect of any break in the news is a break in the reader's concentration.

What are the most likely alternative actions readers may take when confronted by an insert? They may notice the insert, ignore it temporarily and, upon finish-

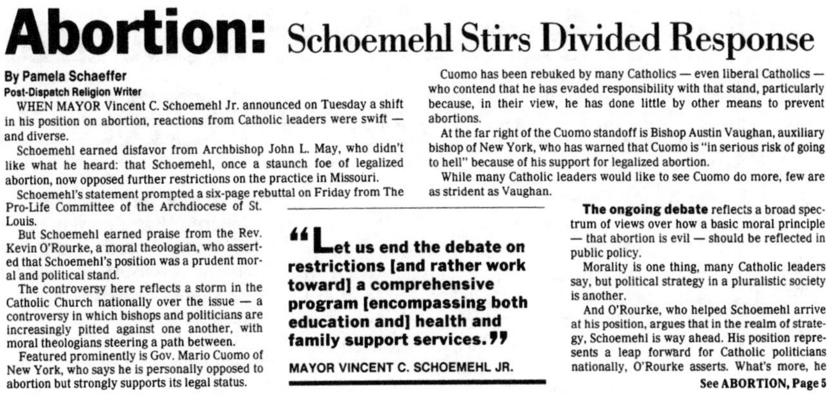

**Abortion:** Schoemehl Stirs Divided Response

**By Pamela Schaeffer**
**Post-Dispatch Religion Writer**

WHEN MAYOR Vincent C. Schoemehl Jr. announced on Tuesday a shift in his position on abortion, reactions from Catholic leaders were swift — and diverse.

Schoemehl earned disfavor from Archbishop John L. May, who didn't like what he heard: that Schoemehl, once a staunch foe of legalized abortion, now opposed further restrictions on the practice in Missouri.

Schoemehl's statement prompted a six-page rebuttal on Friday from The Pro-Life Committee of the Archdiocese of St. Louis.

But Schoemehl earned praise from the Rev. Kevin O'Rourke, a moral theologian, who asserted that Schoemehl's position was a prudent moral and political stand.

The controversy here reflects a storm in the Catholic Church nationally over the issue — a controversy in which bishops and politicians are increasingly pitted against one another, with moral theologians steering a path between.

Featured prominently is Gov. Mario Cuomo of New York, who says he is personally opposed to abortion but strongly supports its legal status.

**"Let us end the debate on restrictions [and rather work toward] a comprehensive program [encompassing both education and] health and family support services."**

MAYOR VINCENT C. SCHOEMEHL JR.

Cuomo has been rebuked by many Catholics — even liberal Catholics — who contend that he has evaded responsibility with that stand, particularly because, in their view, he has done little by other means to prevent abortions.

At the far right of the Cuomo standoff is Bishop Austin Vaughan, auxiliary bishop of New York, who has warned that Cuomo is "in serious risk of going to hell" because of his support for legalized abortion.

While many Catholic leaders would like to see Cuomo do more, few are as strident as Vaughan.

**The ongoing debate** reflects a broad spectrum of views over how a basic moral principle — that abortion is evil — should be reflected in public policy.

Morality is one thing, many Catholic leaders say, but political strategy in a pluralistic society is another.

And O'Rourke, who helped Schoemehl arrive at his position, argues that in the realm of strategy, Schoemehl is way ahead. His position represents a leap forward for Catholic politicians nationally, O'Rourke asserts. What's more, he

See ABORTION, Page 5

*Figure 13–18 An insert set in the interior of a story.*

ing the article, return and read it. They may ignore it entirely. Or they may stop to read it and then try to pick up the thread of thought in the remainder. Even if readers stop to read the insert, they may not read much more of the story. But no matter which option they choose, the insert may break the flow of reading, if only for an instant, and is therefore undesirable (see Figure 13–18).

Nevertheless, editors continue to use inserts, including perhaps at least one in every edition. Too many inserts on one day make the page appear to be full of spots that are unattractive and uninviting. Even one insert stands out as a spot on the page. In such situations, the layout editor interrupts the reader with rules and even a small headline within the insert. That one spot, then, may hamper the efforts of the editor to design a pleasing page because inserts tend to call attention to themselves.

What should be done with the material that may have been used in inserts? One answer might be to place that material at either the beginning, the end or the top center of a story (see Figures 13–19, 13–20, 13–21). Perhaps it can be incorporated into the body of the story and not be obtrusive. Finally, by careful analysis of the news, the editor may treat additional editorial material as another story and place it adjacent to the story in question. If there is reference to a picture, then this may be set in lightface italics and incorporated into the story.

## Boxed Stories

The use of boxed stories in contemporary modular layout is radically different from that in traditional layout. In traditional layout, a short human-interest story or an insert might have been placed in a box. Rarely was a long story boxed.

# Farmers in Midwest affected most by delay of CRP checks

**By Alissa Rubin**
*Eagle Washington bureau*

WASHINGTON — Nearly $1.6 billion in payments to farmers with land under contract to the government will be delayed at least one month because of the government's failure to pass spending bills for 1991, Agriculture Department officials said Thursday.

The land is under contract through the conservation reserve program, which pays farmers to take land out of production for 10 years to protect erodible or fragile land.

The delay hits hardest farmers in Midwest and Plains states, where thousands of farmers were scheduled to receive nearly $1 billion in early October. The states with the most land in the program are Kansas, Texas and Iowa where farmers are due $154 million, $156 million and $148 million, respectively.

"We do not have any appropriated funds so we cannot send out any checks," said Keith Bjerke, chief of the Agricultural Stabilization and Conservation Service, which administers most of the farm program.

Congress has delayed passage of spending bills, under which payments are released from the Treasury, because legislators have been unable to agree how to cut the federal budget deficit.

"As soon as somebody tells us how much money we've got to spend, then we know who the people are who are due the checks and what they've got coming," said Bjerke.

**"We do not have any appropriated funds so we cannot send out any checks."**
*Keith Bjerke, ASCS*

None of this is much comfort to farmers, many of whom rely on checks to make loan payments on land and farm machinery.

"Their cash flows are such that they are expecting that check the first week of October — and the machinery dealer, banker, the seed dealer, the fertilizer dealer, they have to be paid. They have to hope their bankers will bridge them," Knight said.

The only good news is that, contrary to earlier reports, farmers will be paid — albeit a little late — the full amount even if automatic budget cuts go into effect, Bjerke said.

*Figure 13–19 Another insert set in the interior of a story. Check Figure 13–18. Which is best?*

Contemporary layout seeks to dramatize a story or the design on a given page, often through the use of large boxed stories. There is, of course, a danger in using too many such stories on a page. But if only one is used per page, it may liven that page considerably.

An editor boxes a story because he or she considers it significant. Perhaps it is not as significant as the top two stories on the page, but it is still of major importance. The procedure, then, is to place the entire story in a boxed rule. The story must, of course, be squared off so that it fits neatly into the box. A photograph may accompany the story. But the keys to making this box look attractive are the use of

'Many who attend law school will never hang out a shingle. Legal training will be used in another area — it will always stand them in good stead.'

# Good lawyers are his goal

By Maureen Milford

Staff writer

J. Kirkland Grant admits there's a national glut of lawyers, but that's not going to stop him from cranking out hundreds — maybe thousands — of attorneys from the Delaware Law School of Widener University.

The new dean of Delaware's only law school believes quality, not quantity, is the real issue.

"It all depends on what kind of lawyers you're training. Lawyers who are morally and ethically upright, as well as competent, will always have enough work. The competition will be rough, but there's always room at the top," said Grant, who took over as dean in July.

The lawyer-professor said he hopes to get the Delaware legal community to rally around the law school. His figures show that 20 percent of the lawyers in the state are Delaware Law School graduates.

"It behooves the lawyers in the state to be very interested in the law school. After all, it is the only law school in a corporate state," Grant said.

Grant, who at age 37 is one of the youngest law school deans in the country, has his work cut out for him. Many people in Delaware's legal community are shaking their heads at the number of people coming out of law school every year. As one lawyer put it, "People are going to law school the way they used to go into real estate." Another attorney expressed concern that lawyers will take cases that have little or no merit on a contingency basis simply to have work.

The new dean is aware of these concerns, but he says the Delaware Law School graduates have proven themselves top notch in the First State. "The greatest strengths of the law school are its students and graduates," Grant said.

Delaware Law School was a familiar name to Grant when he was offered the job as dean. In 1978 he served on the American Bar Association inspection team for accreditation of the school.

See DEAN — B2, col. 5

J. Kirkland Grant

Staff photo by Pat Crowe

*Figure 13–20 Insert set in a box and before the main headline.*

# Galileo didn't recognize Neptune

By Warren E. Leary

Associated Press

WASHINGTON — The famous astronomer Galileo apparently did not recognize a major planet when he saw it and thus missed out on adding the discovery to his list of credits.

'When he saw Neptune, he was looking for satellites of Jupiter.'

The 17th century genius, using one of his early telescopes, saw the planet Neptune without knowing what it was. This left actual discovery to scientists hundreds of years later, an astronomer said yesterday.

Charles T. Kowal of the California Institute of Technology said he re-examined records of Galileo's observations and calculated that the Italian physicist saw the planet at least two times while working in Florence.

Galileo saw Neptune on Dec. 28, 1612, and Jan. 28, 1613, but erroneously thought it was a "fixed star," even though he noted that it moved in relation to another star, Kowal reported to the National Science Foundation, which finances part of his work. Galileo's observations were 234 years before the distant planet formally was discovered.

Galileo, the first to put the telescope to practical use, discovered that the moon was marked by valleys and mountains, and reflected light instead of generating its own. He also discovered the four major moons of Jupiter and noted the peculiar form of Saturn, which was later recognized as being caused by its rings.

"When he saw Neptune, he actually was looking for satellites of Jupiter," Kowal said in a telephone interview. "It's also somewhat surprising that there are only two observations. Neptune should have been seen for the entire month of January."

Neptune, second only to Pluto as the solar system's most distant discovered planet, is 15 times more massive than the Earth. Very little is known about this cold, gaseous planet because of its distance from the sun — 2.8 billion miles — and the fact that no spacecraft has ever visited it.

Neptune was discovered on Sept. 23, 1846, by a young German astronomer, Johann Galle. Other scientists, notably John Couch Adams of Britain and Urbain Leverrier of France, calculated that an unknown planet had to be in a certain position because something was disturbing the motion of the planet Uranus.

Galle looked at that calculated position and found Neptune with ease.

Kowal said he is looking for early observations of Neptune because its orbit around the sun still is not well defined. The planet takes 165 years to circle the sun and has not yet completed one full revolution since its discovery in 1846, he noted.

*Figure 13–21 Insert set into a story. The technique avoids impediment to reading that would occur if it cut into one column only.*

type rules and more than an ordinary amount of white space inside the box. The only function of the type rule is to set the story apart from all other stories on the page. Very heavy rules usually do not look good. And a fancy border of any kind will call attention to itself, not to the contents of the box. The white spaces, especially between the rules and the body type, are the framing devices that, with the hairline rules, call attention to the story and make it easy to read. Headlines within a box also may be set smaller and in lighter-faced types than those normally used because the rules and white space framing the box make a larger type unnecessary. There is little competition from headlines outside the box.

Boxes, therefore, should be no less than two columns wide, and preferably larger. Some editors place at least one such box on every page where possible as a means of adding dramatic impact. The position of such boxes depends on the sizes and weights of other elements on the page. When other headlines are large and bold, a boxed story should be placed at the opposite side to balance the page. Often they look good at the bottom of the page. In some cases, they might well be used in place of the number-one story (upper-right side).

## Design of Jump Heads

Enough evidence is available through research studies to prove that stories that have been jumped or continued to other pages lose a great deal of readership (see Chapter 2). Nevertheless, the practice of jumping stories cannot always be avoided, and papers should have a policy regarding the design of such heads.

Two related problems arise in the design of jump heads: (1) how to make them easy to find on a page and (2) how to keep their design consistent with both the page and overall newspaper design. It may seem easy to resolve the first problem by setting the headlines in larger and bolder typefaces than other headlines. But if the type is too large or bold, it will call attention to itself and tend to make the page look unattractive. If it is too light, readers may not be able to find the heads. Both problems may be resolved if the following guidelines are observed.

- A jump head should consist of at least one word from the original head to make the transition easier.
- The typefaces and style of arrangement should be consistent with the headline schedule used for other headlines.
- The number of lines and sizes of type used for jump heads should be the same as if the jumped portion were a separate story. The length and importance of the story should be considered in making this determination.
- A contrasting typeface may be used to help the head stand out. If Tempo has been used for most other headlines, a Bodoni Bold italic may provide the necessary contrast.

- Stars, bullets or asterisks, if they are not too obtrusive, may precede the first letter of the jump head to serve as attention-getting devices.
- Ben Day screens may be used in the background for such heads. This is easy for a newspaper printed by the offset technique. When letterpress is being used, the page number from which the story originally started may be placed in a screened background. If several page numbers are screened and kept in logo form, they can be inserted easily under jump heads.

## Tombstoning in Contemporary Design

Almost every editor, from the largest metropolitan newspaper or from the smallest high school paper, knows that tombstoning should be avoided. *Tombstoning* (also called *butting heads)* is a practice of placing two headlines next to each other in adjacent columns, both of which are set in the same type size, weight and style. They also have the same number of lines. But tombstoning was considered a poor layout practice in an age when only 6-point hairline rules were used between columns. Because some newspapers have used hairline rules of even less than 6 points (2- and 3-point rules), there was more danger that a reader might read across the column into the adjacent headline and be confused (see Figure 13–22).

In contemporary design, the space between columns is at least 9 points and as much as 2 picas, and tombstoning may not be objectionable. There may be so much

**Fired Perkins offers no excuses**

Ex-coach believes his plan would have paid off for Tampa Bay

By FRED GOODALL
AP sports writer

**TAMPA, Fla.** — Ray Perkins said yesterday he's confident his efforts to build the Tampa Bay Buccaneers into a winner would have paid off if he had been allowed to complete his five-year contract.

The fired coach declined to discuss specifics about his inability to produce a winning record in four seasons, but did accept responsibility for the failure of his program.

"I'm very disappointed that we have not accomplished some of the things that I felt strongly that we could and that I felt strongly that we would," said Perkins, who was dismissed Monday and replaced by Richard Williamson.

"Being the head coach, being the guy that makes all the decisions, I have let our football players down. I have let our owner down. Needless to say, I understand the workings of the National Football League. I understand those things that go on behind closed doors, and I understand if you don't win that you get fired as a head coach."

The Bucs were 19-41 under Perkins, who sensed last week that he was finished as Bucs coach, regardless of whether the team pulled out of a tailspin that saw Tampa Bay lose seven of eight games.

He said he had no quarrel with the timing of owner Hugh Culverhouse's decision — one day after a victory over Atlanta stopped a six-game losing streak — because the decision already had been made not to bring him back in 1991.

"That's not what Mr. C said yesterday, but that's the fact," Perkins said, disputing the owner's contention he didn't decide to fire the coach until Monday afternoon.

"I felt very strongly on Friday that the decision had already been made. That was my gut feeling," Perkins said. "I had no basis for that except 49 years of 100 percent being right gut feelings."

Perkins, 49, left Alabama in December 1986 to take over a team that lost 28 of 32 games under Leeman Bennett. Using his unchallenged authority as vice president-football operations, he overhauled the team and began this season with 42 players who were not on Tampa Bay's roster when he arrived.

He feels he's leaving behind a good nucleus of talent, including his four No. 1 draft picks — Vinny Testaverde, Paul Gruber, Broderick Thomas and Keith McCants.

"I would like to stay throughout the remainder of my contract to show that it would work, because I still believe my plan will work," Perkins said. "But we'll never know.

"I have no excuses, no alibis," he added. "There are always a lot of reasons for failure ... There are a lot of reasons for us to start out strong and then go into a six-game losing streak. I won't go into them, but I won't deny that I'm part of those reasons. I was, and I am."

Perkins made a seven-minute statement and accepted no questions during a farewell news conference at the Bucs' training complex. He later met with a small group of reporters and said he hopes to coach again, probably on the college level.

"I'm sure I'll have opportunities, and I'm not sure I could do without football," said Perkins, adding he would not accept an offer from Culverhouse to remain with the Bucs organization during the final year of a contract that pays him about $800,000 annually.

"I've got a deep burning desire to be in football, but be in a position where I can make a difference. I don't just want to coach football. I don't want to just draw Xs and Os. I want to be in a position to contribute more to somebody else's life than just being a coach. That could be with or without football."

Early speculation about a possible successor centers around several big names, including former San Francisco 49ers coach Bill Walsh and current NFL coaches Joe Gibbs, Buddy Ryan, Sam Wyche and Lindy Infante.

San Francisco offensive coordinator Mike Holmgren, Giants defensive coordinator Bill Belichick and Dolphins quarterback coach Gary Stevens are pro assistants considered possibilities. And from the college ranks, the names of Louisville's Howard Schnellenberger and Notre Dame's Lou Holtz have surfaced.

**Baseball salaries average $597,537**

By RONALD BLUM
AP sports writer

**NEW YORK** — The average baseball salary rose by a record $100,000 in 1990 to $597,537, according to final figures issued yesterday by the Major League Baseball Players Association.

The dollar increase was the largest in baseball history and the 20.2 percent increase was the steepest single-season rise since the 1982 season.

Although they failed to repeat as World Series champions, the Oakland Athletics finished first in the payroll race with a team average of $804,643. They were followed by Boston ($777,683), the New York Mets ($758,575) and the New York Yankees ($725,872).

The Baltimore Orioles were the poorest-paid team at $279,326. The Seattle Mariners ($388,649) were the only other team below $400,000.

The union's survey was presented to its executive board last night by its staff. The executive board began a four-day meeting in Orlando, Fla., and details of the report were made available to The Associated Press.

EDDIE MURRAY
. . . highest at first

25 to 40.

The biggest jump of any team was the Milwaukee Brewers, who rose from 20th place in 1989 ($363,517) to 10th place in 1990 ($678,581). New multimillion-dollar contracts to Robin Yount and Paul Molitor contributed heavily to that

*Figure 13–22 An example of tombstoning. The three-column headline at the left may confuse readers who also read the story at the right.*

white space between the columns that the reader can't be confused into reading a headline in the adjacent column. Then the only objection to tombstoning may be that there is not enough variety shown when two headlines of the same size, weight and number of lines are placed next to each other. This is a design consideration, and one of the heads may be changed to provide more type variety on the page.

## Banner Headlines

Many modern newspapers have abandoned daily across-the-page banners and replaced them with spread heads without readouts. A *banner* is a large-type headline that runs the full width of a page. A *spread head* is a smaller head running more than two columns and less than full width. But when a story is assigned a banner headline, it is assumed that the body copy will be placed in the extreme right-hand column. Even though the first, or left-hand, column is the most important position on the page, the right-hand column enables the reader to continue reading the story without returning to the left side of the page. In contemporary design, a number of questions arise affecting the reader's ability to continue reading the story smoothly. The answers to these questions lead to principles of handling banner headlines and readouts.

Can the story be continued in any other column to provide a change of pace and variety in makeup? There are few occasions when the story can be continued elsewhere. Least effective is to continue the story to an inside column. Indeed, it should not be placed anywhere but in the extreme right-hand column (with exceptions that will be noted), because otherwise the reader would have to search for it. Although one may argue that the reader will not have to search very long, any time—even a fraction of a second—is too long. That fraction of a second may be just the time necessary for the reader to switch to some more interesting story. After all, the reader may already know the material in the banner headline and perhaps isn't interested enough to continue when there are any impediments to reading. This is the problem of all layout devices. None should slow the reader more than a fraction of a second. Considered individually, each layout device may seem to be effective. But when there are many such devices on a page, the even rhythm of reading may be broken and reading becomes a troublesome rather than a pleasant experience.

When stories are continued from a banner headline to any column other than the extreme right, the cutoff rule is used to separate the banner from unrelated stories. The assumption is that the only column not carrying a cutoff rule must be the one where the story is continued, and it will be obvious at a glance. Indeed, it is obvious at the right-hand side, less obvious when placed at the left-hand side (first column) and almost obscure when placed inside. Cutoff rules help, but not much. It is difficult to find the column where the story is continued.

Another question arises when using a banner headline. Can more than one story relating to the banner headline be arranged so that the banner reads into each story successfully? This might work in a situation when, say, an election story breaks and one political party wins both gubernatorial and mayoral races. The banner headline may therefore refer to both stories. When the readout headlines are only one column wide, they are often hard to find. But when they are two or more columns wide, they are relatively easy to find.

A final question concerns whether a banner headline should lead into a multiple- or single-column headline. One of the older layout rules was that when a large type was used in a top headline, the reader's eyes would have difficulty adjusting to smaller types in the decks or lead paragraphs. Therefore, the reduction in size was supposed to be at least 50 percent. If a 120-point headline was used for the banner, it should read into a headline of no less than 60-point type. The 60-point type would then read into a 30-point deck, which in turn might read into an 18-point deck and from there into a 10-point lead paragraph. There was never any valid evidence that readers had difficulty adjusting their eyes to the changes in type sizes. Therefore, the only reason for reading from a banner into a multiple-column headline is simply to provide more details in the headline than could have been included in the banner. The only trouble with multiple-column headlines is that they usually lead into a single-column story, leaving the space underneath to be filled in the best way possible. If a headline for an unrelated story is placed underneath a multiple-column headline, the effect might be to confuse the reader. At times a picture is placed underneath. But neither of these solutions looks attractive.

In contemporary layout, banner headlines often lead into a single-column headline, a simple device. Or, if a banner is used, it may lead into a three-column headline, and a story may be wrapped for three columns underneath by squaring off the bottom. A final alternative is to limit the use of banner headlines to rare occasions. When used with multiple-column readouts, they won't look awkward because the news is so sensational. Too many multiple-column headlines, however, make the page spotty because they appear dark (being set in bold typefaces and relatively large types). They become centers of interest because of their weight and may be difficult to balance.

## Headline Decks, an Old Idea for Modern Papers

One technique of headline treatment that was used in the early 1900s and fell out of favor has been rediscovered. It is the headline deck. A deck is a small headline placed underneath a main headline to provide additional information. Because top headlines usually are set in relatively large type, they can use only a limited number of

*Figure 13–23 Examples of decks.*

**Measure 5 may force 2 sessions**

Marion leaders urge early adjournment

**SUPER BOWL**

Players say focus is divided between war and big game

words to communicate the full impact of a story. But a deck is usually set in type that is about 50 percent smaller than the main head. Therefore, a headline using 48-point typeface may require a 24-point deck, and a 24-point top head may require a 12- or 14-point deck. The additional words found in a deck may be just what are needed to tell readers what they want to know about a story and help them decide whether to read or not (see Figure 13–23).

Decks also add white space around headlines, much as in magazines. They can therefore help make the total page design more attractive as long as they are neither overused nor too dominating. Most often decks are set in light- to medium-weight typefaces and used on special pages such as the front page, sports or business.

## THE USE OF DESKTOP COMPUTERS IN PAGE LAYOUT

### Uncluttering Pages

Desktop computers, through programs such as PageMaker and Quark XPress, have had a major effect on page layout. Most noticeably, they have helped layout editors produce cleaner, less cluttered pages. That is possible because the computer lends itself to the production of squared-off rectangular modules that lend an aura of keen organization to a page. Perhaps the most outstanding characteristic of contemporary layout is the use of various types of rectangular modules, both small and quite large, arranged both vertically and horizontally. All these variations are necessary to enliven the page.

Although the layout person need not necessarily be restricted to rectangles, he or she will most likely take advantage of the opportunities that modules (and the computer) provide. Please note, however, that a modular page is not necessarily easier to read. It *is* easier to look at because of its orderliness. This orderliness also makes it relatively easy for readers to find stories they might miss on "busy" pages with a complex structure. Desktop layout therefore provides great opportunities for achieving page simplicity, one of the main objectives of good newspaper layout.

## Seeing an Entire Page While Planning the Layout

Desktop computers also allow the layout person to see how each element will look on the page. Such a bird's-eye view is usually helpful because the planner can see relationships that might otherwise have gone unrecognized.

Laying out pages on a tiny screen may be difficult for some people, whereas a larger screen may provide a more realistic view. However, the small screen usually is an advantage rather than a hindrance. Even smaller are the thumbnail sketches of pages that some computer programs offer. By seeing thumbnail sketches of all pages at one time, the layout person can check the style for an entire newspaper section at one viewing.

## Computers Offer Many Layout Alternatives

Another advantage of computers is that many desktop programs allow a variety of useful layout alternatives. Among these are the ability to incorporate illustrations into the text, to set type by following picture contours, to turn type lines to an angular position and to add stylistic page devices such as type rules or type boxes.

With desktop computers, layout editors can design pages quickly. It does not take long to decide where to place stories on a page. Many alternative designs can be explored until a particular one looks right.

## A New Language to Page Layout

Unfortunately, some of the programs for page layout have created a new language for the layout person to learn. Type boxes, for example, do not mean the same thing they do in the printing industry. Some program designers call them *frames.* When new layout editors get on the job, they may find that the terminology is not exactly the same as they learned in school.

## TAKE A LOOK AT FULL-PAGE NEWSPAPER LAYOUTS

Newspapers in the United States are usually attractively designed (see Figures 13–24, 13–25, 13–26, 13–27 and 13–28). Most are very readable, and editors take special pains to avoid errors. More time and effort is taken these days to do a good job, and the computer is the designer's best friend.

In studying these examples, note the effect of one particular layout and design technique: dividing an entire page into two *uneven* major segments. On vertical layouts, one column may be chosen as the most significant page divider, such as between the second and third columns, and the column space may run from the very top to the bottom of the entire page. Such a page division is usually attractive, but it also has a value of bringing order to the page.

This space also may be used horizontally, instead of vertically, dividing the page into two-thirds at the top and one-third at the bottom. There also are other page divisions that may look attractive, but these two are the most outstanding.

The most important part of laying out and designing a newspaper is realizing that creativity underlines these two activities. No hard-and-fast rules can control these two functions. It is time to develop more creative people who do layout and design.

What is needed now is to relate research on newspaper reading with the practice of design. This, along with learning as much as possible from daily newspapers, will help designers meet the challenge of other media, especially the Internet. Computers have helped in the past, but the best alternative methods have yet to be discovered.

Chicago Tribune

**INSIDE SECTION 7**



# Business

SUNDAY, AUGUST 27, 1995

N

**AN EARLY COMEBACK**

*Some consumers are coming out in favor of mortgage prepayment penalties.*

**See Jane Bryant Quinn, Page 3**

**BINARY BEAT**

**Microsoft quickens antitrust suspicions**

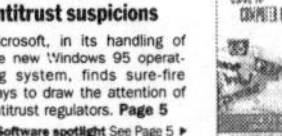

Microsoft, in its handling of the new Windows 95 operating system, finds sure-fire ways to draw the attention of antitrust regulators. **Page 5**

**Software spotlight** See Page 5 ▸

**YOUR MONEY**

**Defending against airborne dollar**

The currency's sudden jump is spurring some investors to put money to work in certain foreign companies. **Page 3**

**Lots of companies playing summer stock**

Season's high-tech hype offers glimpse of future in the computer-technology sector. Andrew Leckey, **Page 3**

Daniel Barbier/Image Bank

**WEEK IN REVIEW**

- ABC News apologizes for reporting that cigarettes were "spiked" with extra nicotine.
- Billionaire Warren Buffett snaps up rest of insurance firm GEICO for $2.3 billion.
- Media moguls Barry Diller and John Malone make ripples in a calm pool. **Page 2**

AP photo

Waste Management Inc. co-founder and former Blockbuster Entertainment Corp. chief H. Wayne Huizenga is now focusing on building and diversifying Republic Waste Industries Inc.

## 'Mr. H': Billionaire with Midas touch

### Huizenga again trying to turn garbage into gold

**By Casey Bukro**
TRIBUNE STAFF WRITER

Billionaire H. Wayne Huizenga's startling re-entry into the garbage business last May jolted the industry, and admiring investors cast votes of approval with millions of dollars.

The former Blockbuster Entertainment Corp. chief stepped in as chairman and chief executive of Republic Waste Industries Inc., leading a group of investors that sank $126 million into the Atlanta-based firm.

Republic Waste stock zoomed 766 percent, from $3 to $26 a share, before leveling off at $20.

The man with the Midas touch, analysts said, made the trash business smell sweeter. Observers prophesized that Huizenga would build a waste-management empire.

Actually, Huizenga—the only man in America to control three professional sports franchises—is far more ambitious than that, which is in keeping with his reputation as an architect of companies that grow like gangbusters.

"We will be a diversified-service company," says Huizenga, 57, his voice soft but edged with enthusiasm. "We plan on being in five of six different industries and build all of them simultaneously. We will not be a pure play, which some people consider a negative.

"We will grow by acquisition and internally in each of those five or six industries, so growth is more rapid and it will be a more exciting company," says Huizenga, who heads Huizenga Holdings Inc. in Ft. Lauderdale, where his staff refers to the boss as "Mr. H" or "Wayne."

Aside from his interest in Republic Waste and the $30 billion solid-waste disposal industry, Huizenga reveals that the electronic-security business for commercial establishments is the next target for his multifaceted service company. He declines to name others, fearful of tipping his hand.

What are his chances of success? Consider this:

A businessman who inspires envy and close attention, Huizenga made his first fortune by co-founding Waste Management Inc. in 1968. Huizenga (pronounced HIGH-zenga) left the Oak Brook-based company, now called WMX Technologies Inc., in 1984 as vice chairman with an annual salary of $427,680.

Three years later, he took control of Blockbuster Entertainment. By 1994, Blockbuster had grown from 27 stores to more than 3,000. Last September, Huizenga sold Blockbuster to Viacom Inc. in a $7.6 billion stock merger, producing his second fortune.

While still at Blockbuster, he bought the Miami Dolphins football team, Florida Marlins baseball team and Florida Panthers hockey team for $254 million. Financial World Magazine estimates that the franchises now are worth $325 million, giving him his third fortune.

But once Blockbuster was sold, Huizenga was without a job. That's when opportunity knocked, in the way it often does for highly successful—and alert—business people.

As Peter Huizenga, Wayne's cousin, tells the story:

"Events gave him an opportuni-

SEE MIDAS, PAGE 2

## Entertainment superstar of future supermarkets

### Innovation playing at a store near you

**By George Gunset**
TRIBUNE STAFF WRITER

Juvenal, dead these 1,900 years, thought the fellow Romans of his day had grown soft and sought only two things: *panem et circenses*.

Bread and circuses are just as important in our times. Man still doesn't live by bread alone and there is growing evidence that Americans want to be entertained when they shop for bread and other food items.

In 10 years, or maybe five, the U.S. supermarket could be remade, putting in place more changes than have occurred in the last 40 years.

Customer ordering by computer and more home delivery seem certain. But the real wave of the future, it appears, is new store formats and concepts to attract and retain customers.

Already, a few food stores have live piano performances. Fine dining restaurants also are being tried.

The best example, said supermarket industry consultant W.R. Bishop Jr., is Byerly's, a 10-store operation based in Minneapolis that soon will open a store in Deerfield.

"Byerly's has gained such a good reputation for its wild rice soup, that it is selling the product to other stores," Bishop said.

Some stores, such as Price Chopper, a regional chain based in Schenectady, N.Y., are even featuring "singles" nights where the younger crowd apparently can meet, shop and date at the same time.

Retailers "are concerned with finding ways of standing out from their competitors," Bishop said. And those competitors are legion.

Grocery stores today number more than 130,000. Supermarkets themselves total 29,000. Then there are the convenience stores. And the gas stations. Hundreds of supercenters also have sprung up, many by retailing heavyweights Wal-Mart Stores Inc. and Kmart Corp., which combine traditional discount stores with grocery items and have the advantage of cheap labor and warehouse prices.

"Entertainment and information are keys to the supermarkets of the future," said Glen Terbeek, managing partner of Andersen Consulting for food and consumer packaged goods. "The shopping experience will become much more important."

Stores that adapt will thrive, but they must add value to the shopping experience, he warns. "The value has to be innovative, exciting, selling solutions, and removing uncertainty."

Innovation in the $400 billion-a-year industry had not exactly been a byword in the last couple of decades. The market hit a saturation point in those years, as mass marketing, merchandising and production resulted in perhaps too many stores and too many products but, because of flat population growth, no new customers.

Before the mid-1970s, the industry had a reputation for efficiency and sophistication.

"In many ways, the ideal of modern capitalism is not the factory, or even the stock market," said John P. Walsh, assistant professor of sociology at the University of Illinois at Chicago.

"From the consumer's point of view, the supermarket is the embodiment of the basic American values of consumption and affluence, rivaled perhaps by the shopping mall," said Walsh, who has chronicled changes in food stores over the last 40 years in his book "Supermarkets Transformed: Understanding Organizational and Technological Innovations."

Technological change within the industry is not just a scientific or economic process, he said.

"Innovations upset the established social relations within an organization and other organizations in its network," he said. "One cannot easily change one part of the system without accounting for its effects on the rest of the system."

For example, Walsh pointed out that when price scanners were introduced, workers, customers, grocers and computer firms each argued for adoption of a system favorable to their interests before a compromise was reached.

Most supermarkets are built around central buying, distribution and self-service, Terbeek said, so the customer in the old corner store "became a consumer. We lost

SEE STORE, PAGE 13

Glen Terbeek of Andersen Consulting poses at the firm's Smart Store research center. Besides pasta, this display features breads, spices and even wines.

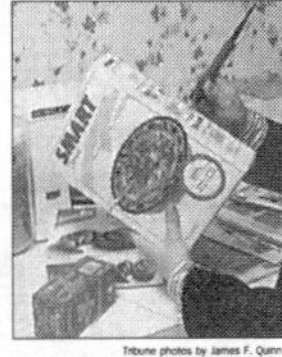


Tribune photos by James F. Quinn

A scanning wand is demonstrated at the center. A customer would use the wand at home to order fresh supplies.

## Microsoft uses soap-marketing secrets to clean up in software sales

**By Marianne Taylor and James Coates**
TRIBUNE STAFF WRITERS

Can you sell a computer the way you would sell soap?

More precisely, can Microsoft Corp. sell computer software by using the tried-and-true principle of "branding," which is how consumer-product companies build loyalty to a particular kind of laundry detergent?

In a word, yes.

That's precisely what Microsoft has set out to do.

Behind the megahype of this week's rollout of Microsoft's new computer operating system, Windows 95, is a quieter campaign to build Microsoft's products into brand names as recognizable to consumers as Procter & Gamble Co.'s Tide

Important in any business, brand-name recognition is key to commanding the vast consumer marketplace that Microsoft wants to dominate with its Windows 95 operating system and, more important, its computer-software products.

Microsoft equips roughly 9 out of 10 personal computers sold to consumers with its own operating systems, but the real prize—now and in the future—comes from selling the accompanying software, the programs that allow you to do the tasks you need and play the games you want on your home computer.

Microsoft has a great deal of competition in this field from players such as Novell Inc., Oracle Systems Corp., International Business Machines Corp., Apple Computer Inc., Lotus Development Corp., Borland International Inc. and many others.

And this is where it wants to build brand-name loyalty.

Until now, Microsoft has largely operated in a traditional business-oriented customer base, concentrated its advertising there and allowed a somewhat fragmented approach in its marketing worldwide.

The unbelievably profitable Microsoft, however, sees even higher sales and profits in the consumer realm. Sales to consumers now account for roughly 20 percent of Microsofts' $4.6 billion in annual revenues. But they are expected to exceed the company's sales to businesses by 1998 because of the boom in home computers and on-line services, a top Microsoft executive said in an interview earlier this year.

That's where the growth in the business will be, Microsoft Chairman Bill Gates has said, as computers increasingly become the delivery vehicles for the new

SEE MICROSOFT, PAGE 4

*Figure 13–24 This well-designed business page is easy to read because it is orderly. The design is contemporary and not dull. It is a mixture of vertical and horizontal display.*

Sunday News Journal, Wilmington, Del. Oct. 15, 1995 · Section B

# Local

## Panel endorses school election registration

### Task force: Control by Del. would cut opportunity for voter fraud

By JAMES MERRIWEATHER
Staff reporter

To avoid the potential for voter fraud, the state's 19 public school districts should hand over the responsibility for board elections to the state Department of Elections, a legislative task force has recommended.

Legislation recently filed supports the task force's recommendation. Some school officials say they'd like to end their oversight of school elections. However, others fear a proposal to combine the general and school elections could politicize school elections. Two bills that call for changing the school election process will be pending in the House Education Committee when the General Assembly reconvenes in January.

The seven-member task force, appointed to root out potential for voter fraud, said voters should be required to register to participate in school board elections. But it recommended waiting until 2002 to take that step.

The head of a New Castle County citizens' group says that time lag keeps the recommendations from being little more than a few good first steps.

"Without question, registration is what's needed," said Allen H. Kemp of Wilmington, president of Citizens for Fair School Taxes.

"To really have safeguards, we need for people to register to vote like they do for all other elections."

> "Without question, registration is what's needed. To really have safeguards, we need for people to register to vote like they do for all other elections."
>
> Allen H. Kemp

The task force acknowledged as much, but was put off by the time and money needed to put a registration system into place. Without it, the recommendations offer only a short-range "Band-aid approach" to shoring up procedures for determining voter eligibility, said state Elections Commissioner Thomas J. Cook, the task force chairman.

Currently, school district information — which would show, for instance, which street addresses are in which school districts — is not included in the state's voter registration system. Estimates are that it would take 18 months and $250,000 to implement a system including such data — thus the recommendation to wait until population figures are updated by the 2000 census.

As things stand now, the task force determined, school districts use a hodge-podge of procedures to determine voter eligibility. In some cases, poll workers merely take the word of walk-in voters that they live in the appropriate district and that they have not voted previously at other polling places.

"Unfortunately, there is a possibility for fraud the way it's set up now," Cook said. The Department of Elections would be expected to bring uniformity, as well as expertise, to the process.

State law requires potential voters be asked for their names and addresses, whether they are U.S. citizens and whether they already have voted in the election. The task force, authorized in February by a joint resolution signed by Gov. Carper, proposes to require that their responses be

See ELECTIONS — B4

## Making an impression

The News Journal/GARY EMEIGH

Ten-year-old Kyle Price (left) watches as Dover policeman Ralph Taylor Jr. fingerprints his 6-year-old brother, Ryan, during Child Safety Awareness Day Saturday at the Dover Mall. Taylor works in drug education and maintains children's fingerprint records.

## Fingerprinting, photos help keep kids safe

By ESTEBAN PARRA
Sussex bureau reporter

DOVER — B.J. May was a little jittery Saturday afternoon. After all, this was the first time the 7-year-old had been fingerprinted.

"You nervous?" Dover policeman Ralph L. Taylor Jr. asked the Fairview Elementary first-grader.

"Yes," May responded, looking at the black ink covering the fingertips of her right hand.

B.J. was one of more than 100 children fingerprinted Saturday at Boscov's department store in Dover Mall. Boscov's event kicked off the YWCA's Week Without Violence, a national effort to make communities a safer place to live, work and learn.

"It's been a positive showing," Taylor said. "The parents have been very supportive."

Fingerprinting is one way to help parents and police find missing children. Taylor, whose four children are fingerprinted, said keeping current photographs or videos of children is another way to help.

But educating children is the best way to prevent many potential dangers, the four-year police veteran said.

"I would say the best thing for a parent is to open a line of communication," he said. "Establish guidelines, something that they can go by for the rest of their lives."

B.J.'s mother, Michelle Kuntzman, said events like this help parents as well as kids.

She said she learned about child abuse while listening to child abuse expert Yvonne Kirksey, with Delawareans United to Prevent Child Abuse, made sure parents were around to listen and learn when she quizzed children on personal safety.

For instance, she said, many parents are not aware there are four private parts children should not allow others to touch: the buttocks, the groin area, chest and mouth.

"The mouth is the one parents usually don't think about," she said.

Other participants included the city of Dover Electric Department, Delaware State Police and the Department of Public Safety. McGruff the Crime Dog and Curly the Clown also appeared, and Halloween candy collection bags were given to children.

**MORE EVENTS**

YWCA Week Without Violence Events continue today as the Kent County Community Justice Center hosts a Conflict Resolution Training session from 9 a.m. to 6 p.m. at the Tatnall Building.

Statewide events are scheduled through Oct. 22. For information, call the New Castle County YWCA at 658-7548 or 366-1673.

## Downpour brings welcome relief

### But it won't wash away drought

By ANN MANSER
Staff reporter

DROUGHT '95

Anyone who braved the downpour Saturday at the University of Delaware's Homecoming or got caught in the day's string of rain-induced traffic accidents may not believe it, but the drought of 1995 isn't over. Still, October is off to a good start.

New Castle County Airport was pelted with 1.69 inches of rain by 10 p.m. Saturday, bringing the total for the month to almost 4 inches, said the National Weather Service in Mount Holly, N.J. In a typical October, only 1.3 inches would have fallen by now. The rain fell even harder north and west of Wilmington, with up to 3 inches falling.

"But, drought-wise, we still need a lot more help," Meteorologist Anthony Gigi said Saturday. "This rain today won't ... end the drought."

And rain isn't likely for the next several days. The weather service forecast shows only a 30 percent chance of showers today and predicts fair skies from Monday through Wednesday.

This summer's drought, the third-worst in a century according to the weather service, severely reduced stream flows in northern Delaware and forced officials to impose restrictions on water use.

The weather service calculates rainfall per "water year" — a designation that begins on Oct. 1, when ground water reserves usually begin replenishing. In the water year that just ended, 31.5 inches of rain fell in northern Delaware, compared with an average yearly total of 41.38 inches.

Yet, the year ended well with 5.17 inches falling in September — 1.58 inches above normal.

Thanks to the September rains and to a drenching storm Oct. 5, officials last week eased a few of the drought-related restrictions imposed last month north of the Chesapeake & Delaware Canal.

## United Methodists approve ministry plan

By RHONDA B. GRAHAM
Staff reporter

► Members criticize model B6

More than 500 United Methodists gathered in Dover Saturday to vote on a plan that could lead to a major restructuring of how the denomination conducts its work in Delaware and on Maryland's Eastern Shore.

A majority approved The Lost Coin report, which calls for establishing "Mission Areas" as a model for ministry and an expression of Methodism founder John Wesley's belief that "the world is my parish."

The model calls for clergy and lay members to be appointed to teams in mission areas based on geographic and demographic boundaries.

Bishop Susan M. Morrison will name a transition team to work out details with the 477 congregations in the Peninsula-Delaware Annual Conference.

The team's mandate is a strategic plan that focuses each of the 97,440 United Methodists in the region on missions, stewardship and teamwork. "Those are the things that lead to church growth and effective ministry," said the Rev. Susan Keirn Kester, the Wilmington District superintendent.

Kester was part of the Lost Coin group, which spent 18 months researching and meeting with members of the conference.

They learned there are at least 15 different worship styles in the conference, but 65 percent of the members do not attend the principal weekly Sunday service.

The strategic plan will be voted upon in June 1997.

After listening to the hour-long debate, Marion Handy of Union Church in Delmar changed her mind about voting against the resolution. Her church had instructed her to vote it down because it did not feel the report included its views.

She said, "I think the feeling is that we were being forced to align with the white churches. They need us and we need them, but don't jam us down their throats."

## Planners propose major remodeling of parts of U.S. 13

### Walkways and bike paths would be added to areas of the road that were bypassed.

By JEFF MONTGOMERY
Dover Bureau reporter

U.S. 13 would become a slower, gentler "boulevard" with walkways and bike paths, according to a draft transportation plan scheduled for public workshops in Kent County next week.

The Dover-Kent County Long Range Transportation Plan outlines dozens of future projects, studies and strategies aimed at meeting the region's transportation needs thro.. h 2020.

Developed by the Dover-Kent County Metropolitan Planning Organization, the local plan would become a chapter in Delaware Department of Transportation's statewide long-range transportation plan. The statewide version is now due in early 1996.

Among the features:

■ Conversion of bypassed portions of U.S. 13 in Dover and Smyrna into city boulevards, with walkways, bike paths, landscaping and lower speeds.

■ Development of U.S. 13 south of State Street into a limited access highway — a proposal bitterly opposed by highway businesses in 1993 and shelved under pressure by DelDOT.

■ Construction of a Camden truck bypass, and studies of several other projects, including creation of a west Dover bypass, partial Delaware 1 interchange at Delawa-e 8 and a new connector from Saulsbury Road to U.S. 13 south of the city.

■ Establishment of a commercial large aircraft maintenance center at Dover Air Force Base, and expansion of civilian aviation activities at the base.

The planning group is governed by representatives from Dover, Kent County, county municipalities, DelDOT and the Delaware Economic Development Office.

Federal law requires DelDOT to accept the group's recommendations in spending decisions involving federal aid within the region.

"We haven't discussed speed limits on Route 13 yet, but I think the speed would come down from 40 and 45 mph to 35 mph," said Dover-Kent MPO Executive Director Juanita Wieczoreck. "The highway has been bypassed, and now its purpose has changed. It's become more of a local street."

South of a planned new connector between U.S. 13 and Delaware 1, Wieczoreck said, U.S. 13 serves interstate purposes.

"Once Delaware 1 is finished, 13 will end up acting as a de facto north-south interstate highway," Wieczoreck said. "It needs to be upgraded to handle the purposes it's serving. We need to preserve land along the corridor. But when we get down to places like Harrington, that's going to create questions about how it will affect the town."

The plan recommends preserving land along the U.S. 113 and Delaware 1 corridor for future needs, in addition to U.S. 13. Tighter controls on entrances and intersections were recommended for all three roads, as well as Delaware 8 between Dover and the Maryland line.

See U.S. 13 — B4

**IF YOU GO**

WHAT: Dover-Kent County Metropolitan Planning Organization public workshops on a long-range transportation plan for Kent County.

WHEN AND WHERE:

4 to 6 p.m. Oct. 23 at Smyrna Town Hall, Market Street Plaza.

4 to 6 p.m. Oct. 24 at Milford Public Library, 11 S.E. Front St, Milford.

3 to 6 p.m. Oct. 25 at Dover City Hall, Loockerman Street.

NEW CASTLE COUNTY
KENT COUNTY
Dover
New walkways, bikeways, landscaping and lower speeds proposed for U.S. 13 "boulevard"
Proposed limited access highway
The News Journal

*Figure 13–25 An inside section with somewhat the same design as the page in Figure 13–24 except this page has two columns on the right and four on the left. Also, there is more horizontal display on this page. Note the use of tone borders at the top and bottom, which provide a frame and sharpen the appearance of order on the page.*

**A12** LEXINGTON HERALD-LEADER, LEXINGTON, KY. ■ SATURDAY, NOVEMBER 4, 1995

# EDITORIALS

## Improved, not perfected

KERA test has helped writing in public schools . . .

Opponents of Kentucky's school reforms are using perfection as a weapon. If a portion of the school reform act isn't working perfectly, it is considered prima facie evidence that the entire law should be used for kindling.

We saw perfection wielded against progress again this week. Opponents of the Kentucky Education Reform Act released selective criticism of the state's testing program. Experts were quoted as saying the test was "seriously flawed" and that using it to decide which schools receive monetary rewards puts the program "in jeopardy of a successful lawsuit."

Once contacted, the experts cited by the Family Foundation of Kentucky all said their criticisms were taken out of context. Penney Sanders, director of the Kentucky Office of Education Accountability, is right when she says the foundation's efforts "serve only to inflame." The state, Sanders says, is "caught in a regrettable anti-education furor . . ."

True enough. And worse. Critics of KERA — like the foundation and Republican gubernatorial candidate Larry Forgy — have made perfection the measure of success. A new test is "seriously flawed," and that is reason for throwing it out.

That's reactionary. KERA demanded a new test because the old tests were meaningless. Who, after all, has a job answering multiple-choice tests? Real work demands thinking, problem solving and communicating. The KERA test was meant to measure what a student could do in the real world. The KERA test asks students to perform in realistic situations.

The problem was (and is) that no such test existed. Kentucky could have stayed with the old-style "multiple-guess" test, or it could build its own system. Kentucky decided to build.

That task was enormously complicated. And it was guaranteed to have problems.

But so what? The state needed a new test to spur innovations in teaching. That's happened. A group of Harvard University professors, reporting recently to the Brookings Institution, has found that "as a result of the assessment's emphasis on writing, more students are writing in Kentucky than were before." Writing has improved and teachers are responding to a test that asks students to perform, not just regurgitate.

The Kentucky test may be "seriously flawed" as an accurate, lawsuit-proof measure of student achievement. But it is working to improve classroom studies.

That's a darn good start. Because reform is about continuous improvement, not perfection. It is about finding better ways every day to teach, learn and test.

### . . . While others say testing is excessive

Americans' obsession with school tests is a mystery to educators in other countries.

We visited recently with Manfred Ehringer, head of the local school inspectorate in Stuttgart, Germany. "We think you test to a very great excess," Ehringer told us. "We have turned away from that."

In order to put Ehringer's comments in perspective, you have to understand that Germany's primary and secondary schools are remarkably efficient. Everyone learns, and that is no exaggeration. Companies say they have no reason to test applicants who have completed German schools. A diploma is evidence enough of academic competence.

The German dissatisfaction with testing lies in the realization that tests are wildly imperfect tools. "Testing is always an unusual situation," Ehringer said. "It is a small basis for making a decision."

What do the Germans do? They rely on the judgment of classroom teachers in the early years. "We only test when necessary," Ehringer said. "The final say is the teacher's." And in higher grades, German students are asked to perform, to do real-life work that is then graded, subjectively, by panels of experts.

Just like KERA.

## Writing and reading

Applause for authors — of fiction and better laws

They weren't quite enough to fill Rupp Arena, but the 600 or so people who turned out for an evening with four Kentucky authors a week ago did exceed the capacity of The Carnegie Center for Literacy and Learning, forcing the event to move to Christ Church Cathedral. And while the authors seemed a tad uncomfortable speaking from a pulpit, their words did have a spiritual quality.

The purpose of the event, sponsored by The Writer's Voice, was to pay homage to the late Wallace Stegner, who was a mentor to each while they were writing fellows at Stanford University. Their reminiscences of Stegner vividly demonstrated the enormous influence one person's life can have on another — a heartening message.

At a time when many are deploring the quality of programming on television and the relentless violence in movies, it's also heartening that so many people think spending an evening listening to Kentucky writers qualifies as Saturday night entertainment.

■ ■ ■

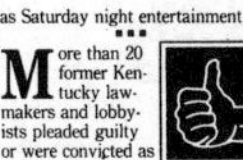

More than 20 former Kentucky lawmakers and lobbyists pleaded guilty or were convicted as a result of Operation BOPTROT. The body count might have been higher, according to former U.S. Attorney Joe Whittle, if federal anti-corruption laws didn't have so many holes.

Because of inadequacies in the corruption statutes, BOPTROT prosecutors relied mainly on laws involving extortion, wire fraud and mail fraud. In some cases, Whittle said recently, those laws did not apply to the wrongdoing uncovered by federal investigators.

U.S. Sen. Mitch McConnell has introduced a bill to rectify the problems with the corruption laws, and to give federal officials broad authority to prosecute government corruption and election fraud. Similar legislation has passed the Senate four times, only to be killed by the House. This time around, the bill deserves full congressional approval.

JOHN BRANCH, San Antonio Express-News

# LETTERS

## Environment needs more attention in the schools

I have discovered that our educational system has drawn away from the environmental subjects they are supposedly teaching. For instance, how many times in a school year are students allowed to go out and plant trees or gather cans and other recyclable wastes? This is probably done once a year, or in some schools, not at all. It should be a monthly event. I'm not saying that the entire school should go out each month, but at least one class should have the opportunity to help our Earth.

Not only will the students be learning, but they will also be improving the future of our environment. We need to raise children who are aware of their surroundings. Not all people are meant to be mathematicians or historians. Some people have hopes of being conservationists and biological experts. These outside studies will teach children to appreciate the good things our Earth holds. If we don't do something today, there will be no tomorrow.

PEGGY ALLEN
LONDON

## ORVs do damage

Responsible four-wheelers — Not!

It is hard for me to conceive of letter writer Kendra Cook's "responsible" use of her off-road-vehicle in Washington, D.C. (Letter, Oct. 2) There might be some responsible ORV users in Kentucky, but they are certainly not the lowest common denominator of this group.

How can this group claim to be responsible when its members routinely drive their vehicles where it is unlawful for them to go? Their sport is to drive straight up steep hillsides causing severe and permanent erosion damage. ORV users do not have this right!

Second, I am curious. To which "organized four-wheeling community" does Cook belong? It would be my guess that Honda, Suzuki and Kawasaki would like to join, if they are not already members. And I am surprised at the Herald-Leader. I thought the use of this forum by organized lobbyists is not advocated.

If Cook's organization is doing anything in Kentucky, spending any money to repair damage caused by ORVs, it is not only insignificant, it is invisible. They should consider doing much, much more.

DOUG EPLING
LEXINGTON

## In defense of tourism

As president of the Greater Lexington Hotel-Motel Association, I would like to voice our organization's strong disapproval of the Sept. 20 editorial, "Not another Branson, please." Comparing Lexington with Branson or Kentucky's tourism industry with that of Missouri's is unfair and unrealistic. Branson may well have become an overnight boom-town suffering the problems associated with extraordinary success, but Lexington's strong leadership and clear vision would certainly not repeat the mistakes made there.

"Tourism makes for a stinking economy," your editorial stated. Lexington's economy remains quite healthy, and we enjoy one of the lowest unemployment rates in the state. I dare say this would not be true if the tourism industry and its financial contributions were no longer a part of our community.

Your editorial also stated, "It is low-pay, low-skill, part-time work." This must be the reason travel and tourism is the second largest private employer in the nation, and why 145,000 Kentuckians work in our industry. I can assure you the majority of these people do more than "make beds, take tickets and wash dishes," and usually work more hours each week than most full-time employees in other businesses. Please check your facts before you attack another industry that contributes so many positive aspects to the city in which you live.

MALCOLM JENNINGS
PRESIDENT
LEXINGTON HOTEL-MOTEL ASSOCIATION

## 'Sick' tax must go

You and your readers should aware of the severe inequity in the Kentucky health-care system referred to as the provider tax or "sick" tax.

This tax is assessed to health-care professionals and providers to secure more federal matching money to provide health-care services for the uninsured. However, this money goes into the state's General Fund with no required redistribution to the health-care system. It is interesting to note that the state had a $130 million surplus in the General Fund this past fiscal year.

Since neither Medicare nor Medicaid covers the cost of the tax, this tax adds as much as 10 percent to the costs of sick people who are commercially insured or pay privately for their health care. The tax is money that would otherwise be available for direct patient-care use. Also in jeopardy are jobs in the health-care market.

This tax needs to be eliminated and a mechanism devised to provide money for uninsured citizens of Kentucky. Since health care for the uninsured is a societal issue, funding for these services should be borne by society as a whole, not just one sector.

Many of our legislators in Frankfort want to eliminate this tax. As citizens who vote, we should demand that our next governor do the right thing and work with our legislators to abolish this tax.

People should not pay a tax just because they get sick. Let the gubernatorial candidates know that this tax is misguided and wrong, and they must take a stand against it!

KENNETH R. UNGER
CHIEF EXECUTIVE OFFICER
SCOTT GENERAL HOSPITAL
GEORGETOWN

## A low level of discourse

I am a newcomer to Kentucky and to voting. I moved here just in time to register to vote and I am finally of the age that I may "officially" take part in the decision-making of our nation. Once registered, I began searching out sources of information so that I might make a more informed decision as to whom I would favor. So far I have learned much about the alleged characters of the candidates, but try as I might, I know next to nothing about where they stand on the issues or what the issues are for that matter.

Is this the breadth of our political discourse? Will I be forced to vote intuitively for lack of any substantive debate? I recognize that the character of a candidate can be viewed as an important factor, but is it so crucial as to be given such a disproportionate amount of coverage, or are we the people viewed as being so ignorant that our elections have degenerated into little more than Geraldo-style finger-pointing and petty argumentation?

I am told that the majority of Americans do not even vote. How can I be surprised when I can barely read the comments of our candidates without becoming bored with their shallowness. Is this what I have to look forward to?

For our own benefit as well as that of a generation just entering this political system (a generation widely perceived as apathetic enough already), please let's all get together and see if we can't demand more of our leaders and those in the best position to ask them the tough questions. If pettiness is all we seek, it can be found easily enough on our television screens at much less risk than the fate of our nation and future of our children.

JASON FULTS
MARIBA

## Bad public policies

Why are candidates Larry Forgy and Paul Patton, as well current public office-holders, so willing to support public policies that result in Americans killing Americans?

According to the Centers for Disease Control more than 400,000 Americans die every year from the poisonous, addictive effects of tobacco products. Annual sales of tobacco products in this country amount amount to $50 billion. One American dies for every $125,000 of tobacco product sales. Most people would say that sacrificing one life for this amount of business is valuing human life too cheaply. Yet Kentucky's members of Congress, our governor, most of our state legislators, the two gubernatorial candidates, etc., all publicly profess strong support for the well being of the tobacco industry. Why can they not acknowledge that the tobacco industry is a health disaster and work on a way to rapidly end the production and sale of tobacco? Every voter should ask this question of all the people in and running for public office.

Candidates Forgy and Patton are in favor of legalizing the carrying of concealed guns. Since guns can only be used for destructive purposes, it is appalling that they take such an irrational position. Close to 40,000 Americans are killed with guns every year. About 80 percent of these deaths involve members of gun-owning households, their friends, neighbors and co-workers. Our public officials should be proposing policies to reduce the irresponsible use of guns.

STEPHEN SENFT
LEXINGTON

## GOP harms health care

We've all read about the Republicans' proposal in which they claim will "save" Medicare. This rescue, however, could prove catastrophic to Medicare and Medicaid recipients.

Cutting physician payments while doubling patient costs and encouraging recipients to join managed-care programs will prevent many people in desperate need of care from receiving it. People with disabilities whose medical, and often everyday, needs are dependent on Medicaid will pay the price of the Republicans' so-called reforms. Necessities such as wheelchairs or in-home care may become luxuries that patients with spinal chord injuries can no longer afford.

Hospitals providing rehabilitative care receive much of their income from Medicaid. How will they continue to provide their services without cutting quality or the number of patients they serve? Will they be forced to serve only wealthy patients or will they not provide the time for patients and their families to learn the skills necessary to cope?

For those who still believe that this Republican proposal is what we all need, think again. The people who will be affected are Alzheimer's patients, spinal-cord injury and nursing-home patients and people with other disabilities. They are your parents, your children, your neighbors.

TOM GRAVITT
LEXINGTON

LEXINGTON HERALD-LEADER

LEWIS OWENS,
President and Publisher

TIMOTHY M. KELLY,
Editor and Sr. Vice President

DAVID HOLWERK,
Managing Editor

PAM LUECKE,
Editorial Page Editor

BILL BISHOP,
Associate Editor

The Lexington Herald-Leader is a Knight-Ridder newspaper, published by the Lexington Herald-Leader Co., 100 Midland Avenue, Lexington, Ky. 40508. Editorials reflect the opinion of the editorial board. Members are Lewis Owens; Timothy M. Kelly; Pam Luecke; Bill Bishop; editorial writers Bob Campbell, Larry Dale Keeling, Audrey Lee and Art Jester, and cartoonist Joel Pett.

***Figure 13–26** An editorial page that also uses solid two-point borders top and bottom. This page is much lighter than front pages but is still interesting. Editorial pages can use horizontal display, but this one is all vertical.*

Reprinted with permission of The Lexington Herald-Leader.

Daily Camera
TUESDAY
October 17, 1995

# TODAY

Tell us what you think of today's paper. Call the Comment Line, 473-1615

EDITORIALS ■ 2
OBITUARIES ■ 3
CLASSIFIEDS ■ 4
**B**
Editors: Coleman Cornelius, Kevin Kaufman

### ■ DOUBLE ROYALTY
**Twins take crowns at Fairview homecoming**

All homecoming week, **Corry** and **Clay Judd**, seniors at **Fairview High School**, thought maybe their friends weren't joking when they all said they were voting for the Judds — a pair of 17-year-old twins — for homecoming king and queen.

"It was so weird when my brother's name was called," **Corry** said about the announcement that **Clay** was king. "And then I thought, 'Oh no, not me.' "

But it was **Corry**, selected by seniors from five queen candidates. And **Fairview** had its first twin king and queen.

The two, who attended **Douglass Elementary** and **Platt Middle** before **Fairview**, are used to sharing things.

They both are athletes — **Clay** hopes to play college soccer at Boston University or CU, and **Corry** was a state tennis champ last year — and they serve together on the Fairview Student Council.

### ■ DON'T MISS IT

**School board forum:** The Interfaith Council of Boulder will quiz the candidates at a forum called "Troubled Times in our Public Schools: Can we find common ground?" from **7:30 to 9:30 p.m. Thursday,** at **St. Andrew Presbyterian Church,** 3700 Baseline Road, Boulder.

**School board forum II:** The Boulder Valley School District Parent Council is sponsoring a school board forum from **7 to 9 p.m. Oct. 24** at **Centaurus High School**, 10300 South Boulder Road, Lafayette.

### ■ GOLD STARS

✓ To commended students in the 1996 National Merit Scholarship competition from **Centaurus High School: Ashley Johnson, Conor Merrigan** and **Erica Vanbuskirk.**

✓ To **Fairview High School's** semifinalists and commended students in the National Merit Scholarship competition: **Paul Covell, Peter Daw, Jennifer Diebel, Andre Houssney, Nicole Itano, Emily Mankin, Tim Pepper, David Sloss, Nathaneal Smith, Laura Stoffer, Isaac Andres, Scott Barkley, Jay Brasch, Jacqueline Decker, Curtis Dozier, Katherine Henneman, Amritha Ibrahim, Colette Meremonte, Brett Phillips, Dana Schneider, David Sparkman, Brian Stone, Katherine Walline, Greg Yamada** and **Trent Yang.**

✓ To Advanced Placement Scholars from **Fairview High School: Helena Broad, Adam Brod, Jane Burkett, Ryan Chuang, Katherine Doherty, Jed Burslek, Lisabeth Hall, Nicole Itano, Dana Kaersvang, Ryan Kirkpatrick, John Lavinsky, Emily Mankin, Stacey Morris, Tim Pepper, Ilana Poley, Carlos Rodriguez, Alicia Smith, Benjamin Yokell** and **Erin York.**

✓ To National Merit Scholars semifinalists from **Broomfield High School, Jeremy Jaeger, Talitha Voth** and **Daniel Yost.**

✓ To Advanced Placement Scholars with honors from **Broomfield High School, Andrew Callaway, Traci Close, Christopher DeLange, Julie Marshall, Jennifer Bourne, Charles Fuller, Katie Milan, Kelly Nebergall** and **Simne Ng.**

### ■ LIFE ON THE ROOF
**Principal makes pact, spends day outside**

**Southern Hills Middle School Principal Mike DeGuire** hedged a bet that paid off in thousands to his school and sent him outside on one of the most gorgeous days of 1995.

As an incentive to the school's 460 or so students, **DeGuire** said if the students sold $50,000 worth of magazine subscriptions he would spend a day on the roof.

The kids sold $51,000, breaking a school record and a district record for the school's size. And they get to keep $20,000 for the school — most of it will go to pay for special teaching projects. Seventh-grader **Ruth Schauble** sold $992 on her own and broke another record for individual sales.

And DeGuire got a lot of paperwork and grounds supervision done. Only those who really wanted to talk to him climbed the steps through the custodian's office.

"It was actually more productive than I thought it would be," DeGuire said.

## Group home fears to be aired

### Angry neighbors to take concerns about facility for teens to council

By JAMES BURRUS
Camera Staff Writer

Fear and outrage are prompting future neighbors of the Halcyon Residence home for youths under mental health care to bring a list of grievances to tonight's Boulder's City Council meeting.

Specifically, neighbors of the new facility at 23rd Street and Hawthorne Avenue in North Boulder are upset they were not notified the teens were being moved there. Residents also worry the background of the teens living there cannot be verified to make sure they pose no threat to neighbors' safety, as required by city law.

"We're not opposed to this group home coming here, it's just that they don't report acts of aggression in the facility because they are teenagers," said Pam Hoge, who lives nearby on Iris Avenue. "There is no way to find out if they have a criminal background because their files are closed because they are juveniles. If there is no cause for fear, and they (the city) had been straight with us about this from the start, I don't think any of us would have a problem."

About 100 neighbors of the facility, slated to open Friday, expect to attend the City Council meeting in a show of solidarity. The meeting will begin at 6 p.m. in the council chambers of the Municipal Building at 1777 Broadway.

Boulder County Mental Health Director Phoebe Norton, whose organization oversees the home, said there is no need for the neighbors to be afraid because "the boys have not been a threat to other people."

But Norton said she understands the neighbors' concern and plans to work with them to allay any fears.

"We hope they will join with us to form a neighborhood advisory committee to get ongoing input from them on how to have a well-run program that doesn't interfere with the neighbors so we can be an asset to the neighborhood," she said.

The Halcyon Residence is where students of the Halcyon School live. The school is a year-round special education facility in Martin Acres for emotionally disturbed children that also provides group and family therapy and skill-building activities.

Norton said the facility will be home to as many as eight teens, aged 11 to 15, who have problems such as attention deficit disorder, severe depression and schizophrenia.

The city approved the move Aug. 31. There is no requirement to notify neighbors that such a home is moving in nearby, according to Bob Cole, city director of project review in the planning department.

Emilie Young, who lives next door to the future facility, said she is not against the home, but is worried it might not be run properly and there will be nothing she can do about it.

"There is no level of accountability for this facility," she said.

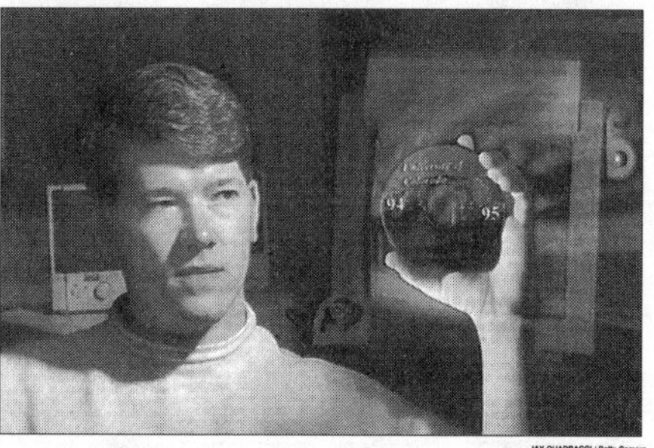

JAY QUADRACCI / Daily Camera

**PALM-SIZED YEARBOOK:** Troy Surratt, a University of Colorado May 1995 graduate, single-handedly has produced a yearbook for the 1994-95 school year on a CD-ROM computer disk after the "Coloradan" went defunct.

## 'Coloradan' lives on in CD-ROM

### CU biz grad gathers photos, video clips of '94-95 school year

By CAROL CHOREY
Camera Staff Writer

The "Hail Mary" pass, winning the Heisman Trophy and controversies concerning University of Colorado President Judith Albino and hiring new head football coach Rick Neuheisel are just some of the events that made the 1994-95 school year memorable for CU students.

But instead of flipping through a traditional yearbook to remember the past, students and alumni this year need to flip the switch on their computer and load up a CD-ROM.

Since CU's yearbook, the "Coloradan," folded, CU graduate Troy Surratt of Broomfield has been trying to rekindle interest with a yearbook of some 100 photographs, three video clips and a 3½-minute camcorder tour digitalized on a CD. The printed yearbook was last published for the 1993-94 academic year.

Although some high schools offer yearbook supplements on CD-ROM, Surratt said the electronic yearbook he developed to remember his senior year may be the first college yearbook of its kind.

The project cost him $10,000 — requiring him to sell 500 disks just to break even — but the 23-year-old business graduate said he didn't want his class to be the one that let the yearbook die.

"To me, it was my senior year and this was such a good idea, I couldn't not have done it," Surratt said. "These things happen only once and it's important to record them for history."

After all, CD-ROMs are riding the wave of computer technology, so why not use one for a yearbook?

"Yearbooks are tangible and traditional and that's important, but CD-ROMs have video, audio and the ability to search for things, which you can't do with a book," said Surratt, who is self-employed in computer graphic arts.

Although his first attempt at a CD-ROM yearbook doesn't have search capabilities, the potential is there if he or current students decide to do another one.

Ironically, Surratt thought of the idea after applying and being turned down for a job at the "Coloradan" his sophomore year. After the "Coloradan" folded in late January, he and four other students in a business entrepreneurship class wrote a business plan for a computerized yearbook. Surratt alone decided to invest in the project and carry it through.

> '''To me, it was my senior year and this was such a good idea, I couldn't not have done it. These things happen only once and it's important to record them for history.''
> — Troy Surratt

Though it doesn't have the usual stock of team and organizational photographs, one thing it does have that is unusual in a traditional yearbook are games. In one, Ralphie, the CU mascot, is stolen from a football game and players must find clues around Folsom Field to find him.

Surratt, who produced the yearbook in two months last spring, had hoped to get it out in time for May graduation. But passing muster for CU licensing — required for use of the CU name and logo — took all summer, which means the yearbook just recently became available.

(See 'COLORADAN', Page 3B)

## Airline standards tightened

By LINDA CORNETT
Camera Staff Writer

The Jefferson County Airport Authority has proposed tighter financial constraints on the Texas entrepreneur who wants to ferry passengers through Jefferson County Airport, although the changes fall short of the restrictions opponents have demanded.

The proposed changes in the requirements that would be placed on charter airlines at Jeffco will be the subject of a meeting of the airport authority at 2 p.m. Thursday at the airport, on the south side of the Turnpike across from Broomfield.

Although the regulations would apply to all future commercial airline operations, John Andrews and his investors in Centennial Express Airlines are first at the gate with their proposal to fly 30-passenger propeller-driven planes from Jeffco and Centennial Airport south of Cherry Creek Reservoir in Arapahoe County.

The cities of Westminster, Louisville and Superior have joined forces to fight the proposal and enraged residents of Rock Creek and other residential neighborhoods beneath the airport's flight pattern have organized Citizens for General Aviation. The group, claiming to have 400 active members, has collected the signatures of 1,500 people opposed to adding passenger flights at Jeffco.

Thirty-eight aviation companies operate 430 daily flights from the airport, making it the fourth busiest in the state. Centennial Express would start with five daily flights, increasing to 20 a day.

Among the changes proposed:

■ The application to run a charter operation at the airport would be $10,000 instead of $500;

■ Liability on aircraft would be $50,000 instead of $10,000;

■ General liability would be $50,000 instead of $1,000 per passenger seat per aircraft (about $30,000 in the planes Andrews is planning for Jeffco);

■ Flights would have to depart and land during normal hours of airport operation;

(See AIRLINE, Page 3B)

## Getting to know you: Park East neighbors meet

NEIGHBOR TO NEIGHBOR
News & notes from over the backyard fence

Residents of the Park East neighborhood in south-central Boulder will gather Wednesday to "make sure we are no longer a neighborhood of strangers," in the words of the 29-year resident who got the movement started.

Jim Burkepile, a retired Ball Aerospace engineer and a resident of Park East since 1966, said, "This is something we should have done a long time ago."

What brought it to a head was fights involving juveniles armed with guns at Park East Park and Aurora 7 Elementary School in August. "I happened to be the one to pick up the phone first," Burkepile said. He called Mayor Leslie Durgin, "an exceptional lady," and soon several city departments were involved, neighborhood meetings were called and a block party brought together 225 neighbors.

Many residents of the neighborhood, which is bounded by Baseline Road, Colorado Avenue, Gilpin and Morgan drives, are already involved, and "we are hoping for many, many more" after Wednesday's meeting, Burkepile said. "We decided it was time we organized to make sure we have a cohesive neighborhood and no one could take it away from us. We're going to look out for each other."

The purpose of the meeting is to elect officers, begin drawing up by-laws, organize working committees and sign up block captains for a Neighborhood Watch.

Beyond the usual functions of keeping residents informed about activities that could affect them, "More important than that is to foster a feeling among the people that live in the neighborhood that it's a neighborhood they are glad to be a part of and feel safe in," Burkepile said.

In addition to the neighborhood association, residents are organizing a citizens' patrol to walk the neighborhood.

**WHEN & WHERE:** 7 p.m. Wednesday, Aurora 7 School, 3995 E. Aurora Ave., Boulder.

■ **NEIGHBOR LABOR:** Volunteers are needed to work on the city of Boulder Conference on Neighborhoods, Nov. 17 (potluck dinner) and Nov. 18 (all day discussion of neighborhood issues). Call Molly Dessonville, neighborhood liaison, at 441-3155.

■ **MOVIN' ON:** Boulder's most transient community is . . . University Hill, right? No, according to Subcommunity Profiles, a collection of statistical information about nine subcommunities in the city of Boulder, prepared by the city's Center for Policy and Program Analysis. Gunbarrel, whose residents have spent an average of 7.9 years in Boulder, was the most restless, followed by East Boulder with 9.4 years. Residents of the CU subcommunity, the area around the University of Colorado, have dug in for an average of 12.6 years. The group with the longest tenure in Boulder live in Palo Park — 17.3 years.

*Figure 13–27 A nicely designed inside page that is interesting and easy to read. It is orderly and inviting. The top and bottom borders are in different colors.*

Reprinted courtesy of The (Boulder, CO) Daily Camera.

Sunday News Journal, Wilmington, Del. Oct. 15, 1995 • Section D

# Sports

SCOREBOARD HIGH SCHOOLS FOOTBALL BASKETBALL

**Brian White**

## Hens regain a key ingredient: solid defense

NEWARK — Jim Reid was being complimentary, but at one instance he worried that he wasn't sounding nice. His 12th-ranked Richmond football team had just come up empty in the Mud Bowl that used to be Delaware Stadium, and Reid didn't hesitate in picking a key to the game.

"I'll tell you," Reid said, "Delaware is playing really good defense and, uh, . . ."

He stammered for a second.

". . . And that's something that, uh, normally they, uh, are not as adept at. They're usually behind the offense, is what I'm trying to say."

What he was trying to say is what nearly every coach who has been beaten by the 6-0 Hens has tried to say: that this isn't the same old Delaware, where offense ruled and defense induced gray hairs.

It used to be that one good defensive stand might be enough for a 58-55 victory. It used to be that the Wing-T offense had to work perfectly for the Hens to win. It used to be that the joke was this:

Q. What's the easiest way for a UD student to score on the weekend?

A. Line up against the UD defense.

After the Hens' 15-0 victory Saturday, those jokes were history. In recording Delaware's first shutout against a Division I-AA team (translation: not West Chester) since 1985, the Hens proved again to be tough, energetic and opportunistic. A driving rainstorm played a part in limiting offenses, but, in the end, Delaware did the better job of surviving.

Defenders, again joining wholeheartedly in the celebration of a victory, were being asked for the second straight week how it feels to bail out a sputtering offense.

Saturday's rain turned a once-beautiful field into slop, but that was OK for the Hens' defense for two reasons. One is that the Delaware defenders have renewed confidence. The second, lineman Mark Hondru said, is that he and his lineman friends revel in those elements because, "We're like a bunch of pigs.

"We're just nasty people," he said. "We knew today it would be in-the-trenches football. No passing. Boom, boom, boom — down the field. It shows how much guts you have."

Delaware won at Boston University in large part because of a late goal-line stand. The Hens beat Youngstown State on a day when the vaunted UD offense could do little. And they won Saturday because they allowed exactly zero big plays.

There was a scary moment, though. With the Hens leading 8-0 in the third quarter, the Spiders crawled to the Delaware 24, and on third-and-4 fullback Elio Imbornone appeared to slip unnoticed through the Hens' defense. Jason Gabrels lofted a pass to the end zone, but safety Kenny Bailey stormed in, went airborne and intercepted. The play, another in a long line of big plays this year by the junior out of Lake Forest High, stunned Richmond.

"Geez, did he leap up there?" Reid said. "What's that saying: taller than a three-story building? I wanted to rip his jersey off and see if there was an 'S' under it."

Reid, former head coach at UMass and defensive coordinator at Boston College, said this is the most-athletic Hens defense he has seen. He said a once-porous secondary is now so fast that it causes play-calling problems.

"And," he said, "the linebackers are hard and tough and take pride in playing hard and tough."

The architect of the unit is Bob Sabol, who is in his fifth season as defensive coordinator. Sabol's defense started poorly last year because of frequently changing personnel, but he kept the unit working, improving and gaining experience. By the end of '94, Sabol said, the Hens had upgraded into an "average" defense.

Now it is much more than that, and it is having fun and winning games. That group that was tossing a shutout and rolling around in the mud like little kids Saturday was just showing us that it has grown up a lot since last year.

▶ Brian White is assistant sports editor of The News Journal.

# Hens better mud-slingers

## UD wins battle of unbeatens

By TOM TOMASHEK
Staff reporter

NEWARK — Faster than you could say 'Here's mud in your eye," the University of Delaware scored on a first-period flurry, then spent most of the afternoon maintaining its footing on the soggy Delaware Stadium turf.

Whatever, the Blue Hens weathered the storm in a 15-0 victory over the University of Richmond in a battle of nationally-ranked unbeatens before a rain-soaked homecoming crowd of 18,926.

| | |
|---|---|
| Delaware | 15 |
| Richmond | 0 |

Delaware, (6-0, 4-0 Yankee Conference) and No. 8 nationally, led by only eight points with 8:05 left in the game. No. 12 Richmond (5-1, 3-1) was camped on the Blue Hen 14 but Spiders quarterback Joe Elrod fumbled, the Blue Hens recovered and quarterback Leo Hamlett directed a game-clinching 85-yard touchdown drive with only 6:12 left.

The victory extended Delaware's unbeaten streak to 12 games, longest nationally in Division I-AA. These Hens became only the second Delaware team since 1974 to open with six consecutive victories.

▶ Hens' Scott a valuable weapon in second half .... D6

The shutout was Delaware's first since 1991, when it blanked Division II West Chester, and the Hens' first over a I-AA opponent since 1985 against Boston University.

It was the lowest-scoring game in Delaware Stadium since 1977, when Temple beat Delaware 6-3. However, the events of the inclement afternoon inspired accolades from both sides.

"It was a great football game played between two potentially great football teams," Blue Hen

See HENS — D6

The News Journal/ FRED COMEGYS

Delaware's Eddie Conti (above) eludes Richmond's Mike White, but then slips on the wet turf during a punt return Saturday. Conti later caught a touchdown pass from Leo Hamlett. Defense was the key for the Hens as Denis Hulme (right) tackles Richmond quarterback Jason Gabrels. Delaware allowed the Spiders just 166 total yards.

| Team | Conf | All |
|---|---|---|
| James Madison | 5-0 | 6-1 |
| Delaware | 4-0 | 6-0 |
| Connecticut | 3-0 | 6-0 |
| Rhode Island | 4-1 | 5-2 |
| Richmond | 3-1 | 5-1 |
| William & Mary | 3-1 | 5-2 |
| Massachusetts | 1-3 | 2-3 |
| Villanova | 1-3 | 2-4 |
| Boston U. | 1-4 | 2-4 |
| New Hampshire | 1-4 | 2-4 |
| Maine | 0-4 | 2-4 |
| Northeastern | 0-5 | 1-5 |

**Saturday's scores**

Wm. & Mary 48, Penn 34
James Madison 23, N. Hampshire 19
Villanova 27, Northeastern 24
Connecticut 31, Maine 30
Rhode Island 22, Boston U. 19

## Colonials roll over Concord

By KEVIN TRESOLINI
Staff reporter

BRANDYWINE HUNDRED — There was no inclination to search for hexes or sinister forces beyond Concord High's control, or to feel like it just was not meant to be.

| | |
|---|---|
| Wm. Penn | 35 |
| Concord | 13 |

When it was done, and Concord's football game against William Penn had ended same as the previous 15 against the Colonials, there was only stark reality.

"They just took it to us," Coach George Kosanovich said.

And that was that.

No. 1-ranked William Penn prevailed 35-13 over No. 5 Concord because it ran and passed better and because it defended the run and the pass better. Nothing is more basic than that for a William Penn team that has built a 5-0 record on the ability to do mundane things, such as control the football, in spectacularly consistent fashion.

In the rain at Concord Saturday, William Penn turned a 22-13 halftime cushion into a lopsided win by running 34 plays in the second half. Concord ran just 16, but managed only 12 total yards from that, with zero first downs.

Previously unbeaten Concord (5-1) had outgained William

See COLONIALS — D8

▶ No. 6 Dickinson romps D9

The News Journal/ BRIAN BRANCH-PRICE

William Penn's Rodshik Edwards slips the tackle of Concord's Griff Burk.

## Braves finish off sweep of the Reds

By CHRIS SHERIDAN
Associated Press

▶ Tribe gets even ...... D10

ATLANTA — The Atlanta Braves, baseball's best team in the '90s, made it to the World Series again — this time with ease.

The Braves got a three-run homer from MVP Mike Devereaux and a combined three-hitter from four pitchers to beat Cincinnati 6-0 Saturday night and finish a four-game sweep.

| | |
|---|---|
| Braves | 6 |
| Reds | 0 |

Atlanta, which reached the World Series in 1991 and 1992 with seven-game wins over Pittsburgh, has six days off before starting the World Series at home against Cleveland or Seattle.

This Braves team is considered their best yet in the '90s, and it showed against the Reds. Atlanta trailed only once in the four games, won twice in its last at-bat and outscored Cincinnati 19-5. The Reds hit just .103 with runners in scoring position.

Aside from starting pitcher Pete Schourek, the Reds didn't put up much of a fight in their last game under Manager Davey Johnson, who will be replaced by Ray Knight next season. Cincinnati struck out 10 more times and hit into another three double plays.

The only hits Avery allowed were a line single in the second by Benito Santiago and a grounder that Barry Larkin beat out in the sixth. Ed Taubensee also had an infield single in the eighth off Alejandro Pena.

Atlanta got its first run in the third when Rafael Belliard led off with a single. Avery moved him to second with a line drive off Schourek's left wrist and Mark Lemke hit a two-out single.

**INSIDE**

**College football**

Florida State 72, Auburn 13
Nebraska 57, Missouri 0
Florida 49, Auburn 38
USC 26, Washington St. 14
Tennessee 41, Alabama 14
Oklahoma 24, Texas 24
Penn State 26, Purdue 23
Notre Dame 28, Army 27
Fla. A&M 24, DelState 21
Wesley 62, Curry 0

**Articles, D4-5**

AP/ TODD ROSENBERG

**George to the rescue**

Ohio State's Eddie George heads downfield on a 51-yard touchdown run in the fourth quarter of Saturday's game against Wisconsin. George rushed for 141 yards and three TDs as the No. 4 Buckeyes rallied for a 27-16 victory.
**Article, D4.**

The News Journal/ BOB HERBERT

**Viera gets win No. 600**

University of Delaware volleyball coach Barb Viera notched her 600th career victory Saturday night as the Blue Hens defeated Hofstra 15-11, 15-7, 15-9.
**Article, D6.**

*Figure 13–28 A sports page with action photographs dominating the page. The page design is not as simple as other pages shown, but it is not hard to read. Sports page readers will probably enjoy this page.*

# CHAPTER 14

# AN INTRODUCTION TO MAGAZINE AND NEWSLETTER DESIGN

## WHY MAGAZINE AND NEWSLETTER DESIGN IN AN EDITING BOOK?

Editing is a journalistic activity that helps make words and pictures simpler and more understandable. Any time that words are written anywhere, some editing is necessary. Its most obvious use is in newspaper publications, where time doesn't permit lengthy editing analysis and corrections, but editing proves very helpful nonetheless. But it is by no means limited to that medium. It is also a necessity for modifying words and pictures written for magazines, even though there is often considerably more time available to produce the average magazine issue. Magazines are also like another print medium that is growing in popularity, size and importance in the United States: newsletters.

There is a great deal of similarity between the editing techniques of magazines and newsletters, as there is between graphic design in both media. So it is reasonable to include them in this chapter, which will cover graphic design editing and its effects on readability.

## DESIGN IN MAGAZINES AND NEWSLETTERS

Magazines and newsletters provide great opportunities to enhance words with pictures, which in turn enhances both of their meanings. In general, what makes this possible is simply that there is more space available for that purpose than there is in newspapers. When space is limited, as in newsletters, there is very little room left for design to affect readability.

Furthermore, since the editorial material of magazines is often emotional and dramatic, rather than simply a presentation of facts, good graphic design better fits the communication environment than elsewhere. A deeply moving story with full-color illustrations, printed on enameled or glossy paper, is a magazine's stock in trade. Such stories stand out and become desirable to a reading public.

Newspapers, because they are published daily, are more restricted in using graphic design for the same purpose, no matter how valuable it is. In addition, new equipment in print media seems to be using less and less quality-grade paper stock for such things, while magazines and newsletters are using better paper.

The design of magazines, then, becomes a criterion for helping readers to choose one magazine over another, or to choose any other medium. A magazine that uses outstanding graphic design tends to attract advertisers whose advertisements also are beautifully designed. Thus, most magazine advertisements harmonize with the basic idea of providing good design, making the entire publication consistently attractive.

## The Values of Design

Several elements make outstanding graphic design valuable:

- **Variety of design keeps magazines interesting.** Reading magazines is not only easier and faster than many other print media, but more pleasurable (see Figure 14–1). The choice of typefaces in magazines does not necessarily have to conform to all the other typefaces used throughout the same magazine, as they traditionally tend to do in newspapers. Type can be chosen that precisely fits the nature of a story, especially for headlines.
- **Better paper makes better picture detail possible.** In most magazines, the halftone screens for photographs are finer than in other media (over 100 lines to the square inch, and as high as 133+ lines). What makes that possible is the quality of paper stock on which magazines are printed. Photographs appear snappy and sharp, so that details that might have looked fuzzy with lesser screen sizes now stand out clearly (see Figure 14–2).
- **Varying art fits the nature of a story.** Illustrations in magazines do not necessarily have to be halftones. They may be in almost any art form there is, such as scratchboard, water color or oils, giving an aura to a story and magazine that the editor wants.
- **Changing column widths better fits page design.** Column widths need not be restricted to a standard width but may be varied to allow each story to have its own width with distinctive connotations.
- **Bringing order to a page.** Some page designs have too many column widths or are too complex (see Figure 14–3). Limiting the page to one column helps control the design and brings order to the page.

# The Net 50

A YEAR AGO ONLY A HANDFUL of computer cognoscenti had heard of Netscape's Marc Andreessen or Sun Microsystems' James Gosling. Now they're considered to be computerdom's hottest visionaries. Where to find the Big Thinkers of tomorrow? Start here, with NEWSWEEK's list of the 50 People Who Matter Most on the Internet. Many of them aren't household names, but they're supplying the vision, the tools and the content that are getting millions of people to turn on their modems.

**Halsey Minor**
NET PUBLISHER
In 1995 the 31-year-old founder of c|net, "The Computer Network," launched a cable-TV show and two Web sites, and lured big-name investors like Microsoft cofounder Paul Allen to help finance his burgeoning multimedia business. Minor's company is also at the forefront of new technologies that will let advertisers target their customers online. To trumpet his Web site, he's blanketed New York City buses with advertisements. C|net doesn't yet deserve the name "network," but it might if Minor realizes his real goal: an "all computers, all the time" version of the Cable News Network.

**Lawrence Landweber**
INTERNET GUARDIAN
Landweber was one of the founders of CSNET, an early network for computer scientists. Today the University of Wisconsin professor is president of the Internet Society, and he's rallying companies around the globe to maintain technical standards for the Net. His goal: to ensure that your computer will be able to talk to others. His trade organization, made up of 5,000 Internet professionals from 120 countries, trains Third World engineers to create networks in their own countries, publishes a bimonthly magazine and is working to protect the term Internet from companies that are trying to copyright it.

**Srinija Srinivasan**
ONTOLOGIST
She may be the best-kept secret at Yahoo!, the company that produces the wildly popular Web search engine. Trained in library science, Srinivasan is the one who decides how the thousands of Web pages submitted to Yahoo! should be categorized and classified, making it as intuitive, expandable and maintainable as possible.

**Bart Decrem**
CYBER ROBIN HOOD
A native of Belgium and graduate of Stanford Law School, Decrem is bridging the gap between the low-income neighborhoods of East Palo Alto and the information-rich companies next door, in Silicon Valley. His nonprofit Plugged In provides a vast computer lab with Internet classes, open 70 hours a week to kids and adults from the community. He has also acquired a $20,000 grant from the U.S. Commerce Department and recruited companies like Intel and Apple Computer to help make personal computers and online access available to low-income people.

**Nicholas Donatiello**
INTERNET STAT-MAN
Sen. Bill Bradley's former campaign manager is now the chief executive officer of Odyssey, a San Francisco-based market-research firm dedicated to tracking America's online behavior. His Homefront Survey, produced twice each year, provides a rare and reliable glimpse into the technology habits and hangouts of cybersurfers everywhere. Odyssey's list of clients includes blue-chip companies like Microsoft, AT&T, all the online services and six of the seven regional phone companies.

**Omar Wasow & Peta Hoyes**
ONLINE INNOVATORS
Co-administrators of New York Online, a self-described "jazz

42 NEWSWEEK DEC. 25, 1995/JAN. 1, 1996

***Figure 14–1** This page uses innovative design, such as a shortened headline set in very large type, a heavy type rule at the top, a lead paragraph set full page width (in larger type), good body-type leading and bold subheads.*

*My garden's brightest springtime yellow comes from eared coreopsis. Both it and frothy, white foamflower thrive in filtered sun.*

# Gold for the Garden

You've heard of gold in them thar hills. But what about gold in that thar garden? Here are three easy, low-growing plants that shower the garden floor with yellow blossoms each spring and summer.

Eared coreopsis *(Coreopsis auriculata)* gets its name from the pair of lobes that sticks out from the base of each leaf. The foliage stays green all year and slowly spreads to form mats. Bright-yellow blossoms atop 18-inch stems appear from spring through early summer. The flowers of a dwarf selection, called Nana, stand only about half this tall. Eared coreopsis likes well-drained, woodsy soil and filtered sun. Plant it at the edge of rock gardens, woodland gardens, or natural areas, where it won't be shaded by taller plants.

*Bright-yellow blossoms sparkle among the handsome foliage of this ground cover known as green-and-gold. It makes an excellent filler plant between other wildflowers and perennials.*

*Middendorff sedum benefits from the strong sun and excellent drainage atop a low, stacked wall.*

I combine it with foamflower *(Tiarella cordifolia)* and blue phlox *(Phlox divaricata)* and praise my genius every April.

Green-and-gold (*Chrysogonum virginianum*) shares much in common with eared coreopsis. It, too, is a creeping, evergreen ground cover featuring sunny blossoms and deep-green foliage, and it also likes rich soil, partial sun, and good drainage. But it's distinctly prostrate, barely rising 6 inches tall. Keeping fallen leaves and pine straw from smothering the foliage is essential. Green-and-gold makes a nice filler plant between other wildflowers and low-growing perennials. I also like to plant it on small spring bulbs, such as dwarf narcissus, Spanish bluebells, crocus, and dwarf crested iris.

No one will scoff if you haven't grown eared coreopsis or green-and-gold. But what gardener worth his Epsom salt hasn't tried a sedum once in his life? There are about a zillion kinds, most of which creep along the ground, sport yellow or pink flowers, and are easier to grow than the national debt. The one pictured here is middendorff sedum *(Sedum middendorffianum)*.

As you can see, this low, mounding succulent with needlelike leaves looks great at the edge of a stacked, stone wall. You can also tuck it into pockets between garden walls or paving stones, or place it at the front of a mixed border. Middendorff sedum isn't fussy about soil and doesn't need much water. But if it doesn't receive full sun and excellent drainage, kiss it goodbye.

*Steve Bender*

***Figure 14–2*** *This is a simple and attractive page design. It was printed in full color. The picture at the top provides design variety. The other photographs are nicely placed, and the type has been given adequate leading.*

# Read On...

February/March 1996

**News and Information from the Mount Prospect Public Library**
10 South Emerson Street • Mount Prospect, IL 60056 • 708/253-5675 • TDD 253-5685

## BE Creative and eXpress Yourself!

How do you eXpress yourself?
Do it through poetry or prose! and
you might win both money and recognition!

If you like to eXpress yourself through poetry or prose (fiction or non-fiction), you should enter Mount Prospect Public Library's Creative Writing Contest, promoting lifelong learning and recognizing National Library Week, April 14-21, 1996.

*Funded by the Friends of the Library.*

Entries may be made in two categories: poetry or prose (fiction or non-fiction); they *must* be unpublished works. Each person may submit only one entry in each category.

Qualified judges will consider creativity, originality and quality of writing in selecting the winning entries. Winners will be notified by telephone and announced at the final party for National Library Week.

First place grade school winners in each division and category will receive $50; second place winners will receive $25.

First place high school student and adult winners in each division and category will receive $125; second place winners will receive $50.

***All first place entries will be published.***

Creative writing forms may be obtained at Mount Prospect Public Library or through the schools beginning February 1. Entries must be received by Mount Prospect Public Library by April 5.

For more information, call Community Services at Mount Prospect Public Library at 847/ 253-5675. ▲

> A good free library is a great boon to any town. It makes every farm worth more and puts it in the power of every citizen to grow more intelligent. But it is of no use to have it unless you're going to use it.
>
> -- *Old Farmer's Almanac, 1880*

**Contest Divisions:**

(maximum 500 words)
Grades 1-2, Grades 3-4,
Grades 5-6, Grades 7-8

(maximum 1000 words)
High School Students
Adults

Turn
Back Time
Through MPPL's Winter
Multimedia Programs


*January 8 through March 15!*

✔ ***T'where the Magic Begins*** (Ages 2-12)
✔ ***Retro Read*** (Ages 11-18)
✔ ***Those Were the Days*** (Age 18 and Over)

***Read books, watch videos and listen to books on tape and music to gain great chances to win some fabulous prizes!***

***Figure 14–3** An example of a typical newsletter. This shows the need for a simpler, and more dramatic page design that can organize and display many different kinds of news.*

The potential for producing a readable and enjoyable story is more likely a reality for large-circulation magazines. Magazines with smaller production budgets cannot produce what larger ones can, but they may apply some design principles within their budgets.

To see where graphic design is done most and how, or why, some editors do a better job designing, this chapter comments on samples of designs shown in Figures 14–1 to 14–9.

## MAGAZINE CLASSIFICATIONS

It is important to remember when discussing magazine design that magazines can be classified in many different ways.

### Consumer Magazines

Consumer magazines are the largest general category of publications, distinguished by their coverage of a broad range of subjects that interest different groups in the general population (see Figure 14–4). The differentiation is based primarily on the special interests of certain population segments. A small percentage of these magazines could cover a sport, such as football, golf or tennis, or provide specialized stories on investing, gourmet cooking, child care or fashion (see Figure 14–5). The segmentation is based on different needs, frequently within male and female groups. Some more-specialized interests might be the subject of other magazines that seek to fill a clear niche.

Consumer magazines use the most elegant graphic designs and are usually recognized because of generous amounts of white space. Some consumer magazines are in highly competitive fields, and the graphic designs there tend to have less white space (see Figure 14–6). The pages in these magazines need to be a bit smaller, to get as much information as possible in a smaller space. The use of colored innovative graphics may be quite restrictive in some of those magazines.

### Business Magazines and Newsletters

A very large group of periodicals fall under the heading of business magazines, which in turn have a large number of categories (too many to discuss all their designs here). Here are a few:

- Magazines that cover retail and wholesale selling, such as *Chain Store Age.* Products are often distributed to middlemen who sell to retailers and who in turn sell to consumers. This is a very active group of businesses.
- Another category is written for those in raw materials businesses such as mining and manufacturing (see Figure 14–7).

PHOTOGRAPHS COURTESY OF PRINCESS CRUISES

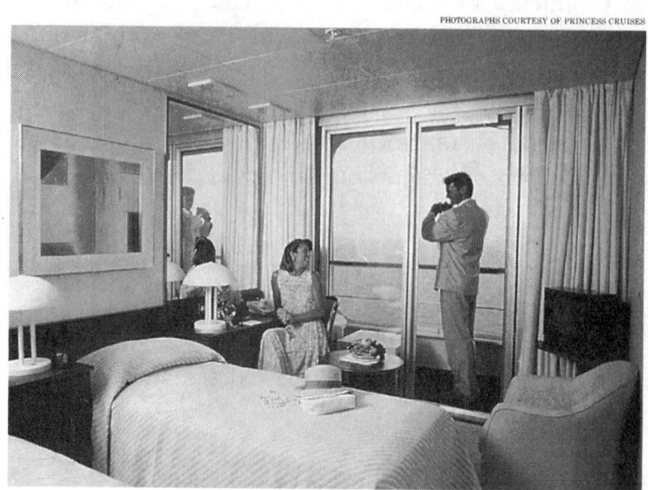

*Cabins are spacious, though only the most expensive have verandas.*

# A Traveler's First Cruise

Bands played. Paper streamers scribbled colorfully over the water. Passengers waved goodbye to friends, and to their everyday, landlocked selves. Ahead lay the mysterious, enticing blackness of the ocean, filled with possibilities.

It was the start of my first cruise, and it was magical. I felt like Fred Astaire in those old movies, setting off on a grand adventure. Any second now, I would bump into Ginger Rogers, and we would go waltzing across the deck.

I never did find Ginger. Instead, a few days later in the main lounge, I found myself watching well-dressed, mature adults stuffing large, shiny spoons down the fronts of their dresses and pants. The spoons were tied to pieces of string, and two teams of audience volunteers were racing to stitch themselves together while the orchestra played the theme from *The Flintstones*.

I spent most of my cruise somewhere between those two extremes of the sublime and the silly. But what I remember is the sublime.

I was aboard Princess Cruises' *Crown Princess*, which at 70,000 gross tons is typical of the megaships that have dominated the cruise market in recent years. This floating condominium can hold 1,590 passengers.

The boxy shape may sacrifice the sweeping angularity of the old transatlantic liners, but it does maximize interior room. My cabin was bigger than some hotel rooms I've inhabited, and it even had a tiny veranda. I propped the door open at night so I could smell the ocean's tang and hear the muted thunder of the ship shoving its way through the water.

Being at sea can invest even the most mundane moments with romance. Lunching by the pool one day, trying to keep my salad from blowing overboard, I realized that the tanned gentleman who had just walked past was Tony Bennett, who was performing that day in the main lounge. And you haven't seen the moon until you've seen it from a ship's rail. Painting the waves with a dazzling stripe of molten silver, the moonlight leads right to your feet.

Neither duty-free shopping nor the packaged shore excursions held much appeal for me, so I wasted most of my time in our first couple of ports. I wandered around aimlessly, hoping in vain to stumble across some spectacular restaurant or hidden pocket of local charm.

Finally, in Antigua, I got it right. I hired a pleasant fellow named Jonathan Teague to give me a personal tour in his taxi. As we jounced over the bumpy roads in his right-hand-drive Datsun, he told me about the island's history and politics.

Most of Antigua (pronounced an-TEE-gwah) resembles a ragged green carpet tossed over a pile of rocks. Jonathan told me that all the trees had been cut down centuries ago, using prison labor, to clear land for long-departed sugar plantations. But the coast is gorgeous. Antigua boasts that it has 365 beaches, one for each day of the year, and they're all public.

After we stopped to pick up his daughter from school, Jonathan

*One of the ship's pools includes a full-service bar.*

24 SOUTHERN LIVING

***Figure 14–4** A nice page design with balanced pictures. However, if one of the photographs were larger or shaped differently, it would have helped balance the page better.*

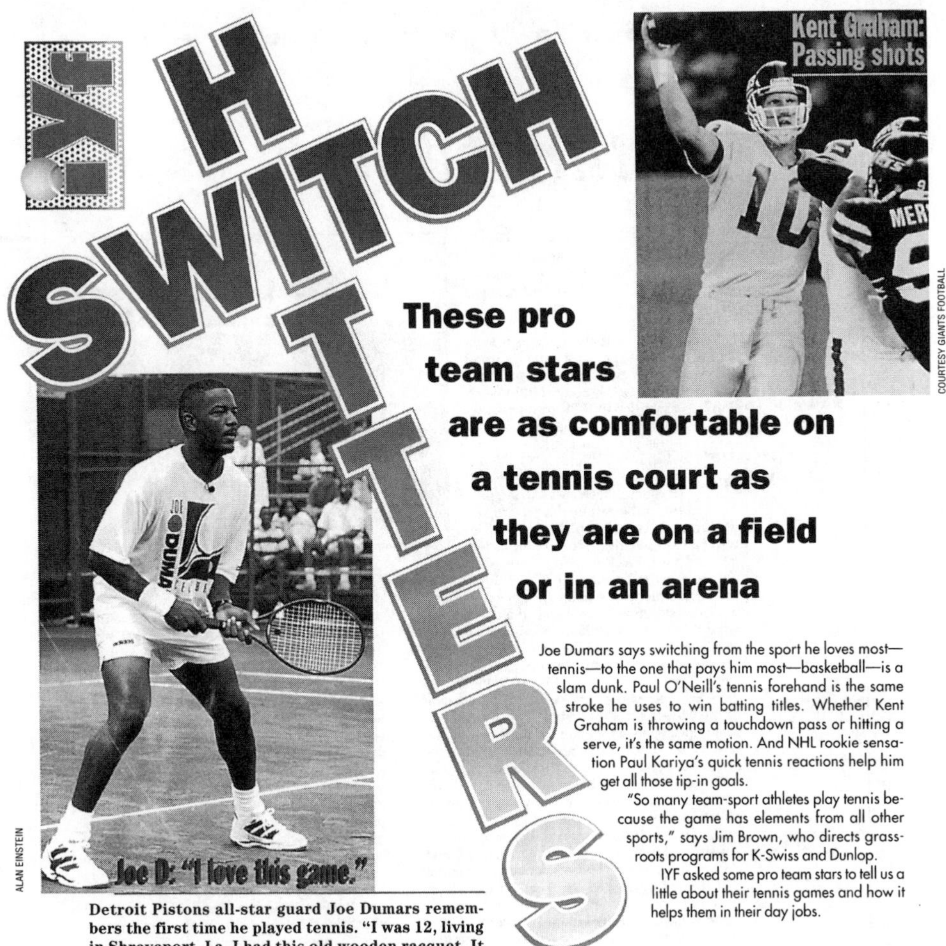

IYF

SWITCH HITTERS

Kent Graham: Passing shots

COURTESY GIANTS FOOTBALL

These pro team stars are as comfortable on a tennis court as they are on a field or in an arena

Joe D: "I love this game."

ALAN EINSTEIN

Joe Dumars says switching from the sport he loves most—tennis—to the one that pays him most—basketball—is a slam dunk. Paul O'Neill's tennis forehand is the same stroke he uses to win batting titles. Whether Kent Graham is throwing a touchdown pass or hitting a serve, it's the same motion. And NHL rookie sensation Paul Kariya's quick tennis reactions help him get all those tip-in goals.

"So many team-sport athletes play tennis because the game has elements from all other sports," says Jim Brown, who directs grass-roots programs for K-Swiss and Dunlop.

IYF asked some pro team stars to tell us a little about their tennis games and how it helps them in their day jobs.

**Detroit Pistons all-star guard Joe Dumars remembers the first time he played tennis. "I was 12, living in Shreveport, La. I had this old wooden racquet. It was pure joy. I still get that same feeling every time I pick up a racquet."**

**Joe picks up a racquet a lot. During the off-season, he tries to get in at least one match a day, and each summer he runs the Joe Dumars Celebrity Tennis Classic in Michigan, which features his tennis-playing friends such as Pistons teammate Grant Hill, Steve Smith of the Miami Heat, Tim Cheveldae of the Winnipeg Jets and tennis pros Aaron Krickstein and Carrie Cunningham.**

**Joe D says his best move on the tennis court is his serve, but his approach shots need work. "I'm pretty good once I get to the net," he laughs.**

**Joe sees a lot of similarities between basketball and tennis. "Playing tennis as much as I do really helps to build my endurance in basketball," he says. His basketball skills cross over into tennis, especially the 1994 Dream Teamer's defensive skills. "In basketball, you use your body to force a player to move to your strength. In tennis, it's the same thing—you force your opponent to hit shots that you know you can get to. I use the one sport to help train for the other."**

**Regardless of the court, whether he's guarding Michael Jordan or trading volleys with his friends MaliVai Washington and Todd Martin, Joe's game plan is to compete.**

**"Tennis is really my favorite sport. I just love to play the game." —*David McConnachie***

70 • TENNIS • AUGUST 1995

*Figure 14–5 This design gets readers' attention. Getting readers to read the text is more difficult. The subject matter here has many devotees, and it is hoped they will read on.*

Courtesy of The New York Times Company Magazine Group.

A U D I O / V I D E O F I L E

AUDIO/VIDEO FILE 17 NORTH SHORE/JANUARY 1996

# Picture perfect

**Sony's latest TV offers optimum viewing**

A rose is a rose is a rose. Unless you view it on most color TVs. On TV, red roses fade to pink or shift toward orange. But the new Sony KV-32XBR100 reproduces a bouquet of red, red roses worthy of Valentine's Day. It also remains true to mums, petunias, pansies, asters, carnations, snapdragons and sunflowers. In fact, if Sony could invent a way for this TV to reproduce fragrance, it might put the local florist out of business.

Whimsy aside, this $3,300 set produces the best direct-view (meaning that you view the picture tube rather than a projection screen) television picture available today. Competition from companies such as Toshiba, Panasonic and RCA pushed Sony to develop the KV-32XBR100—the ideal centerpiece for a home theater system with limited space. Sony minimized the cabinet size around the set's 32-inch screen, and shaped the cabinet to fit neatly into a corner, if you desire.

**High-tech wonders work behind Sony's revolutionary 32-inch screen.**

Sony integrated many technological advancements to create the KV-32XBR100. It researched more accurate phosphors to paint on the inside of the TV screen. When struck by an electron beam, these phosphors glow and create the picture. Making the TV screen flatter than ever before reduced internal picture distortion and external reflections. Sony also increased the power of the electron gun, causing the phosphors to glow more brightly. Sophisticated computer circuitry monitors the picture, continuously making subtle adjustments to keep it clear and accurate. The set contains Dynamic Picture for optimum contrast, video noise reduction, and automatic color correction.

One boon to viewers who insist on the truest, most natural picture, is the ability to select the National Television Standards Committee (NTSC) color temperature reference standard of 6,500 degrees K. While this sounds mighty technical, it means that you see natural whites and warm reds. More than 95 percent of the other TVs on the market come adjusted for excessively higher color temperatures. This shifts white to blue, which is only useful for Cheer detergent commercials. The tradeoff is that at the lower reference standard the picture is not as bright as some people prefer. So Sony offers three Trinitone color temperature settings for people who prefer brightness to accuracy.

The Sony set comes in two pieces. A separate "feature box" contains all of the inputs and outputs, and controls the TV via the remote. The box includes five audio/video inputs with "S" jacks, eliminating the need for external switchers. This allows you to connect two VCRs, a laserdisc player, and a digital satellite system (DSS) receiver, and still have an input left over for your camcorder. The box connects to the TV with a single cable. Sony hides the stereo speaker system deep in the TV cabinet with the sound radiating from invisible grilles at the sides of the screen.

This lavish Sony incorporates every feature ever built into a TV set. Not only does it have picture-in-picture (PIP), but also picture *and* picture (P&P), which Sony calls Twin View. The latter permits you to view two 13-inch diagonal pictures side by side. Since this set contains two TV tuners, you can watch Leno and Letterman simultaneously. Or you can fill the screen with nine mini-pictures continuously scanning available channels and instantly select one for viewing with the remote control.

The versatile universal remote control also stands out. It controls not only the TV and multiple VCRs, but also cable boxes and DSS receivers. A mini-joystick harnesses much of this power, making it easy to use.

The KV-32XBR100 proves that even a 32-inch TV can draw you into the full home theater experience. While projection TVs produce bigger pictures, none produces a better one. Since Sony builds most of these sets in America with mostly American parts, you might call this rose an American Beauty. N

Y R I C H W A R R E N

*Readers can contact Rich Warren via e-mail at rwarren@prairienet.org.*

*Figure 14–6 This page has a great deal to communicate, so there are many words on the page, yet it is superbly organized. The spacing of elements is also generous and helpful for the reader. However, adding subheads may have brought more light into the interior of the story.*

North Shore Magazine, Winnetka, IL.

PLANT ENGINEERING
MAINTENANCE
MANAGEMENT

# Mon Valley Works Wins Highest Honor in Maintenance

RICHARD L. DUNN, Editor

**U.S. Steel's giant complex rises to world-class excellence on a total approach to continuous maintenance improvement**

The ghosts of Big Steel still haunt the Monongahela River valley southeast of Pittsburgh. The graves of the once-huge plants are vacant fields with skeletal buildings or abandoned blast furnaces for headstones. Here is where the American steel industry was born. And here the names of the dead — Homestead, Duquesne, National — conjure only memories.

But here, too, U.S. Steel's Mon Valley Works rises like a Phoenix, a reborn industrial giant growing stronger with new ideas and technology. Once again a world-class producer, the Mon Valley Works has not only survived, it's now winning awards — including the prestigious 1995 North American Maintenance Excellence (NAME) Award.

The ghosts of Big Steel Past are constant reminders of the work needed to ensure the success of Big Steel Present and Future in Mon Valley. This awareness, coupled with the leadership of management and unions alike, has forged a new attitude and new work paradigm. As one of the shop managers expresses it, "For the first time in perhaps 100 yr of historical conflict, there is a growing realization that the destiny of the employees and that of the company is the same." Or, as one hourly worker explains, "We're talking security. We're talking jobs in the future. We're talking my future."

Old relationships are indeed changing, so much so that today one finds an amazing integration all the way from the plant manager's business plan to the daily maintenance operations. That's one reason why Mon Valley won the NAME Award. The teamwork between production and maintenance is exemplary.

Here's a quick look at some of the programs and processes that are making Mon Valley's maintenance operations world class.

**Mon Valley Tomorrow**

The Mon Valley Tomorrow program is the Works' formalized process for continuous improvement (CI). Using a system of teams, this CI process taps the knowledge and ideas of the plant floor personnel and pushes decision making to the lowest levels possible.

Teaming is relatively new to Mon Valley, but the supporting organizational structure is in place, and new teams are now building on the successes of the original three pilot areas.

The structure includes four levels of

**U.S. Steel's Mon Valley Works comprises two plants along Pennsylvania's Monongahela River southeast of Pittsburgh. The Edgar Thomson Plant, shown here, produces the basic steel slabs that are rolled into finished sheet in the Irvin Plant.**

*Figure 14–7 This page from an industrial magazine shows an interesting design. It has generous white space, but the type rules may distract readers.*

- Still another category is for professional persons such as physicians and professors.
- The last category is newsletters and house organs, which can vary from one page (front and back) to eight- or 16-page publications.

Professional magazines have been showing more attractive graphic designs lately, although they are often restricted to a scholarly style of writing, which in turn tends to limit designs. At one time professionals felt that only the type and tables of data were needed to keep readers informed. Design had little place in their periodicals. But today, almost all professional publications pay a great deal of attention to attractively designed charts, graphs and maps that make statistics readable.

Newsletters usually have very restricted graphic designs, not because their editors disapprove of design, but because their production budgets are so small, with few exceptions (see Figure 14–8). However, when newsletters are part of a public relations program of a company, they tend to be more attractive.

### CONTROLLED-CIRCULATION MAGAZINES

Many business magazines use a controlled distribution technique, which means the publisher gives the magazine away free to a controlled segment of the business or professional population. Advertisers like to buy advertising in these publications because they know the magazine is going to precisely defined kinds of people who tend to buy the products advertised. Therefore, there should be no wasted issues.

Controlled circulation makes a lot of sense, but it may adversely affect the readership of a magazine because a person who agrees to take the free publication is under no obligation to read it. On the one hand, readers may be more likely to read the magazine because their interests are the same as those in the particular business or industry. On the other hand, especially when professionals are too busy, tired or hurried, they may not read it at all.

This is where good graphic design may help. Good graphic design of controlled magazines may influence readers by enhancing the pleasure and enjoyment of reading.

## PRINCIPLES OF NEWSPAPER DESIGN ALSO APPLY TO MAGAZINES

Although magazines offer more opportunities for creating readable and attractive page designs, problems arise in these same areas. One problem is that the competition has been so keen and aggressive that each magazine cannot fully stand out from the others. Even though each has a distinctive design, many readers fail to notice or appreciate this in their favorite publication. Also, new magazines usually try to start with eye-catching designs but often overdo it, which may turn off readers. The search is still on for *more* readable and *more* attractive magazines.

The basic principles of design (balance, contrast, proportion and unity) apply in magazine planning the same as they do in newspaper design—sometimes more. This is because there are fewer restricting features in magazines than in newspapers, and fewer restrictions mean more planning.

Magazine pages usually do not have to cram a great deal of news into very small, highly structured pages. Most consumer magazines are 8½-by-11 inches (in trimmed page size), and there are larger-size magazines that give the page designer even more options in creating an attractive design. There is also more white space available in key stories in magazines than there is in newspapers.

More options come in the form of different ways magazines arrange stories. Some arrangements could not be done easily in any other medium, for example:

- Magazines that run each story from start to finish in the same vicinity so readers do not have to stop in the middle and search through many pages for the continuation. Graphic design is now concentrated around each individual story and may be changed if the nature of some story demands a different design.
- Magazines that concentrate all editorial pages in the center, and advertisements are congregated either in the front or back. Here, editorial design would usually be the same in front and back, with the center having a different design.
- Magazines that start each story somewhere near the front and then jump (or continue) in the back. Graphic design should be the same throughout the entire magazine in this case.

## Goals of Page Designers

The primary goals of magazine page designers are as follows:

- To create an environment where readers are likely to find an assembly of stories in which they would be interested.
- To attract readers to specific stories as readers casually thumb through pages of a magazine (see Figure 14–9).
- To hold readers' attention in each story until they have finished. Designers do this by following the principles of readability discussed in Chapter 12.
- To use design devices that dramatize the key issues of each story. This is a matter of creating a competitive edge so that readers remember that the magazine is superior to others in the same category.
- To make design maximize readability of each story or page.

***Figure 14–8** These two facing pages are from a newsletter created in a magazine format. The design is eye-catching because of its dramatic illustration, a unique type arrangement (with headline and deck above, in center) and excellent leading. It is very readable.*

**Carol Simpson Stern, the Graduate School's first female dean, steers the school on a people-first course.**

# AT THE STERN

**By Clare La Plante**

A colleague calls Carol Simpson Stern "one of the country's leading administrators," and Stern herself allows that she has "an appetite for administration." But while many people equate administration with a diet of numbers, the first female dean of Northwestern's Graduate School wins respect for focusing her bottom line on people.

"Carol brings her humanity to the Graduate School," says Leila Edwards, the school's senior associate dean. "Sometimes leadership gives too much importance to numbers crunching, forgetting that there are people behind the numbers. It's important in training graduate students that the dean be a humane person."

Stern (G64, 68) made her bottom line known in one of her first initiatives after becoming Graduate School dean in 1993. Concerned that graduate students are well versed in their disciplines but perhaps not well enough in how to teach, she spent countless hours seeking a grant to develop new approaches to educating future professors. Northwestern was awarded a $170,000 "Preparing Future Faculty" grant from a program funded by the Pew Charitable Trusts. Seventy universities competed, and Northwestern was one of only five to receive full funding.

C. William Kern, vice president for research and graduate studies, admits that his reaction was tepid when Stern proposed applying for the grant. "I must say that I was reluctant to see her get entangled in it," Kern says. "The monies weren't that large, and it would take a lot of her time. The complexities of the project I found quite formidable. But Carol really insisted. She sees this as a long-term way to foster and promote graduate education."

"I saw this project as a vehicle to promote reflection on the aims of graduate education," says Stern. "Are we offering the appropriate kinds of courses to train our students, not simply in the knowledge base of the discipline, but as future professors? Graduate faculty have not given much attention to these questions outside of schools of education. I also thought we could give students a greater understanding of the nature of the academic profession that many of them will enter."

In two years as dean, Stern already has moved the Graduate School significantly along toward her goals of improving the diversity and the quality of graduate students and giving them more financial aid and better training for teaching.

She has aggressively looked for ways to increase minority enrollment, seeking outside money and supplementing awards with Northwestern funds. The school successfully competed for a federal grant to provide fellowships for minority teachers interested in doctoral degrees and careers in higher education. As a result of such efforts, minority enrollment has significantly increased. Minorities now make up 7.3 percent of new Graduate School students, compared with 2.8 percent in 1989, an increase of 165 percent.

Stern has also, notes provost David Cohen, "increased the Graduate School's visibility on a national level" by being elected to the board of the Council of Graduate Schools, a national organization of 410 universities that grant nearly all of the doctoral degrees in the country. She also testified before Congress about Title IX aid for graduate students.

"I think the University has been well served by Carol's deanship," says Kern in an understatement.

Even before ascending to the deanship, Stern had been serving the University well for nearly two decades. A professor and former chair of the performance studies department, she is responsible for several innovative programs and academic reforms and is a respected scholar and teacher as well as administrator.

Northwestern hired Stern in 1974 as an associate professor in the interpretation department (later

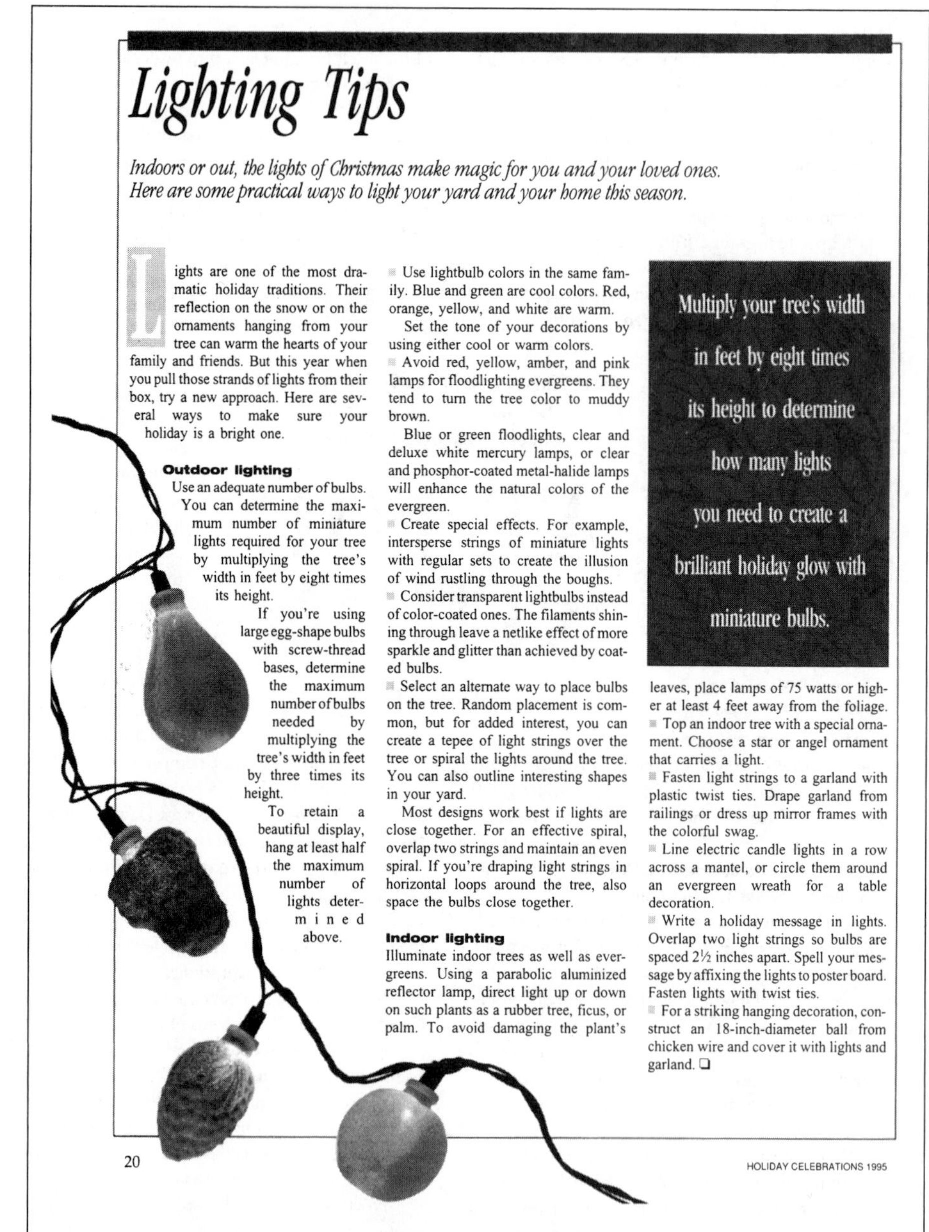

# *Lighting Tips*

*Indoors or out, the lights of Christmas make magic for you and your loved ones. Here are some practical ways to light your yard and your home this season.*

Lights are one of the most dramatic holiday traditions. Their reflection on the snow or on the ornaments hanging from your tree can warm the hearts of your family and friends. But this year when you pull those strands of lights from their box, try a new approach. Here are several ways to make sure your holiday is a bright one.

**Outdoor lighting**

Use an adequate number of bulbs. You can determine the maximum number of miniature lights required for your tree by multiplying the tree's width in feet by eight times its height.

If you're using large egg-shape bulbs with screw-thread bases, determine the maximum number of bulbs needed by multiplying the tree's width in feet by three times its height.

To retain a beautiful display, hang at least half the maximum number of lights determined above.

- Use lightbulb colors in the same family. Blue and green are cool colors. Red, orange, yellow, and white are warm.

Set the tone of your decorations by using either cool or warm colors.

- Avoid red, yellow, amber, and pink lamps for floodlighting evergreens. They tend to turn the tree color to muddy brown.

Blue or green floodlights, clear and deluxe white mercury lamps, or clear and phosphor-coated metal-halide lamps will enhance the natural colors of the evergreen.

- Create special effects. For example, intersperse strings of miniature lights with regular sets to create the illusion of wind rustling through the boughs.
- Consider transparent lightbulbs instead of color-coated ones. The filaments shining through leave a netlike effect of more sparkle and glitter than achieved by coated bulbs.
- Select an alternate way to place bulbs on the tree. Random placement is common, but for added interest, you can create a tepee of light strings over the tree or spiral the lights around the tree. You can also outline interesting shapes in your yard.

Most designs work best if lights are close together. For an effective spiral, overlap two strings and maintain an even spiral. If you're draping light strings in horizontal loops around the tree, also space the bulbs close together.

**Indoor lighting**

Illuminate indoor trees as well as evergreens. Using a parabolic aluminized reflector lamp, direct light up or down on such plants as a rubber tree, ficus, or palm. To avoid damaging the plant's leaves, place lamps of 75 watts or higher at least 4 feet away from the foliage.

> Multiply your tree's width in feet by eight times its height to determine how many lights you need to create a brilliant holiday glow with miniature bulbs.

- Top an indoor tree with a special ornament. Choose a star or angel ornament that carries a light.
- Fasten light strings to a garland with plastic twist ties. Drape garland from railings or dress up mirror frames with the colorful swag.
- Line electric candle lights in a row across a mantel, or circle them around an evergreen wreath for a table decoration.
- Write a holiday message in lights. Overlap two light strings so bulbs are spaced 2½ inches apart. Spell your message by affixing the lights to poster board. Fasten lights with twist ties.
- For a striking hanging decoration, construct an 18-inch-diameter ball from chicken wire and cover it with lights and garland. ❑

20 HOLIDAY CELEBRATIONS 1995

***Figure 14–9** An excellent design that is readable and has a free-form element (the lights and cord). The insert in reverse type is a bit strong but provides page balance.*

Reprinted from Better Homes and Gardens® Special Interest Publications, *Holiday Celebrations.*

- To avoid page designs that are so powerful that readers spend too much time admiring the design and lose their desire to read a particular story.

## GRAPHIC STYLING ALTERNATIVES

There are so many ways to lay out a magazine page that it is difficult to define any one particularly outstanding style. But the easiest way to lay out a page is by a columnar technique in which each page is divided into a certain number of columns (usually three or four). The trend today is to vary column widths to maximize readability. The designer usually determines how many columns will be used and implements the idea unless there is some good reason for changing it. A 7-by-10-inch page (one of the standard sizes used in U.S. magazines) would usually be divided into three columns. Four columns may be used occasionally; five columns are rare.

The first page of any story usually has better design techniques on it to make it more interesting than or different from ordinary pages. The technique used is the designer's choice. The following sections describe some of the more popular techniques.

### Bleed Pages

A *bleed* page, or a bleed illustration, is one that prints off the edge of a *trimmed* page. Different illustrations may bleed on different sides of the same page. Illustrations of all kinds are candidates for a bleed treatment. Some pages have bleed on all four sides (with the inside gutter bleed disappearing into the binding). In order to make a bleed illustration, the designer extends it into the white space on the side where the bleed is to appear. When the magazine is trimmed (one of the later steps in production), the entire page is cut to make all pages appear even.

Bleed pages attract readers' attention and make the page attractive at the same time. Most pages have a white margin around the outer edge, so when a bleed page is seen for the first time, it seems to have eye-catching power.

### Body and Headline Typeface Styles

If you examine the typefaces in major magazines, you will find a wide variety of typefaces and a wide variety of styles. However, there are some similarities, too. Most common is the use of serif as opposed to sans serif body type. Serif type is generally agreed to be the most widely used typeface in the United States, even though the specific names of typefaces vary a great deal. But here the variance is mostly unnoticeable by readers, who can't tell the differences among Baskerville, Caslon, Goudy and a hundred typefaces that look alike. Typographers who have developed a keen eye usually can tell the difference, and that is how they decide which typeface to use.

The differences are actually very slight. The one thing that readers can agree on is that they are all equally readable. Check the body typeface found in three different magazines. Even though each may differ, they look very much alike.

Headlines are also set in a wide variety of type styles, as the graphic designer attempts to be truly innovative in choosing typefaces. Not only are computer typefaces used, but many are drawn, or painted or designed especially for the main story. The typefaces used tend to have connotations related to the action or concepts of a particular story. These special typefaces are reserved for the first and presumably the most important headline, while subheads or minor heads are usually selected from the many typefaces that the typesetter has available.

## Line Spacing (Leading)

Line spacing varies a great deal because research evidence shows that more space is needed between lines as the type size and line widths are increased.

There are usually differences of opinion about how much line spacing is optimal, but the amount found in the most readable magazines is generous. Almost every magazine uses line spacing as a design technique, and almost all of it is excellent. If there are errors in line spacing, it is probably because some magazines have overdone it, and there's too much.

## Type Rules

Type rules are lines that print wherever needed in varying widths and styles and can be used to set portions of a story apart so that readers will not miss them. The thinnest hairline, to widths as large as 24 points, can be used, either in black and white or in color. If the designer wants a rule that is distinguished, the editor may have an artist paint it in a distinguished style, exactly as wanted.

A rule can be added at the top, bottom or anywhere in the middle of a page. The objective is to set that page apart from other pages.

## Boxes

Boxes are usually type rules placed around a paragraph or more of type that sets it apart on the page. The weight of the rules may vary from hairlines to six points in width. The key to using attractive boxes is to make sure that the space inside the box (between the type and rules) is adequate. Some well-designed magazines have shown boxes with less than 12 points of space inside the box on each side, and that usually is unattractive. In this contemporary world, at least 12 points of space is needed on all four sides. An attractive box used inside a story can be charming, setting off a piece of the story that the author wishes to emphasize, without harming readership of the entire story.

## Type Inserts

One of the most popular design techniques in magazine pages is the use of *inserts.* An insert is usually a phrase or sentence from the story, set apart as a free-floating headline to attract attention. In older magazines a story's headline might have a deck underneath the main headline, or perhaps two decks or more. Each deck contained more information about the story. Today inserts usually contain some spectacular statement by one of the principles or the author.

Placement of inserts is up to the designer. They often break into a column of type, or they break into the middle of two adjacent columns, and the story is wrapped around the inserts. They may also be placed at the head of a column, at the head of a story or just about anywhere the graphic designer wants, as long as it does not halt the reader's thoughts (which it can do at times).

## Tint Blocks

A tint block is usually a tint of some color (or a color mixed with white) used as a background for type. Tint blocks are most often prepared in a rectangle, but they may also be free-form.

The words used in a tint block are usually taken verbatim and represent some unusual aspect of a story. They are placed in the block (or rectangle), and the block is placed in one or two columns. Since the words chosen are pertinent to the story, they tend to attract continued interest in a story when, perhaps, interest has waned a bit. Readers see a particular block and may be encouraged to continue reading. (See Figure 14–9.)

Common colors for tint backgrounds are light red, tan and light blue, but they may be almost any color. The typeface used in the block is usually medium to heavy. If the tint block color is very light and the typeface is quite delicate, the entire idea may not work. A danger in using tint blocks is that the contrast between words and background is not sharp enough. If the tint block and type on top isn't seen by a large percentage of the audience, then it hasn't fulfilled its function.

Another danger is in the tint block being used as an insert. In this case it may attract so much attention that readers stop reading at that point and miss the end of the story. That should be avoided.

## Picture Styles and Uses

One of the key graphic techniques of magazine designing is the use of essential parts of a picture to communicate ideas. Pictures can give instant insight into who the protagonists are in a story. Of course, words can also do this, but pictures chosen for their dramatic impact usually can do it better.

Some techniques that provide insight are:

- Use large head shots, either painted or drawn photographically, or drawn/painted in color with backgrounds that help set the scene of a story.
- Most readers will remember the key illustrations in a story if they provide drama. The general artistic technique is to have one dominant illustration in a story and a number of relatively minor illustrations that support the large one. When all illustrations in a story are the same size, there is usually a loss of dramatic impact. (See Figure 14–2.)
- Too many illustrations are rectangular, and a change of pace is needed. An outline halftone is a good way to provide change. In the outline technique, all of a picture's background is cut photographically, thus concentrating on the subject in the foreground. This is a popular modern technique.

Illustrations may be photographs, drawings or paintings. Each has its own value, but what is most important is that these illustrations provide much needed *contrast* to make a story interesting and make it stand out from the others.

Photographs usually are excellent for attracting attention and working with words to tell a story. The most popular kind are rectangular in shape, in either strong horizontal or vertical shapes. Square pictures tend to be used for small insert purposes. It is very important that every illustration have cutlines, or captions, that explain who or what is being shown. (See Figure 14–4.)

## Use of Color

Consumer magazines are usually ablaze in color because color is seen as having the ability to attract readers' attention and thereby enhance readability. This technique is known as the *four-color process.* The term *process* means that when red, yellow and blue inks are added to each other in different colors (or shades), almost any color in the rainbow can be reproduced.

Most publishers do a credible job in printing colors, but their success may depend on which colors make up the process and how well every illustration is retouched. Good retouching adds printed values and can make one magazine look better in color than others. Color as a design strategy is risky. Some magazines have so much color on any one page that the effect seems overdone, and the value of the color is somewhat diminished.

Color should be used for dramatic effect, but it can reach a point of diminishing returns. Therefore, the use of a judicious amount of color should be a concern for all designers. There is no question that, when used properly, color can be eye-catching.

## Use of Initial Letters for Drama

One of the most often used techniques for adding drama to a magazine page is the use of a large initial letter at the beginning of a story, the beginning of a page inside the story or just about anywhere in editorial material. (See Figure 14–6.)

The use of initial letters is an old idea that has been modernized by graphic designers who have changed some of the unwritten rules of the game. In older days, an initial letter at the beginning of a story was followed by whatever remaining letters were necessary to complete the first word of the story. Therefore, if a story started with the word "America," a large capital A was the initial letter and the remaining letters were also capitalized for readability. The next word and all words thereafter in the first sentence were set in lowercase letters.

Today, however, dramatic changes have been made in the use of initial letters in the following ways:

- The size of the initial letter has sometimes been increased enormously. Whereas in the past the size of letters ranged from about 36 points to 72 points, now they might be as large as the length of an entire page. These huge letters are so large they have to be placed in the margins.

  This effect can be visualized by thinking of a page with an enormous *T* running from the top to the bottom. At the top, next to the *T* is a tiny, almost invisible, 10-point letter *o.* Together they represent the word *To.*

  In no way can this technique of using initial letters be considered as making the text more readable. In fact, this is an example of design that impedes reading. The huge initial letter is eye-catching and interesting, but not readable.

- Initial letters often vary a great deal from story to story. Consistency of style is assumed to have little or no value. This kind of assumption cannot be justified.

  Sometimes initial letters vary so much that the appearance of the magazine is also inconsistent. Sometimes a lowercase sans serif letter is used; sometimes an oldstyle typeface is used; sometimes one initial is a capital letter and the next one is lowercase.

- Today there seems to be no attempt to cut an initial letter so that extra amounts of white space appear in words that start with F, P, T, V or W. It takes only a bit to change the letter on a computer keyboard to undercut and eliminate the gap of white space that these letters create.

Careful design makes the use of initial letters a great idea. Designers must remember that some typefaces look very attractive as initial letters, and some, such as sans serif bold, look unattractive.

### A Final Rule of Thumb About Design

As stated earlier, design exists to enhance readability. Therefore, whenever design becomes the center of attention, it may impede readability. It is possible for design and readability to coexist. Some pages may be so readable that readers return to the material just read. In such cases, good designers have a right to show off their wares with pride. But the good designer also continues on his or her quest to produce material of high readability.

## MAGAZINE COVERS

Magazine covers are usually considered a separate design entity from the body of inside pages. *Therefore, Page One of the magazine is not the front cover.* Page One is the first page of the inside signature (or body of pages). The cover is usually printed on a thicker, glossier, stronger paper stock than the inside pages and is wrapped around one or more signatures.

In order to refer to the four pages that are usually wrapped around all the inside sections (called signatures), a different term is used. The "outside front cover" is the first cover page. The next cover pages in the wraparound style are referred to as "inside front cover," "inside back cover" and, finally, "outside back cover."

The outside front cover gets more attention than any other page because of its premier position but also because of its premier design. It is expected that in most highly competitive magazines, the design will mention two or three leading stories and perhaps show illustrations that dramatize the story. The best covers feature some well-known, outstanding person who attracts attention because of his or her unusual life.

Front covers can become so busy with detail that they dissuade more than they persuade potential readers. The solution is to simplify page structure without losing dramatic appeal. Inside front, inside back and outside back covers are usually reserved for advertising, and each of these three usually has high readership.

# PART 4

# EDITING FOR OTHER MEDIA

# CHAPTER 15

# EDITING FOR THE BROADCAST MEDIA

Most of the techniques suggested for editing in newspapers and magazines apply as well to news on radio and television. Those responsible for news copy for any medium must have good news judgment, a feeling for the audience and the ability to handle the language.

Broadcast news differs from newspaper or magazine news in two major respects. Both radio and television news programs must aim at the majority audience and cannot, as newspapers can, serve the interests of the minority. And because the broadcast newscaster must pack enough items into the newscast to give listeners and viewers the feeling they are getting a summary of the big and significant news of the moment, condensation is required. (Figure 15–1 shows a typical network-affiliated control room.)

A newspaper often offers its readers a 1,000-word story and lets them decide how much, if any, they want to read. The broadcast audience has no such choice. If the newscaster gives too much time to items in which listeners and viewers have only a mild interest, they turn the dial.

Here are two wire service accounts of the same story, one intended for the newspaper members, the other for radio and television stations:

> MASSENA, N.Y. (AP)—Unarmed Canadian police scuffled with some 100 Mohawk Indians today and broke an Indian blockade of the international bridge that goes through Mohawk territory in linking the United States and Canada.
>
> The Indians put up the human and automobile blockade after Canadian government officials refused to stop levying customs duties on Indians—duties the Indians say are illegal under the Jay Treaty of 1794.
>
> The Indians had brought 25 automobiles into line at the center of the bridge linking the United States and Canada, and Indian women had thrown themselves in front of police tow trucks to hinder the clearing of the roadway.

***Figure 15–1 Control room of a network-affiliated television station.***
Photo by Scott Serio.

There were no reports of serious injury. Forty-eight Indians were arrested–including most leaders of the protest–and taken into Canadian custody by police on Cornwall Island.

A spokesman for the Indians called for the other five nations of the Iroquois Confederacy to join the protest Thursday.

The Indians began the blockade after the Canadian government refused Tuesday to stop customs duties on Indians who live on the St. Regis Reservation, part of which is in the United States and part in Canada.

Scattered fighting and shoving broke out among the Mohawks and police when officers tried to move in to clear away the automobile blockade. One automobile and two school buses were allowed over the international span around noon.

A newspaper copy editor might trim as many as 50 words from that lengthy story to make it tighter and to eliminate repetition. The story was pared to about 70 words for the broadcast wire roundup item:

(MASSENA, NEW YORK)–UNARMED CANADIAN POLICE HAVE ARRESTED 48 MOHAWK INDIANS. THE INDIANS HAD FORMED A HUMAN WALL AND BLOCKED THE INTERNATIONAL BRIDGE LINKING CANADA AND THE UNITED STATES NEAR MASSENA, NEW YORK, TODAY.

THE MOHAWKS ARE UP IN ARMS ABOUT CANADA'S INSISTENCE ON COLLECTING CUSTOMS DUTIES FROM INDIANS TRAVELING TO AND FROM THEIR RESERVATION ON THE BRIDGE. THEY SAY IT'S A VIOLATION OF THE 1794 JAY TREATY.

***Figure 15–2** A television assignment editor keeps track of video crews assigned to stories.*

Photo by Philip Holman.

As an item in the broadcast news summary, it was cut even more:

FORTY-EIGHT INDIANS HAVE BEEN ARRESTED BY CANADIAN POLICE NEAR THE NEW YORK STATE BORDER. THE INDIANS BLOCKED THE BRIDGE, WHICH LINKS THE U-S AND CANADA. THEY CLAIM VIOLATION OF A 1794 TREATY. THE MOHAWKS SAY THEY PLAN NO BLOCKADE TOMORROW.

(Although AP style still uses the term *Indians*, the term *Native Americans* is preferred by many as less offensive.)

News is written and edited so that readers will have no trouble reading and understanding it. Broadcasters must write news that they can read fluently and that sounds right to the listeners. Broadcast news style must be so simple that listeners can grasp its meaning immediately. The language must be so forceful that even casual listeners will feel compelled to give the story their full attention.

A reader's eyes may on occasion deceive but not to the extent that the listener's ears deceive. A reader who misses a point can go over the material again. A listener who misses a point likely has lost it completely. All radio-television news manuals

## *Television Newsroom Organization*

Television newsrooms are organized to assemble news rapidly and in multimedia format, one that encompasses the written (or spoken) word, graphics, sound and moving pictures. Combining all those into a coherent newscast is a challenging assignment. Typically, those in television news departments have these titles:

- **News director.** The top news executive usually has this title. The news director manages newsroom personnel and resources and sets newsroom policies. An *assistant news director* or *managing editor* often directs the day-to-day news-gathering operation.
- **Executive producer.** The producer determines the overall look of the station's newscasts, including how video, graphics and live reports are handled, and how upcoming stories are teased before a commercial break. Most stations also have *show producers,* who are responsible for the various newscasts.
- **Reporters.** Like their print counterparts, broadcast reporters do the tough work of gathering the news in the field. But unlike print reporters, broadcast reporters must pay plenty of attention to the demands of the camera. Much of their work involves writing a story on the spot and recording it on tape. Sometimes they work with videographers or field producers to do live remotes from the field. In such cases, they must be exacting in their work; no editor corrects their mistakes.
- **Videographers.** Broadcast camera operators, or videographers, work hand-in-hand with reporters to assemble news reports in the field.
- **Anchormen and anchorwomen.** These lead newscasters work in the studio to tie together the various parts of the newscast. They introduce reports from other reporters or read those that are text only.
- **Desk assistants, assignment editors and other support staff.** Back at the station, others help track the news. These include desk assistants, who assemble news from the wires and coordinate network material to be included in the local newscast, and assignment editors, who may be responsible for assigning news teams to breaking stories. Various technical support staffers are also there to handle live remote transmissions, camera repair and similar chores.

Networks often have additional levels of support. The *correspondents* you see regularly on network news may be backed up by *field producers* or *off-camera reporters* who do much of the work of gathering the news.

caution against the use of clauses, especially those at the beginning of a sentence and those that separate subject and predicate. *The AP Radio-Television News Stylebook* uses this example: "American Legion Commander John Smith, son of Senator Tom Smith, died today." Many listeners may be left with the incorrect impression that Sen. Tom Smith died.

The broadcast message is warm and intimate, never flippant or crude (see Figure 15–3). The tone is more personal than that of the newspaper story. It suggests, "Here, Mr. Doe, is an item that should interest you."

The refreshing, conversational style of broadcast news writing has many virtues that all news writers might study. Broadcast writing emphasizes plain talk.

***Figure 15–3** An anchorman and anchorwoman prepare for a television newscast.*
Photo by Scott Serio.

The newspaper reporter may want to echo a speaker's words, even in an indirect quote: "The city manager said his plan will effect a cost reduction at the local government level." Broadcast style calls for simple words: "The city manager said his plan will save money for the city."

The newspaper headline is intended to capture the attention and interest of news readers. The lead on the broadcast news story has the same function. First, a capsule of the news item, then the details:

> THE F-B-I SAYS THERE WAS AN OVER-ALL 19 PERCENT CRIME RATE INCREASE THE FIRST MONTHS OF THIS YEAR. AND THE CRIME THAT INCREASED THE MOST WAS PURSE-SNATCHING–UP 42 PERCENT....
>
> THE NEW YORK STOCK MARKET TOOK A SHARP LOSS AFTER BACKING AWAY FROM AN EARLY RISE. TRADING WAS ACTIVE. VOLUME WAS 15 (M) MILLION 950-THOUSAND SHARES COMPARED WITH 16 (M) MILLION 740-THOUSAND FRIDAY.

The newscast is arranged so that the items fall into a unified pattern. This may be accomplished by placing related items together or by using transitions that help listeners shift gears. Such transitions are made with ideas and skillful organization of facts and not with crutch words or phrases. Said one wire service editor, "Perhaps the most overworked words in broadcast copy are *meanwhile, meantime* and *incidentally.* Forget them, especially *incidentally.* If something is only 'incidental' it has no place in a tight newscast."

Broadcast copy talks. It uses contractions and, often, fragmentary sentences. It is rhythmic because speech is rhythmic. The best broadcast copy teems with simple active verbs that produce images for the listener.

The present tense, when appropriate, or the present perfect tense, creates immediacy and freshness in good broadcast copy and helps eliminate repetition of "today." An example:

> AN AWESOME WINTER STORM HAS BLANKETED THE ATLANTIC SEABOARD–FROM VIRGINIA TO MAINE–WITH UP TO 20 INCHES OF SNOW. GALE-FORCE WINDS HAVE PILED UP SIX-FOOT DRIFTS IN VIRGINIA, BRINGING TRAFFIC THERE AND IN WEST VIRGINIA TO A VIRTUAL HALT. THE STORM HAS CLOSED SCHOOLS IN SIX STATES.
>
> TRAINS AND BUSES ARE RUNNING HOURS LATE. PENNSYLVANIA AND MASSACHUSETTS HAVE CALLED OUT HUGE SNOW-CLEARING FORCES.

## COPY SOURCES

Copy for the broadcast newsroom comes from the wires of news-gathering associations and from local reporters. The news agencies deliver the news package in these forms:

1. **Spot summary.** A one-sentence item:

   (DENVER)–F-B-I SHARPSHOOTERS HAVE SHOT AND KILLED A GUNMAN WHO KILLED TWO HOSTAGES ABOARD A PRIVATE PLANE AT STAPLETON INTERNATIONAL AIRPORT IN DENVER.

2. **Five-minute summary:**

   (DENVER)–F-B-I AGENTS FATALLY SHOT A GUNMAN EARLY TODAY AS HE BOARDED AN AIRLINE JET IN DENVER, WHICH HE THOUGHT WAS TO FLY HIM TO MEXICO. THE F-B-I SAID THE GUNMAN HAD HIS TWO HOSTAGES WITH HIM AT THE TIME OF THE SHOOTING. HE HAD HELD THEM ON A SMALL PRIVATE PLANE FOR SEVEN HOURS. BEFORE HE LEFT THE FIRST PLANE, THE GUNMAN HAD TOLD AUTHORITIES OVER THE RADIO, "I'LL TELL YOU WHAT. I'M STILL GONNA HAVE THIS GUN RIGHT UP THE BACK OF HIS (THE HOSTAGE'S) HEAD, AND IT'S GONNA BE COCKED, AND IF ANYBODY EVEN BUDGES ME, IT'S GONNA GO OFF, YOU KNOW THAT."–DASH–

   THE CHIEF OF THE DENVER OFFICE OF THE F-B-I, TED ROSACK, SAID 31-YEAR-OLD ROGER LYLE LENTZ WAS KILLED SHORTLY AFTER MIDNIGHT, ENDING AN EPISODE THAT BEGAN IN GRAND ISLAND, NEBRASKA, AND INCLUDED TWO SEPARATE FLIGHTS OVER COLORADO ABOARD THE COMMANDEERED PRIVATE PLANE. NEITHER HOSTAGE WAS INJURED.

3. **Takeout.** This is a detailed, datelined dispatch concerning one subject or event.

4. **Spotlights and vignettes.** Both are detailed accounts, the latter usually in the form of a feature.
5. **Flash.** This is seldom used and is restricted to news of the utmost urgency. A flash has no dateline or logotype and is limited to one or two lines. It is intended to alert the editor and is not intended to be broadcast. The flash is followed immediately by a bulletin intended for airing.
6. **Bulletin.** Like the flash, it contains only one or two lines. A one-line bulletin is followed immediately by a standard bulletin giving details.
7. **Double-spacers.** This indicates a high-priority story not as urgent as a bulletin. The double-spacing makes the item stand out on the wire and calls the item to the attention of news editors:

    HOSTAGES (TOPS)

    (DENVER)–AN F-B-I SPOKESMAN SAYS A GUNMAN WAS SHOT TO DEATH AFTER HE SEIZED A PRIVATE PLANE IN NEBRASKA, ORDERED THE PILOT TO FLY TO DENVER AND HELD TWO HOSTAGES FOR SEVEN HOURS. F-B-I SPOKESMAN TED ROSACK SAYS THE GUNMAN, 30-YEAR-OLD ROGER LENTZ OF GRAND ISLAND, NEBRASKA, WAS SHOT AND KILLED IN AN EXCHANGE OF GUNFIRE WITH F-B-I AGENTS ABOARD A CONVAIR 990 AT STAPLETON INTERNATIONAL AIRPORT.

    ROSACK SAYS THE HOSTAGES WERE NOT HURT.

8. **Special slugs.** These include AVAILABLE IMMEDIATELY (corresponds to the budget on the news wire), NEW TOP, WITH (or SIDEBAR), SPORTS, WOMEN, FARM, WEATHER, BUSINESS, CHANGE OF PACE, PRONUNCIATION GUIDE, EDITOR'S NOTE, ADVANCE, KILL, CORRECTION, SUBS (or SUBS PREVIOUS).

Some local stations broadcast the news in the form they receive it from the news agency. This may suggest that an announcer dashes into the newsroom, rips the latest summary off the machine and goes on the air with it. Although that may have been true in the early days and at the smaller stations, the practice is becoming increasingly rare because news has commercial as well as public service value. Furthermore, the many typographical errors in wire copy force the reporter to read it and edit for errors. Here is a fairly typical example:

A U-S DEPARTMENT OF AGRICULTURE OFFICIAL SAYS IN DENVER HE FEELS INSPECTION REPORTS OF COLORADO MEAT-PLANTING PACKS HAVE BEEN ACCURATE.

How about "packing plants"?

Most broadcast news today is handled by trained reporters who know how to tailor the news for a specific audience. This is done by combining items from several roundups and double-spacers to create the desired format. Increasingly, almost all wire copy is rewritten before it is assembled for broadcast to give the listener some variety in items that may be repeated several times during the broadcast period.

Some radio and television stations subscribe to the national news wire of a wire service as well as to the broadcast news wire. That provides a greater number and variety of stories.

## PREPARATION OF COPY

All local copy should be triple-spaced. Copy should be easier to read in capital and lowercase letters than in all caps, but because reporters are used to reading all-cap wire copy, some prefer the all-cap script style. If a letter correction is to be made in a word, the word should be scratched out and the correct word substituted in printed letters. If word changes are made within sentences, the editor should read aloud the edited version to make sure the revised form sounds right. If the copy requires excessive editing, it should be retyped before it is submitted to a newscaster.

Most television newscasters read stories on the air using a script projection system called a TelePrompTer. The TelePrompTer works by beaming a picture of the script to a monitor mounted on the studio camera. The script image reflects in a two-way mirror mounted over the camera lens, so the newscaster actually can read the script while appearing conversational and pleasant. There is no need to refer to notes, but newscasters have them in case the TelePrompTer fails. Script copy averages only four words a line so the newscaster's eyes do not have to travel noticeable distances back and forth across the page as he or she reads from the TelePrompTer or from the script itself.

All editing of broadcast copy is done with the newscaster in mind. If a sentence breaks from one page to another, the newscaster will stumble. No hyphens should be used to break words from one line to the next.

News editors prefer to put each story on a separate sheet. That enables them to rearrange the items or to delete an item entirely if time runs short. A few briefs tacked near the end of the newscast help fill the allotted time.

Properly edited broadcast copy also should include pronunciation aids when necessary. The most common dilemma for newscasters concerns place names, many of which get different pronunciations in different regions. No newscaster should confuse the Palace of Versailles (vur-SIGH) in France with the town of Versailles (vur-SALES) in Missouri. The editor should add the phonetic spelling to the script, and the newscaster can underline the word on his or her copy as a reminder.

The wire services provide a pronunciation list of foreign words and names appearing in the day's report. The guide is given in phonetic spelling (Gabon—Gaboon') or by rhyme (Blough—rhymes with how; Like-Like Highway—rhymes with leaky-leaky).

PHONETIC SPELLING SYSTEM USED BY WIRE SERVICES

| | |
|---|---|
| A—like the "a" in cat | OW—like the "ow" in cow |
| AH—like the "a" in arm | U—like the "u" in put |
| AW—like the "a" in talk | UH—like the "u" in but |
| AY—like the "a" in ace | K—like the "c" in cat |
| EE—like the "ee" in feel | KH—guttural |
| EH—like the "ai" in air | S—like the "c" in cease |
| EW—like the "ew" in few | Z—like the "s" in disease |
| IGH—like the "i" in tin | ZH—like the "g" in rouge |
| IH—like the "i" in time | J—like the "g" in George |
| OH—like the "o" in go | SH—like the "ch" in machine |
| OO—like the "oo" in pool | CH—like the "ch" in catch |

## BROADCAST STYLE

### Abbreviations

No abbreviations should be used in radio-television news copy with these exceptions:

1. Common usage: Dr. Smith, Mrs. Jones, St. Paul.
2. Names or organizations widely known by their initials: U-N, F-B-I, G-O-P (but AFL-CIO).
3. Acronyms: NATO.
4. Time designations: A-M, P-M.
5. Academic degrees: P-H-D.

### Punctuation

To indicate a pause where the newscaster can catch a breath, the dash or a series of dots is preferable to a comma: THE HOUSE PLANS TO GIVE THE 11-BILLION-500-MILLION DOLLAR MEASURE A FINAL VOTE TUESDAY . . . AND THE SENATE IS EXPECTED TO FOLLOW SUIT—POSSIBLY ON THE SAME DAY.

The hyphen is used instead of the period in initials: F-B-I. The period is retained in initials in a name: J.D. Smith. All combined words should have the hyphen: co-ed, semi-annual. (Spelling should also use the form easiest to pronounce: employee.)

Contractions are more widely used in broadcast copy than in other news copy to provide a conversational tone. Common contractions—isn't, doesn't, it's, they're—may be used in both direct and indirect quotes:

MEMBERS OF THE TRANSPORT WORKERS UNION IN SAN FRANCISCO SAY IF THE MUNICIPAL STRIKE DOESN'T BEGIN LOOKING LIKE A GENERAL STRIKE BY THIS AFTERNOON, THEY'LL RECONSIDER THEIR SUPPORT OF THE WALK-OUT. THE REFUSAL BY DRIVERS TO CROSS PICKET LINES SET UP BY STRIKING CRAFT UNIONS HAS SHUT DOWN MOST TRANSPORTATION IN THE CITY.

Good broadcast writers, however, avoid contractions when they want to stress verbs, especially the negative: "I do not choose to run" instead of "I don't choose to run."

Even in broadcast copy, contractions should not be overworked. Nor should the awkwardly contrived ones be attempted. The result would be something like this:

IT'S POSSIBLE THERE'S BEEN A MAJOR AIR DISASTER IN EUROPE.

A BRITISH AIRLINER WITH 83 PERSONS ABOARD DISAPPEARED DURING THE DAY AND IS CONSIDERED CERTAIN TO'VE CRASHED IN THE AUSTRIAN ALPS.

APPARENTLY NO SEARCH'LL BE LAUNCHED TONIGHT. THERE'S NO INDICATION OF WHERE THE AIRCRAFT MIGHT'VE GONE DOWN.

## Quotation Marks

The listener cannot see quotation marks. If the reporter tries to read them into the script—"quote" and "end of quote"—the sentence sounds trite and stilted. It is easier and more natural to indicate the speaker's words by phrases such as "and these are his words," "what he called," "he put it in these words," "the speaker said." Direct quotations are used sparingly in the newscast. If they are necessary, they should be introduced casually and the source should precede the quotation:

THE SOVIET NEWS AGENCY TASS SAID TODAY THAT SOVIET SCIENTISTS WERE AWARE OF AN IMPENDING EARTHQUAKE THIS MONTH IN CENTRAL ASIA FIVE DAYS BEFORE IT HAPPENED....

THE NEWS AGENCY SAID, IN THESE WORDS, "AT THAT TIME, A CONNECTION WAS FIRST NOTICED BETWEEN THE GAS-CHEMICAL COMPOSITION OF ABYSSAL (DEEP) WATERS AND UNDERGROUND TREMORS."

Quotation marks are placed around some names that would otherwise confuse the reporter:

IN AN ANSWER TO AN S-O-S, THE U-S COAST GUARD CUTTER "COOS BAY" ALONG WITH OTHER VESSELS STEAMED TO THE AID OF THE STRICKEN FREIGHTER. THE NORWEGIAN VESSEL "FRUEN" PICKED UP NINE SEAMEN FROM THE "AMBASSADOR" IN A TRICKY TRANSFER OPERATION IN THE TEMPEST-TOSSED SEAS.

In this illustration the reporter is more likely to fumble "tempest-tossed seas" than the names of the vessels.

## Figures

Numbers are tricky in broadcast copy. "A million" may sound like "eight million." No confusion results if "one million" is used.

In most copy, round numbers or approximations mean as much as specific figures. "Almost a mile" rather than "5,200 feet," "about half" rather than "48.2 percent," "just under two percent" rather than "1.9 percent."

An exception is vote results, especially when the margin is close. It should be "100-to-95 vote" rather than "100-95 vote." The writer or editor can help the listener follow statistics or vote tallies by inserting phrases such as "in the closest race" and "in a landslide victory." Here are some additional style rules for figures:

- Fractions and decimals should be spelled out: one and seven-eighths (not $1\frac{7}{8}$), five-tenths (not 0.5).
- Numbers under 10 and over 999 are spelled out and hyphenated: one, two, two-thousand, 11-billion-500-million, 15-hundred (rather than one-thousand-500), one-and-a-half million dollars (never $1.5 million).
- When two numbers occur together in a sentence, the smaller number should be spelled out: twelve 20-ton trucks.
- Any figure beginning a sentence should be spelled out.
- Figures are used for time of day (4:30 this afternoon), in all market stories and in sports scores and statistics (65-to-59, 5-foot-5). If results of horse races or track meets appear in the body of the story, the winning times should be spelled out: two minutes, nine and three-tenths seconds (rather than 2:9.3).
- In dates and addresses the *-st, -rd, -th* and *-nd* are included. June 22nd, West 83rd Street. Figures are used for years: 1910.
- On approximate figures, writers sometimes say, "Police are looking for a man 50 to 60 years of age." This sounds like "52" to the listener. It should read, "Police are looking for a man between 50 and 60 years old."

## Titles

The identification prepares the ear for the name. Therefore, the identification usually precedes the name: Secretary of State Warren Christopher. Some titles are impossible to place before the name: The vice president of the Society for the Preservation and Encouragement of Barbershop Quartet Singing, Joe Doe. Use "Vice President Joe Doe of the Society for the Preservation and Encouragement of Barbershop Quartet Singing." Use "Police Chief Don Vendel" rather than "Chief of Police Don Vendel."

Some radio and television newsrooms insist that the president should never be referred to by his last name alone. It would be President Clinton, the president or Mr. Clinton.

Broadcast copy seldom includes middle initials and ages of persons in the news. Of course, some initials are well-known parts of names and should be included: Richard M. Nixon.

Ages may be omitted unless the age is significant to the story: "A 12-year-old boy—Mitchell Smith—was crowned winner" and so on. Ages usually appear in local copy to aid in identification. Place the age close to the name. Do not say, "A 24-year-old university student died in a two-car collision today. He was John Doe." Use "A university student died ... He was 24-year-old John Doe."

Obscure names need not be used unless warranted by the story. In many cases the name of the office or title suffices: "Peoria's police chief said," and so on. The same applies to little-known place names or to obscure foreign place names. If the location is important, you may identify it by placing it in relation to a well-known place—"approximately 100 miles south of Chicago." In local copy, most names and places are important to listeners and viewers.

When several proper names appear in the same story, it is better to repeat the name than to rely on pronouns unless the antecedent is obvious. Also, repeat the names rather than use *the former, the latter* or *respectively.*

## Datelines

The site of the action should be included in broadcast copy. The dateline may be used as an introduction or a transition: "In Miami." Or the location may be noted elsewhere in the lead: "The Green Bay Packers and the Chicago Bears meet in Chicago tonight in the annual charity football game."

On the newspaper wire *here* refers to the place where the listener is. Because radio and television may cover a wide geographical area, such words as *here* or *local* should be avoided. Said a radio news editor, "If the listener is sitting in a friendly poker game in Ludowici, Ga., and hears a radio report of mass gambling raids 'here,' he may leap from the window before realizing the announcer is broadcasting from Picayune, Miss."

## Time Angle

In the newspaper wire story almost everything happens "today." Radio copy breaks up the day into its parts: "this morning," "early tonight," "just a few hours ago," "at noon today." The technique gives the listener a feeling of urgency in the news. Specific time should be translated for the time zone of the station's location: "That will be 2:30 Mountain Time."

In television especially, use of the present and present perfect tenses helps downplay the time element:

> SEARCHERS HAVE FOUND THE WRECKAGE OF A TWIN-ENGINE AIR FORCE PLANE IN PUERTO RICO AND CONTINUE TO LOOK FOR THE BODIES OF SIX OF THE AIRCRAFT'S

EIGHT CREWMEN. AUTHORITIES CONFIRM THAT THE PLANE, MISSING SINCE SATURDAY, CRASHED ATOP A PEAK 23 MILES SOUTHEAST OF SAN JUAN.

## TASTE

Broadcast news editors should be aware of all members of their audience—the young and the aged, the sensitive and the hardened. Accident stories can be reported without the sordid details of gore and horror. Borderline words that may appear innocent to the reader carry their full impact when given over the more intimate instruments of radio and television. If spicy items of divorce and suicide are tolerated by the station, at least they can be saved until the late-hour news show when the young are in bed.

The wire services protect the editor by prefacing the morbid or "gutsy" items with discretionary slugs:

(FOR USE AT YOUR DISCRETION)
(RAPE)
MIAMI, FLORIDA–POLICE IN MIAMI REPORT THEY SUSPECT JOHN DOE IN THE CRIMINAL ASSAULT (RAPE) OF AN 18-YEAR-OLD GIRL. DOE–27 YEARS OLD–WAS ARRESTED IN THE CITY MUSEUM AND CHARGED WITH STATUTORY ASSAULT (RAPE).
(END DISCRETIONARY MATTER)

Avoid references to physical handicaps or deformities unless they are essential to the story. Never say "blind as a bat," "slow as a cripple" and the like. Similarly, unless they are essential, references to color, creed or race should not be used.

Wire services handle items involving pertinent profanity by bracketing the profanity:

"GODFREY SAID–IT HURTS (LIKE HELL)."

The practice of including a humorous item, usually near the end, in a newscast has produced some unfunny stories, such as the one about a man breaking his neck by tripping over a book of safety hints. But a truly humorous item lightens the heavy news report. Invariably it needs no embellishment by the editor or reporter.

On many stations someone other than the news reporter gives the commercials. One reason for this practice is to disassociate the newsperson from the commercial plugger. Even so, the director or reporter should know the content of commercials sandwiched in news. If a news story concerns a car crash in which several are killed, the item would not be placed ahead of a commercial for an automobile dealer. Airlines generally insist that their commercials be canceled for 24 hours if the newscast contains a story of an airliner crash, a policy that is likewise applied to many metropolitan newspapers.

The sponsor does not control or censor the news. The story of a bank scandal should never be omitted from a news program sponsored by a bank. Nor should a sponsor ever expect sponsorship to earn news stories publicizing the business.

## ATTRIBUTIONS

Attribution is an important aspect of all news writing. If an error is discovered, the writer has an "out" if the item has official attribution. For example, say, "The State Patrol said Smith was killed when his car overturned in a ditch" rather than "Smith was killed when his car overturned in a ditch." Attribution can also be vital in the event of any court action over a story written and aired by the news staff.

Should identification of accident victims be made before relatives have been notified? Some stations insist on getting the coroner's approval before releasing names of victims. If the release is not available, the tag would be, "Police are withholding the name of the victim until relatives have been notified."

In stories containing condition reports on persons in hospitals, the report should not carry over the same condition from one newscast to another without checking with the hospital to find out whether there has been a change.

## AUDIO AND VIDEO TAPES

All news copy for radio and television should show the date, the time block, the story slug, the writer's name or initials, the story source and an indication whether the story has a companion tape cartridge or a video segment. If more than one tape accompanies a story, the slug would indicate the number of tapes. A tape cartridge is simply a tape recording or audio tape from a news source.

If a tape is used, a cue line is inserted for each tape. Many stations use a red ribbon to type the out-cues or place red quotation marks around the cue line. At the end of the tape, the radio newscaster should again identify the voice used on the tape.

If several tapes are used in one newscast, they should be spaced so that the same voices, or series of voices, are not concentrated in one part. The control room needs time to get the tapes ready for broadcast.

The out-cues of the tape should be noted in the *exact* words of the person interviewed. That will ensure that the engineer will not cut off the tape until the message is concluded. The producer should provide the engineer or board operator with a list of news cartridges to be used and the order in which the producer intends to use them. The same would hold true of telephone interviews, either taped or live.

The broadcast reporter also may have access to audio news services provided by networks, group-owned facilities and the wire services. These feeds, provided to the station on audio tapes, may be voiced reports or recorded natural sound.

## TELEVISION NEWS

Newspapers communicate with printed words, radio with spoken words and television with spoken words and moving pictures. Editing a television news or special event show involves all three. Television news editing is the marriage of words to pictures, words to sound, pictures to sound and ideas to ideas.

Reuven Frank, former president of NBC News, contended that the highest power of television is to produce an experience. Television cannot disseminate as much information as newspapers, magazines or even radio. But in many instances it causes viewers to undergo an experience similar to what would happen if they were at the scene. One can feel sympathy when reading about napalmed civilians or the drowning of a child at a swimming pool. But watching the same thing on a television newscast is a wrenching personal experience that gets people worked up, even angry. Television is an instrument of power, not because of the facts it relates but because of the experience it conveys to viewers.

Words speak for themselves to the newspaper reader. In radio, a newscaster voices the words for the listener. In television, the newscaster is there, talking directly to the viewer about the news. The newscaster is the key actor, and many a station has fallen behind in ratings for its news shows, not because the station did not have good reporters and camera operators or lacked a well-paced news format, but because competing stations had better on-air talent.

In the early days of television, stations hired journalists to report and write the news, then handed over the polished manuscript to a good-looking announcer with mellifluous tones. Today more and more newscasters are men and women with backgrounds who may not sound like movie stars but who know what they are talking about (see Figure 15–4).

News editing, the sorting or processing of the news, requires more time for television than for radio. Producers and writers must spend hours reviewing, sifting and editing all the material available for a single half-hour newscast. They use these criteria in selecting items: how significant the item is, whether it is interesting either factually or visually, how long it is and how well its content complements the rest of the program.

Before local videotape is edited, it must be examined to determine how much to cut. Sometimes a tape may have relatively little news value but is included because of its visual quality. A barn fire might not rate mention on a radio newscast, but the video could be spectacular.

***Figure 15–4*** ***News, weather, sports: These have become the traditional pattern in television news. In this modern studio, those on camera are supported by many behind the scenes. These include producer, director, assignment editor, reporters, photographers, video editors and others.***

Photo by Scott Serio.

Network tapes also are examined. Later afternoon network news stories are reviewed to determine what can be lifted for the late-evening local news program. The networks provide their affiliates with an afternoon news feed for use as the affiliates see fit. This closed-circuit feed from New York consists of overset material not used on the network news. These feeds are recorded on videotape and usually include more than a dozen one- to two-minute video stories from the nation and the world. The producer has to monitor these feeds to decide which stories to use.

Chain-owned stations maintain a Washington or New York bureau that sends member stations daily video reports. These, too, must be reviewed.

In addition to editing these videotape reports, the producer must also go over the vast number of wire service news and pictures, not to mention stories filed by station reporters. Having selected what to use, the producer's next job is to determine how and where it can be used within the few minutes allotted the news show.

Most U.S. television stations today use portable electronic videotape equipment rather than film. These tape minicams, known as ENG (for electronic news gathering), shoot ¾-inch videotape, usually in 15-minute cassettes. The small recorder and battery pack give the ENG camera crew enough mobility to cover fast-paced news stories, and, although the minicam rig is heavier than a film camera, the ENG photographer can see immediately the pictures shot instead of having to wait, often hours, for the film to be developed. That instant image-making, along with the high picture quality and increased editing advantages, make the ENG rig well worth the weight.

It is the ¾-inch videotape editing systems that have transformed television news production procedures most in recent years. The tape editor plays back on one recording machine the pictures shot in the field, selects the shots he or she wants, their length and order, and assembles the edited story by transfer-dubbing those shots to a second tape cassette in another recorder.

The editor also may add a reporter's narrative, natural sound recorded in the field, music and other audio components to the same edited cassette. In most situations, ENG editing is more precise, more versatile and quicker than film editing. Best of all, the editor finishes with a compact videocassette containing all the elements of the story. That simplifies production of the newscast.

In a typical newscast, the anchorman and anchorwoman, sitting in a brightly lighted studio set, read stories into studio cameras. They also introduce edited videotape stories, which beam onto the air when the director instructs an engineer to start the playback machines. A studio camera also may shoot closeups of still pictures in the studio to help illustrate stories.

The minicam's portability has added a new dimension to local television newscasts. Lightweight transmitters allow a reporter and engineer to beam live reports back from remote locations—even from an airborne helicopter—straight to the station and onto the air. Portable editing rigs mean the remote reports can include packaged pieces as well.

## COPY FORMATS

There are almost as many copy formats and scripting styles as there are broadcast newsrooms. Terminology varies, as well. In one newsroom the script designation "voiceover" (or "VO") might mean that a newscaster is reading over videotape; in another newsroom "voiceover" might mean that the reporter's recorded voice is running with video.

Figures 15–5 and 15–6 are typical scripts from KOMU-TV in Columbia, Mo. They are from a 6 o'clock newscast devoted to local news. The robbery arrest story (Figure 15–5) calls for the newscaster to read it all, but after a few lines the picture will change from the newscaster's face to videotape illustrating the story. The script instructions tell the director that the video source is ENG videotape, to begin as the newscaster says "an eyewitness" and to last for 22 seconds. Other designations show

NewsCenter 8 Slug arrest
Newscast 6 Date 9/1 Writer mm Length :28

| | |
|---|---|
| moc: | Columbia police have charged a Moberly man with last week's robbery of the Boone County Bank. |
| ENG :22 | |
| ENG :00-:22<br>Cassette E-100<br>Cut 1 Cued<br>VOICEOVER | An eyewitness identified 23-year old Robert Wilson of Moberly as the man who took almost 10-thousand dollars from the bank Friday afternoon. |
| Key: File Tape<br>(:05-:10) | Acting on a tip, police took Wilson into custody this morning in downtown Columbia.<br>(more) |

NewsCenter 8 Slug arrest ADD
Newscast 6 Date 9/1 Writer mm Length --

| | |
|---|---|
| tape rolling for<br>VOICEOVER: | Wilson denies committing the robbery, but remains in jail under 50-thousand dollar bond.<br>These bank photographs show the robber escaping with the money bag under his arm. |
| back to moc: | Police say they also want to question Wilson about a series of burglaries in the county over the past year. |

*Figure 15–5 Television script indicating use of voiceover. The newscaster continues to read the script while videotape illustrates the story.*

Courtesy of KOMU-TV, Columbia, Mo.

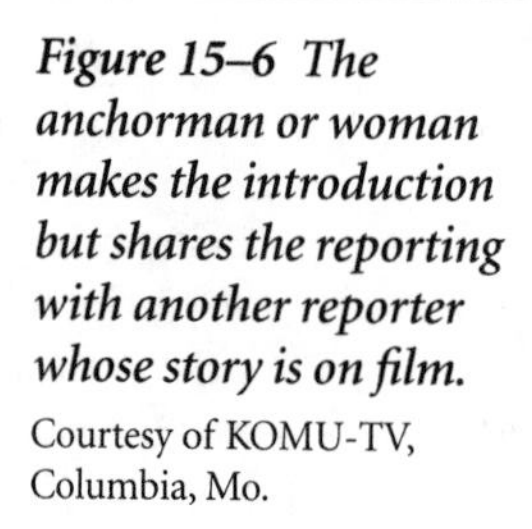

*Figure 15–6 The anchorman or woman makes the introduction but shares the reporting with another reporter whose story is on film.*

Courtesy of KOMU-TV, Columbia, Mo.

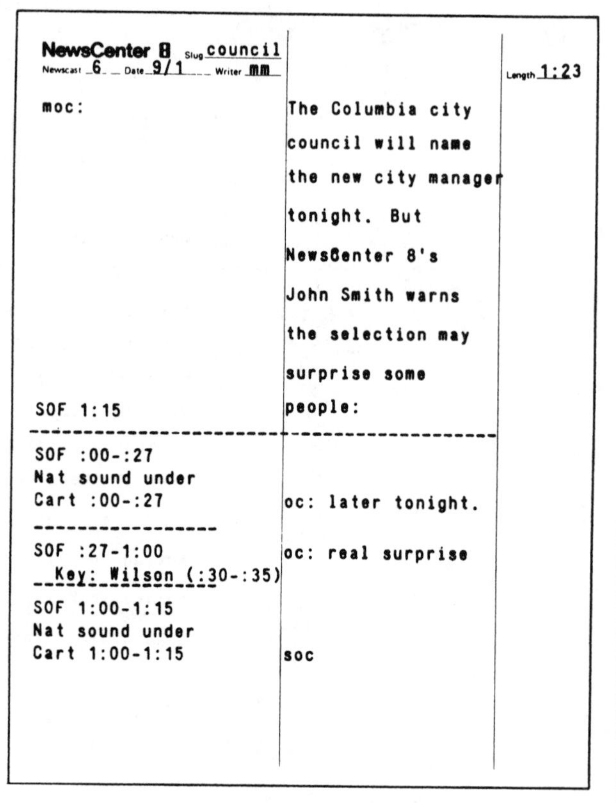

NewsCenter 8 Slug council
Newscast 6 Date 9/1 Writer mm Length 1:23

| | |
|---|---|
| moc: | The Columbia city council will name the new city manager tonight. But NewsCenter 8's John Smith warns the selection may surprise some |
| SOF 1:15 | people: |
| SOF :00-:27<br>Nat sound under<br>Cart :00-:27 | oc: later tonight. |
| SOF :27-1:00<br>Key: Wilson (:30-:35) | oc: real surprise |
| SOF 1:00-1:15<br>Nat sound under<br>Cart 1:00-1:15 | soc |

the number of the cassette, the location of the story of the cassette ("Cut 1 cued") and whatever written information ("key") will be superimposed on the screen and when.

Figure 15–6 shows the other common TV script, the anchorman or woman's introduction to a packaged report. In this story, he or she shares the reading with the reporter. This example involves a filmed story, so the scripting differs from the previous example. The "soc" designation alerts the director that the reporter's narration ends with a standard out-cue such as "John Smith, NewsCenter 8, Columbia."

A good television script should not compete with the pictures. It should prepare viewers for what they are about to see or specify what may be difficult to see, but it should avoid repeating what viewers can see or hear. If the mayor has criticized the city's water supply and his statement is on videotape, the script merely sets up the statement with a brief introduction. Good scripts also direct the viewer's attention to significant details in the video, but they should avoid phrases such as "we're now looking" or "this picture shows."

# CHAPTER 16

# EDITING FOR THE NEW MEDIA

## A NEW SOURCE OF NEWS

The Information Age is here, and with it comes a dramatic change in the way the public gets its news. Today, those with computers and modems in their homes no longer have to wait for the television or radio newscast. Nor do they have to wait for the morning newspaper to get their fill of news. Instead, they simply dial in to one of the public information utilities or the Internet. There they have access to all the news they want, when they want it.

Computer-based information retrieval makes it possible for consumers to serve as their own editors. It's simple to find the up-to-the-minute score of the Minnesota Twins game in progress. It's also easy to get the latest German and English soccer results, something not published in most U.S. newspapers but of great interest to thousands of U.S. soccer fans. Advanced computer programs even make it possible for the computer to retrieve news of interest to you based on a profile you've provided. Each day (or even more often, if you desire) the computer seeks out news on those subjects, then packages it in an easy-to-read format.

As discussed in Chapter 1, the ability to control information resources on an individual basis raises disturbing questions about who sets the agenda for public discourse, traditionally a role of the media. It's now possible for the public to spurn the editor's attempts to set that agenda and go directly to primary sources of news, bypassing the journalist gatekeeper entirely. That option simply hasn't existed until now. Many who view the media as biased relish the idea of cutting editors out of the process.

That makes it more important than ever for editors to make themselves useful. They can do so by helping busy consumers sort through the massive amount of information contained in those computer systems. They also can do so by establishing themselves as honest brokers in the news dissemination process.

Research shows that consumers are busier than ever. They have less and less time each day for perusing the newspaper or watching television news. It follows, then, that new-media editors can perform a useful service by helping readers sort through that mass of information we call news and ranking what's most important. While it's probably true that some will reject that service in favor of wading through the wires themselves, others will *want* help in sorting out what's important. New-media editors who do that as honestly and objectively as possible—those who are seen as not having their own agenda—are most likely to win the public's trust.

Thus, we enter an era in which the editor's role changes rather dramatically. How journalists adjust to that new role will have much to do with whether the new media become a haven for anarchists or the mass medium of the future.

## OUTLETS FOR PUBLISHING

The term *new media* has been used to refer to any number of different attempts to find innovative ways of distributing information electronically. Some include under that heading the *audiotext* services run by many newspapers. With audiotext, users dial a telephone number to connect, then press additional numbers to hear recorded information on a variety of subjects. More often, though, the term new media is used to refer to computer-based sources of information. Three sources have emerged as the most important media distribution services:

- **The public information utilities.** The PIUs (see Figure 16–1) include CompuServe, America Online, Prodigy, The Microsoft Network, Delphi, GEnie and others. All offer news to their customers.
- **The Internet.** The Internet has become the most prominent of all the on-line services. Even the PIUs have been forced to provide connections to the Internet, which arguably has become the world's most important source of information. Some refer to it, with justification, as the world's largest library. Hundreds of newspapers, magazines and broadcast stations around the world have set up shop on the Internet. In addition, the Internet is a repository for massive amounts of information provided by universities, governments and private companies. It's an appealing place to those who have learned to navigate it; where else in the United States can you get today's news from St. Petersburg, Russia?
- **CD-ROMs.** Interactive CD-ROMs, while not suited for breaking news, are fast becoming the most intriguing magazines of the new-media world. *Time* magazine published an interactive CD-ROM on the Gulf War against Iraq, and photojournalists publish the results of the annual Pictures of the Year competition on CD-ROM. There, consumers see not only the winning photos but also get to hear why the judges picked them.

For journalists, new-media operations in all those areas represent new places of employment. Indeed, new-media publishing is probably the fastest growing sector of the media industry. Jobs by the thousands are being created each year.

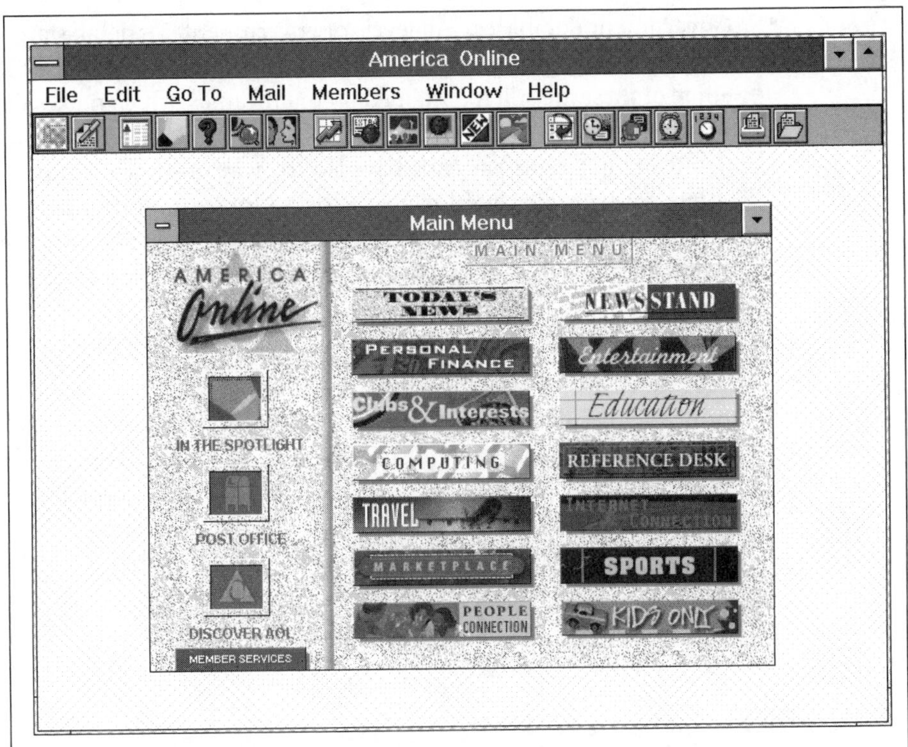


*Figure 16–1 The opening screen of America Online, one of the most popular public information utilities.*

## SOURCES OF INFORMATION

Journalists have two primary interests in computer-based media. As we've already discussed, new media are the newest medium for publishing text, graphics, audio and video. But they also are an invaluable source of information—on-line libraries, of sorts—that can be used to provide background information for stories or to check facts. Obviously, both the PIUs and the Internet are great sources of information for journalists. But some additional sources merit attention as well:

- **Commercial database services.** These services are not intended for the general public but cater to journalists, lawyers and others who need to do background research. The commercial services provide archives of information from wire services, newspapers, magazines and other sources. Lexis/Nexis and Dialog are examples of commercial services.

- **Government databases.** All levels of government—federal, state, county and municipal—maintain extensive databases of information of use to journalists. Some of these are still maintained on mainframes, which makes them difficult to access. Organizations such as the National Institute of Computer-Assisted Reporting in Columbia, Mo., have led the way in showing journalists how to access government data for news stories. Increasingly, though, government agencies are recognizing the usefulness of making that data more readily available. Today, thousands of government databases are accessible through the Internet.
- **Special-interest group databases.** Many private groups maintain databases of information of use to journalists. Usually, these are maintained by nonprofit groups with a cause. Journalists must exercise care in using such databases because of the potential for distortion as these groups champion their causes. Nonetheless, such databases can be quite useful. The journalist may find data there that are not available elsewhere.
- **Other computerized databases.** These include the *electronic libraries* (or *morgues*) maintained by the various media and databases created by reporters themselves as they attempt to document stories.

## THE NEW-MEDIA EDITOR

Editing for the new media has been likened to editing for a wire service. There are no set deadlines; more precisely, there is a deadline every minute. The comparison to wire service journalism has validity in terms of immediacy, but in other respects it falls short. New-media editors face deadlines constantly, but unlike wire editors they must be skilled in the presentation of news in many formats—the written word, graphics, audio and video. That's because the new media increasingly use all those devices to transfer information to consumers. In a real sense, then, new-media editors are *multimedia journalists.*

Indeed, new-media editing requires only a few new skills; instead, it forces editors to learn about editing in all forms—words, photos, graphics, audio and video. Says John Callan, a senior editor at the Microsoft Network News who was instrumental in assembling that service's news staff:

> We knew we needed to bring in people from all the existing media because it was unlikely we would find people who had all the skills we wanted. We brought in newspaper reporters, editors, magazine people, broadcast journalists, photojournalists, graphics artists, even computer programmers. Throwing all those people, with their varying skills, into the same operation is fascinating. We learn from each other every day.

It's unlikely that someone trained as a writer or editor will suddenly become an accomplished graphics artist or photojournalist. But as Callan points out, in the new media it is essential to think in multimedia terms: What device is best to convey this particular bit of information? A story? An information graphic? A video clip? Audio? Or should they all be used? The new-media journalist must be able to think in visual concepts as well as in written ones.

Obviously, the new-media editor must be proficient in using computers, a key tool of the craft. But the medium itself also creates the need for new kinds of journalists. The Microsoft Network News coined the term *linkmeister* to describe the job of one individual in the newsroom. The linkmeister's job? To find sites on the World Wide Web, the user-friendly portion of the Internet to which Microsoft and other PIUs can link. For example, if a news story reports on census trends, it's a simple matter to link to the U.S. Census Bureau site on the Web, where the reader can find plenty of additional information. Doing so provides a service that none of the existing media can match.

Such *hypertext* links to other sites are one of the major advantages of the new media over existing media. It takes just seconds to establish links to other Web sites that may be invaluable in helping the reader understand a story. And for those who are intensely interested in a subject, links are established that provide all the information they want.

Indeed, the new media offer many such advantages over the existing media. It's even possible to provide the user with tools for exploring databases that reside on the Internet. The challenge for journalists is to think creatively about how computers can assist in the presentation of news. As they do so, they redefine the medium.

What Microsoft's Callan and others like him are doing is inventing a new medium, one rooted in the traditions of the existing media but one that rejects their shortcomings as unacceptable. Thus, the new media combine the depth of newspapers and magazines with the immediacy of radio and television. And they combine the eye appeal of television with the enduring quality of magazine reproduction. Those who work in the new media must use their existing skills while learning new ones. They must become multimedia journalists. For those who like challenges and for those who view every day as a new opportunity to learn, new-media operations are exciting places to work.

## THE WEB'S IMPORTANCE

The transformation of the Internet from a computer network for academics and the military into the world's newest news medium occurred almost overnight. It happened during 1994 and 1995 with the widespread adoption of the World Wide Web, a user-friendly front-end for the notoriously unfriendly Internet.

The concept of the World Wide Web was developed by Tim Berners-Lee at a laboratory in Switzerland. What made it popular, however, was the development of a computer program called Mosaic at the University of Illinois in Urbana-Champaign. Mosaic is known as a *browser* because it allows the user to browse the Internet's content with ease.

Mosaic, freely distributed on the Internet, spawned improved commercial versions of browsers, including Netscape, which quickly captured about 75 percent of the market. Today, Netscape sets the standards for browser software. Other browser vendors, including companies as large as IBM and as small as Spyglass, try to keep up as Netscape adds feature after feature. Often, those companies are forced to license Netscape technology. Another key player is Sun Microsystems, whose HotJava technology allows developers to create multimedia objects for Web sites. Other technologies, including virtual reality, add the potential of creating exciting new environments for the presentation of news.

Today, it's fair to say that no journalist can be a serious journalist without a good working knowledge of the World Wide Web. There's simply too much information available there for anyone to ignore it. It's important for journalists to use the Web as a source of information, and it's important for journalists to see what others are publishing on it.

## Tools for the Web

Because of the importance of the World Wide Web in the new media, many journalists are learning how to create Web pages. They do so by learning HTML, or HyperText Markup Language, a simple programming scheme. HTML tags tell a computer how to display text and other items as headings, body text, photos or similar elements.

Figures 16–2 and 16–3 show a page of information with its HTML coding and the page that results, respectively. Increasingly, tools are being developed that make it unnecessary to learn HTML for all but the more complex design chores. With such programs, a click of the mouse inserts the proper coding. Still, anyone headed for a career in the new media should learn HTML. More than one journalism graduate has landed a job at a large newspaper or broadcast station simply because he or she possessed that skill.

The real key to understanding the power of the World Wide Web is understanding hypertext. Within a Web document, links to related information are established. Linked words or phrases appear in a color different from the rest of the text. When a mouse pointer is dragged across a hypertext link, the shape of the pointer changes. Clicking at that spot takes the user to the linked material.

The power of all that is that the linked site may not be on the same computer. Indeed, it may be located in another state or even in another country. With good links, the wide, wide world of the Internet is placed at the user's fingertips.

```
<html><head>
<title>Digital Missourian: First storm is all bluster, no bite</title></head>

<body bgcolor="#EEFFFA">
<a href="../../jpegindex.html"><img src="../images/digmo.gif"  alt="Home"></a>
<a href="frontpage.html"><img src="../images/local.gif" alt="Local  headlines"></a>
<a href="../forum/frontpage.html"><img src="../images/letters.gif" alt="Letters"></a>
<a href="../topnews/frontpage.html"><img src="../images/smtopnews.gif"
alt="Top news"></a><br>
<font size="-1">&copy; 1995, The Digital Missourian</font><p>
<a name="top">December </a>20, 1995<p>


<H1><b>First storm is all bluster, no bite</b></H1>
<ul><li><H4><b>School was canceled, but residents were largely unfazed.</b></H4></ul>
<hr><H5><i>By DAVE RENARD, Missourian staff writer</i></H5><hr>
<a href="nobcut1220.html"><img align=left src="../images/1220smnobite.gif"
hspace=8 vspace=4 border=1></a>
COLUMBIA, Mo. -- Columbia's first winter snow storm of the season didn't
quite live up to its billing.<p>Aside from public schools taking a day
off, it was business as usual Tuesday after 3 inches of overnight
snow.<p>Columbia residents scraped their car windows, cleared their
driveways and went on with the day, as most major roads were cleared for
the morning rush hour and no major accidents were reported.<p>Howard
Wendell made quick work of clearing his driveway at Sexton Road. He said
he enjoys winter weather, but wasn't too fond of last January's 20-inch
snowfall.<p>"If we get that much, I'll get one of those snowblowers,"
Wendell said as he leaned into his shovel.<p>
<a href="slecut1220.html"><img align=right src="../images/1220smsled.gif"
hspace=8 vspace=4 border=1></a> Many Columbia students
weren't expecting school to be canceled, but gladly used their day off to
enjoy some winter activities. Russell Browns, an eighth-grader at West
Junior High School, went sledding at Cosmopolitan Park with several
friends.<p>"I don't know why school was canceled; it didn't snow that
much," Browns said. "I hope it snows more, to cover the hill
better."<p>The Cosmo Park sled hill was worn down to a mixture of slush
and mud by afternoon, with youngsters going down the slope again and
again.<p>"It's usually more crowded, when it snows more," said Browns, who
has been to the hill several times.<p>The hill has a shed near the top
where weary downhill racers can warm up in front of a wood-burning stove,
and a ramp at the bottom where the daring can complete their runs with a
flourish.<p>Because temperatures climbed above freezing today, another
snow storm will probably be necessary to ready area slopes for sledding
again. However, mid-Missouri youths shouldn't hold their breath -- no more
snow is expected to hit Columbia through the weekend.<p><p><a
href="#top">Back to the top</a></body></html>
```

*Figure 16–2 HTML coding shown here creates the document shown in Figure 16–3.*

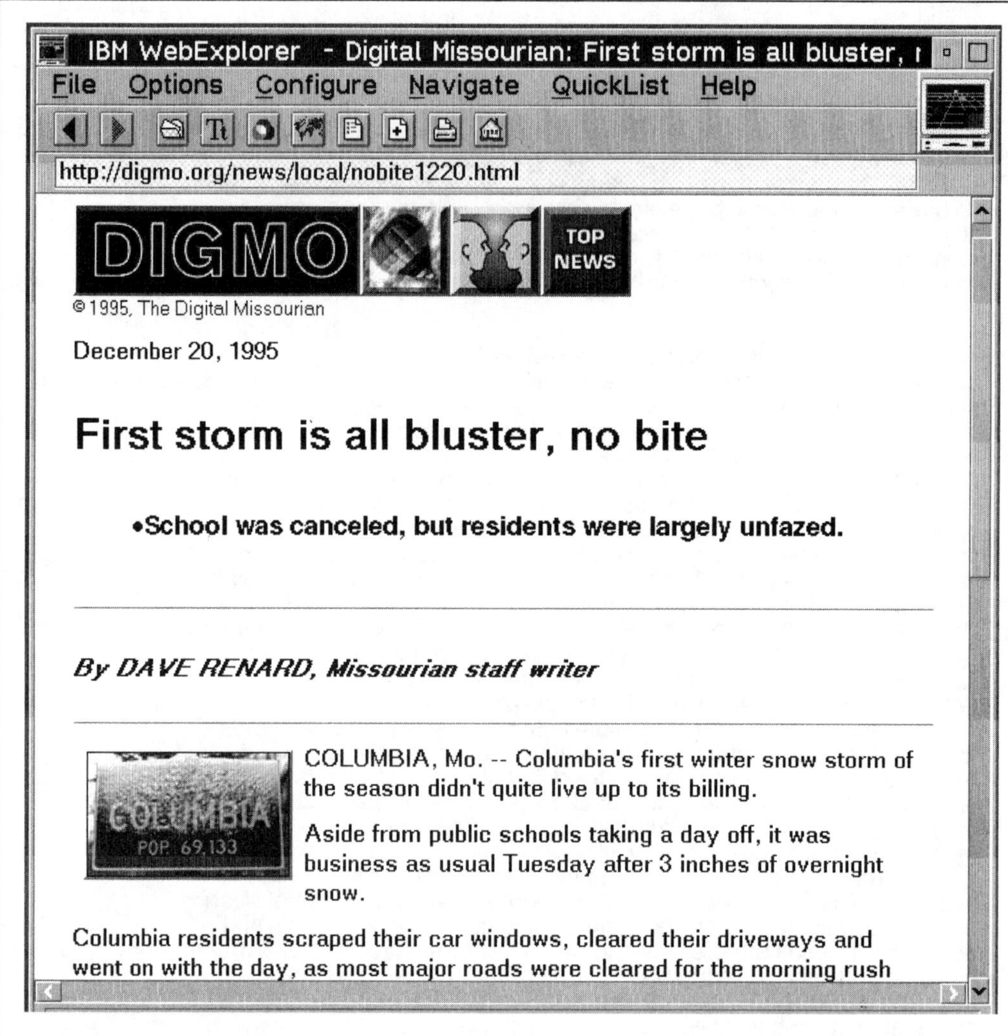


*Figure 16–3 HTML coding in Figure 16–2 produces the results shown here. Only the top of the document is shown.*

All of those pages, and all of those links, are created using HTML tags. Figure 16–4 shows the most common tags that editors use in designing Web pages.

Users also can navigate the Internet by inserting a URL, or *uniform resource locater*, in a space the browser provides (see Figure 16–3). Each site on the Internet has a unique URL. Inserting that address and pressing the computer's ENTER key takes the user directly to that site. For example, to go to the Microsoft Network site on the Web, the user inserts this URL:

http://www.msn.com

All URLs begin with *http,* which stands for hypertext transmission protocol. This alerts the computer that the destination address is a World Wide Web site. Many URL addresses follow with *www,* which is the Web site on the provider's computer. In the preceding example, *msn* stands for the Microsoft Network, and *com* indicates that the site is run by a commercial, or for-profit, corporation. Other common designators are *edu* for educational institutions, *mil* for military sites and *org* for nonprofit organizations.

The World Wide Web makes navigating the Internet a simple matter. Nothing, however, is perfect. Some sites may have slow-speed links to the Internet, which can make the downloading of images, in particular, painfully slow. And because the Internet is a collection of both high-speed and slow-speed computer networks around the world, responsiveness is often less than desirable. Sites also come and go, so today's valid URL may be obsolete tomorrow. Still, the Web is an exciting place for journalists to explore. Becoming familiar with it is almost essential to the practice of modern journalism.

## NEW MEDIA AND THE FUTURE

The phenomenal growth of the new media in general, and the Internet in particular, has fueled much speculation about the future of the existing media. Some even predict an eventual melding of television, radio, newspapers and magazines into a new information medium, or infomedium. That infomedium would be carried on the Information Superhighway, a term widely attributed to Vice President Al Gore. Others think the new media already provide those services and that nothing stands in the way of achieving that goal but cables or telephone lines into homes with bandwidth sufficient for video.

The existing media are betting heavily that a dominant new medium will evolve from the Internet. Major investments in Internet sites by radio, television, newspapers and magazines are ample evidence that the existing media view the new media as both a threat and an opportunity. Instead of fighting the trend, they have chosen to participate. That's the motivation for the media mergers discussed in Chapter 1.

It's clear to most that the Internet is almost surely the Information Superhighway that will carry the infomedium of the future, regardless of its eventual form. No serious new medium could exist without a close connection to the Internet. That's how powerful the Internet has become. And if it is attached to the Internet,

| Name | Opening Tag | Closing Tag | Description |
|---|---|---|---|
| Anchor | <A> | </A> | Hyperlink to a resource |
| Abbreviation | <ABBREV> | </ABBREV> | Enclose abbreviations |
| Abstract | <ABSTRACT> | </ABSTRACT> | Enclose abstracts |
| Acronym | <ACRONYM> | </ACRONYM> | Enclose acronyms |
| Added | <ADDED> | </ADDED> | Enclose added text |
| Address | <ADDRESS> | </ADDRESS> | Format an address |
| Argument | <ARG> | </ARG> | Enclose arguments |
| Array | <ARRAY> | </ARRAY> | Define mathematical matrices |
| Bold | <B> | </B> | Display text in bold |
| Base | <BASE> | No closing tag | Record URL of document |
| Blockquote | <BLOCKQUOTE> | </BLOCKQUOTE> | Include text in quotes |
| Body | <BODY> | </BODY> | Contain the document's body |
| Box | <BOX> | </BOX> | Group mathematical items |
| Line Break | <BR> | No closing tag | Break current line |
| Byline | <BYLINE> | </BYLINE> | Info on document authors |
| Caption | <CAPTION> | </CAPTION> | Table captions |
| Changed | <CHANGED> | </CHANGED> | Mark changed text |
| Citation | <CITE> | </CITE> | Specify a citation |
| Command Name | <CMD> | </CMD> | Set command name |
| Code | <CODE> | </CODE> | Enclose an example of code |
| Definition List Description | <DD> | No closing tag | Description of definition list item |
| Definition | <DFN> | </DFN> | Define instance of a term |
| Directory List | <DIR> | </DIR> | Enclose a directory list |
| Definition List | <DL> | </DL> | Enclose a list of terms and definitions |
| Emphasis | <EM> | </EM> | Emphasize enclosed text |
| Figure | <FIG> | </FIG> | Embed a figure; acts as a paragraph |
| Footnote | <FOOTNOTE> | </FOOTNOTE> | For footnoted information |
| Form | <FORM> | </FORM> | Define form of enclosed text |
| Level 1 Heading | <H1> | </H1> | Enclose Level 1 heading |
| Level 2 Heading | <H2> | </H2> | Enclose Level 2 heading |
| Level 3 Heading | <H3> | </H3> | Enclose Level 3 heading |
| Level 4 Heading | <H4> | </H4> | Enclose Level 4 heading |
| Level 5 Heading | <H5> | </H5> | Enclose Level 5 heading |
| Level 6 Heading | <H6> | </H6> | Enclose Level 6 heading |
| Head | <HEAD> | </HEAD> | Define the head of the document |
| Horizontal Rule | <HR> | No closing tag | Insert horizontal line |
| HTML | <HTML> | </HTML> | Define HTML document |
| HTML+ | <HTMLPLUS> | </HTMLPLUS> | Define HTML+ document |
| Italics | <I> | </I> | Italicize enclosed text |
| Image | <IMG> | No closing tag | Embed an image |

*Figure 16–4 The most widely used HTML tags for coding World Wide Web pages. This table includes both HTML and HTML+ tags.*

| Image | <IMAGE> | </IMAGE> | Embed an image |
|---|---|---|---|
| Input | <INPUT> | </INPUT> | Display entry field |
| Index | <ISINDEX> | No closing tag | Define searchable URL |
| Keyboard | <KBD> | </KBD> | Indicate user-typed text |
| Line Break | <L> | No closing tag | Make explicit line break |
| List item | <LI> | No closing tag | Item of directory list, menu list, ordered list, unordered list |
| Link | <LINK> | No closing tag | Describe relationship between documents |
| Literal | <LIT> | </LIT> | Embed literal texts |
| Margin | <MARGIN> | </MARGIN> | Mark with margin attention label |
| Math | <MATH> | </MATH> | Embed mathematical equations |
| Menu | <MENU> | </MENU> | Enclose a menu list |
| NextID | <NEXTID> | No closing tag | Generate identifier for anchor points |
| Note | <NOTE> | </NOTE> | Bring attention to a point |
| Ordered List | <OL> | </OL> | Enclose an ordered list |
| Over | <OVER> | No closing tag | Divide math boxes into numerator and denominator |
| Paragraph | <P> | </P> (optional) | Define a paragraph |
| Person | <PERSON> | </PERSON> | Embed proper names |
| Preformatted Text | <PRE> | </PRE> | Enclose preformatted text |
| Quotation | <QUOTE> | </QUOTE> | Quote portions of text |
| Render | <RENDER> | No closing tag | Tell browser how to render unknown tags |
| Strike Through | <S> | </S> | Strikes a line through the font |
| Sample | <SAMP> | </SAMP> | Indicate sample text |
| Select | <SELECT> | </SELECT> | Define a set of selectable options |
| Strong Emphasis | <STRONG> | </STRONG> | Strongly emphasize text |
| Subscript | <SUB> | </SUB> | Subscript text |
| Superscript | <SUP> | </SUP> | Superscript text |
| Table | <TABLE> | </TABLE> | Define a table |
| Table Cell Data | <TD> | No closing tag | Define table cell data |
| Textarea | <TEXTAREA> | </TEXTAREA> | Enclose a text area |
| Table Header | <TH> | No closing tag | Define table's row headers |
| Title | <TITLE> | </TITLE> | Define document's title |
| Table Row | <TR> | No closing tag | Define table's row data |
| Typetype | <TT> | </TT> | Display enclosed text in a monospaced (typewriter-like) font |
| Underlined | <U> | </U> | Underline text |
| Unordered list | <UL> | </UL> | Enclose an unordered list |
| Variable | <VAR> | </VAR> | Indicate a variable |

the Information Superhighway becomes a part of the Internet. A separate existence is unthinkable.

In the years ahead, look for the cable television and telephone companies to battle it out to provide your connection to the superhighway. Already, cable modems are being installed that turn cable television lines into high-speed lines for Internet transmissions. Not every cable system is well-prepared to do that, however, and in many ways the telephone companies are in the best position to succeed. They are rushing to provide those connections before the cable companies are ready.

As usual, government will play a key role in how this develops. As government sets the rules for the Information Age, expect less regulation of the media than before. As this was written, Congress was near passage of a new communications act to supersede the Communications Act of 1934. Indications were that the act would greatly diminish regulation of the broadcast media, telephone companies and possibly the cable television industry. The rules Congress sets, or doesn't set, will do much to determine how quickly and in what form the new media proliferate.

For now, however, control of the electronic media and common carriers (as the phone companies are known) is a governmental nightmare. The federal government regulates broadcasting, state governments regulate telephone companies, and local governments regulate cable television companies. Who, then, will regulate this new medium and to what degree will it be regulated? The answers to those questions will have much to do with the speed of its spread into the marketplace.

Through it all, don't expect the existing media to disappear. While it's likely that the number of newspapers will continue to decline, they won't just vanish. Nor will broadcast radio and television. Nor will magazines. For the foreseeable future, we're likely to see fierce competition among them all.

# CHAPTER 17

# EDITING ADVERTISING AND PROMOTIONAL COPY

Many students in schools and departments of journalism who are preparing for careers don't edit news at all. Instead, they write and edit advertising copy or promotional copy such as company manuals and press releases. High-quality editing is just as important in those fields as in the news.

## PRINT ADVERTISING

The media depend on the professionals who create advertising to bring home the revenue that pays the bills. News departments inevitably are drains on company resources; advertising departments, on the other hand, create the revenue to support news operations.

Print advertising specialists, in particular, borrow many techniques from the news department. Key among them is learning to write a headline that sells. Like the news headline writer who tries to lure the reader into the story, the ad headline writer lures the reader to the wares of a paying customer. In both cases, the headline is expected to be compelling.

Writes Alastair Crompton, a British advertising expert, in *The Craft of Copywriting*:

> One point I cannot emphasize too strongly. There can be times when a headline [in an advertisement] does not need an illustration; but there can *never* be a time when an illustration doesn't need a headline. Of course you will see ads that go straight into the body copy from the picture: I call these ads "headless wonders." They are the occasions when the copywriter has copped out. Advice to art directors: Never let your writer make you run an ad without a headline. If he suggests it, tell him not to be so damned lazy. A great picture deserves a great line to back it up; even the painting of the Mona Lisa in the Louvre needs a small brass plate underneath reading "La Gioconda by Leonardo da Vinci."

ONLY YOUR MOTHER IS MORE OBSESSED WITH YOUR SAFETY.
Ford Safety Engineers: Karin H. Przybylo, Steve Pingston, Mike Foster.
Where would we be without our mothers? They take care of us and protect us. So, we're proud to say, when it comes to safeguarding drivers, at FORD MOTOR COMPANY our maternal instinct becomes very apparent. You can feel it in our TRACTION CONTROL system. And in our ANTI-LOCK BRAKES. It's why DUAL-AIR BAGS* are standard in all our cars. And why ROADSIDE ASSISTANCE is available 24 hours a day. We're also developing a Vision Enhancement System — to help drivers when "mother" nature acts up. All this might be considered obsessive. But at Ford Motor Company, we believe such commitments to safety and security will enhance the quality of all our lives. Besides, it's for your own good.
• FORD • FORD TRUCKS • Ford • LINCOLN • MERCURY •
QUALITY IS JOB 1. SM
*Always wear your safety belt.


*Figure 17–1 Sometimes an ad seeks to improve a company's image. Here, Ford tries to convince the public of its commitment to quality.*

The relationship between the picture or drawing in a print ad and the headline is critical to an ad's success. The two must work together, not compete, Crompton says. He warns the beginning ad designer not to let a picture simply illustrate the headline. Instead, he urges, make sure the two key parts of the ad do their own half of the work.

Both the advertising illustration and the headline are designed to catch the reader's attention quickly. That's because most readers spend no more than a second or two scanning an ad. If it doesn't sell immediately, it's unlikely to sell at all. For that reason, the elaborate rules for writing news headlines are abandoned in advertising. Ad headlines often don't have verbs. Sometimes they are prepositional phrases. Occasionally, there may be no complete thought at all. Some examples:

**ONTARIO, CANADA**
*We treat you royally.*

**THE SOUND OF FAST RELIEF. PLOP, PLOP, FIZZ, FIZZ.**

In each case, the ad's purpose is clear. Most readers would instantly recognize the first as a tourist ad for Ontario and the second as one for Alka Seltzer. One of the classic advertising headlines of all time was one simple word below a picture of a Volkswagen Beetle:

**LEMON.**

Readers were instantly drawn to the ad, wondering why Volkswagen would label one of its products with the most derogatory of all comments about cars. In the text block that followed, readers learned why: The car, perfect on the outside, had been rejected by a Volkswagen inspector for a tiny blemish on the glove box. The ad highlighted Volkswagen's commitment to quality. Such eye-catching cleverness is what the advertising creative process is all about.

The lemon headline is a classic example of the *curiosity headline,* one of the three types used in advertisements. The others are *news headlines* and *benefit headlines.* News headlines, as the name implies, do much the same job as those in the news pages: They inform. Benefit headlines help the reader understand why he or she should be interested: **EARN MONEY IN YOUR SPARE TIME.** Curiosity headlines, like the one for Volkswagen, can be fun to write but are risky. If the reader isn't curious, the ad fails.

Getting the reader's attention with a compelling headline and illustration is merely the beginning of the creative process. Once the writer has the reader's attention, it's time to follow up with material to show the reader why he or she should be interested:

HEADLINE: **EVERYTHING FROM A TO SEA.**

FOLLOW-UP: With so much to do at Hilton beach resorts, the only hard part is deciding what to do first. Whether you want to stretch your legs with a game of tennis or just stretch

> out for a few winks. Come discover the ways to play that suit your style. For information and reservations, call your professional travel agent, Hilton's Resort Desk at 1-800-221-2424 or 1-800-HILTONS.

After the ad captures the reader's interest, it builds desire for the product—in this case a resort vacation—and tries to convince the reader to act. Here, the objective is to convince the reader to call Hilton for more information. The four-step process used in the Hilton ad is one known to advertisers as AIDA:

- Attract ATTENTION.
- Build INTEREST.
- Create DESIRE.
- Compel ACTION.

Too often, though, ads fall victim to the same problems that plague the news columns—incorrect grammar, spelling errors and similar unforgivable mistakes. Protecting and polishing the language is just as important in ads as in news. Educated consumers find such mistakes irritating, and as a result they may vow *never* to buy the product. When that happens, the ad has the opposite result from the one intended.

## BROADCAST ADVERTISING

Like new-media journalists, broadcast advertisers must learn to work with all forms of communication—the written word, graphics, audio and video. All combine to produce one of the most effective forms of communication ever devised, broadcast advertising.

Radio ads are a delight to write, edit and produce. It's relatively easy to play appealing tricks on listeners with sound effects and simple devices like British accents. Radio ads have only one dimension, audio. Television ads, on the other hand, become major productions. They are expensive to produce, and they must combine the effective use of audio, video and graphic design. Once it is produced, the television ad competes with dozens of others for the viewer's attention and often is sandwiched between them.

Writing and editing a compelling script is merely the beginning of the process of creating a successful television commercial. Creative video and audio production are critical to its success. Finally, editing it all together is a talent that is much in demand.

Like print advertising, successful broadcast advertising demands attention to detail, including effective communication based on the fundamentals of good grammar and proper spelling. Too often, these are lacking in broadcast commercials.

## PROMOTIONAL COPY

Public relations practitioners and corporate communications specialists are called upon almost daily to write and edit news releases. In most cases, a news release is written in the inverted pyramid format on the assumption that the person who reads it will skim the release quickly to see if anything in it is of interest.

Typically, the audience for which news releases are intended is working journalists—newspaper city editors and broadcast station news directors. The release attempts to provide basic information about the topic. It's understood that the newspaper or broadcast station will not use the release verbatim. If the story is used at all, it likely will become the genesis of the media outlet's own story.

News releases usually come in three varieties:

- Announcements of coming events or actions.
- Information promoting a cause.
- Information designed to build an image.

The best releases give the media the information they need quickly and succinctly. They also tell them where to go for more information. In all cases, they must be accurate and truthful. Editors who feel misled by inflated or overstated news releases soon learn to ignore the next one from the same organization.

It's important for the promotional writer to remember that the media are inundated with news releases. Releases with the best chance of attracting attention localize news, are short and to the point and are clearly written. Promotional writers will also find that it's much easier to get attention from the editors with whom they have developed a working relationship. Like journalists, promotional writers must develop strong principles of professional ethics. Truthfulness is the foundation upon which successful public relations careers are built.

Promotional writers and editors also produce corporate reports, manuals and other materials that will be used both inside the company and by external audiences. In each case, the quality of the publication directly affects improvements or deterioration in the corporate image. Audiences often judge institutions by the quality of the written materials they produce.

IBM is a good example. The company won the Malcolm Baldridge Award for its high-quality products. But its image still suffered because the manuals that accompanied those products were "written backward," in the words of many critics. To find what you wanted to find in an IBM manual, the joke went, you turned to the back of the manual first.

In too many cases, there was truth in that criticism. As a result, the company launched an effort to streamline and simplify its product manuals based on time-tested principles of simplicity in writing and editing. Whether the company has yet

Commerce Bancshares, Inc. **CBSH**

1000 Walnut • P.O. Box 13686 • Kansas City, MO 64199 • (816) 234-2000

Financial News Release

FOR IMMEDIATE RELEASE:
Tuesday, January 16, 1996

**COMMERCE BANCSHARES, INC. REPORTS RECORD EARNINGS FOR 1995**

**Commerce Bancshares, Inc. announced record earnings of $107.6 million for the twelve months ended December 31, 1995, an increase of 12.0% compared to $96.1 million for 1994. For the year, earnings per share were $2.85 compared with $2.72 for 1994, and the return on assets was 1.21% in both 1995 and 1994.**

**Net income for the fourth quarter was $28.2 million, an increase of 16.4% compared to the same period in 1994. Quarterly earnings per share were $.75 compared to $.69 for the same period in 1994, an increase of 8.7%.**

**In making the announcement, David W. Kemper, Chairman and CEO, said, "We are pleased to report record annual earnings for the fifth year in a row. Our fourth quarter results were helped by a stable interest margin and good growth in non interest income. These factors produced a solid increase in net income for the quarter despite higher provision for loan losses than the same quarter last year. Non-performing assets totaled $33.9 million at year-end which represents .35% of total assets."**

**Mr. Kemper continued, "Significant acquisitions in our Wichita and Bloomington markets added major new opportunities for our Company this year. At year end, including all acquisitions, total loans were up 20% compared to a year ago, while deposits also increased 17% over the previous year. Integrating these acquisitions with existing operations has positioned Commerce as the third largest bank in both Kansas and Central Illinois. Also, during 1995 Commerce embarked on a productivity improvement program entitled 'Tomorrow's Bank'. This initiative as well as the successful integration of our 1995 acquisitions will position Commerce for a prosperous 1996."**

**Total assets at December 31, 1995 were $9.6 billion, total loans were $5.3 billion, and total deposits were $8.2 billion. The allowance for loan losses totaled $98.5 million which is 1.85% of total loans and net charge-offs for the year were $16.2 million which represents .31% of average loans. Non-interest expense including acquisitions, was up only 8% and reflects ongoing productivity initiatives plus lower FDIC insurance. Additionally, the loan loss provision increased over the previous year by $8.8 million as a result of both loan growth and higher credit card loan losses.**

*Figure 17–2 The news media are deluged with releases from organizations of all types.*

achieved excellence in editing is questionable, but the improvement is dramatic nonetheless. For some products, manuals were reduced from book-sized documents to mere brochures.

As in news writing and editing, the lesson is simple: Good promotional writing and editing must be based on the four C's: Be correct. Be concise. Be consistent. Be complete.

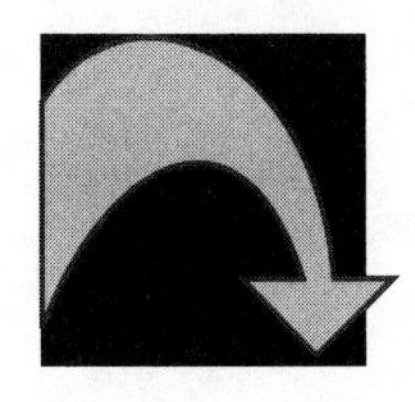

# APPENDIXES

# APPENDIX I

# WIRE SERVICE STYLE

The stylebooks of The Associated Press and United Press International are reference books as well as stylebooks, and they contain much useful information. This appendix is a condensation of the key rules of usage found in those stylebooks. *Webster's New World Dictionary* is the primary source for references not found in the stylebooks. The information herein is used with permission.

## CAPITALIZATION

In general, avoid unnecessary capitals. Use a capital letter only if you can justify it by one of the principles listed here.

### Proper Nouns

Capitalize nouns that constitute the unique identification for a specific person, place or thing: *John, Mary, America, Boston, England.*

Some words, such as the examples just given, are always proper nouns. Some common nouns receive proper noun status when they are used as the name of a particular entity: *General Electric, Gulf Oil.*

### Proper Names

Capitalize common nouns such as *party, river, street* and *west* when they are an integral part of the full name for a person, place or thing: *Democratic Party, Mississippi River, Fleet Street, West Virginia.*

Lowercase these common nouns when they stand alone in subsequent references: *the party, the river, the street.*

Lowercase the common elements of names in all plural uses: *the Democratic and Republican parties, Main and State streets, lakes Erie and Ontario.*

## Popular Names

Some places and events lack officially designated proper names but have popular names that are the effective equivalent: *the Combat Zone* (a section of downtown Boston), *the Main Line* (a group of Philadelphia suburbs), *the South Side* (of Chicago), *the Bad Lands* (of South Dakota), *the Street* (the financial community in the Wall Street area of New York).

The principle applies also to shortened versions of the proper names for one-of-a-kind events: *the Series* (for the World Series), *the Derby* (for the Kentucky Derby). This practice should not, however, be interpreted as a license to ignore the general practice of lowercasing the common noun elements of a name when they stand alone.

## Derivatives

Capitalize words that are derived from a proper noun and still depend on it for their meaning: *American, Christian, Christianity, English, French, Marxism, Shakespearean.*

Lowercase words that are derived from a proper noun but no longer depend on it for their meaning: *french fries, herculean, manhattan cocktail, malapropism, pasteurize, quixotic, venetian blind.*

## Other Key Points of Capitalization

**academic departments** Use lowercase except for words that are proper nouns or adjectives: *the department of history, the history department, the department of English, the English department.*

**administration** Lowercase: *the administration, the president's administration, the governor's administration, the Clinton administration.*

**air force** Capitalize when referring to U.S. forces: *the U.S. Air Force, the Air Force, Air Force regulations.* Use lowercase for the forces of other nations: *the Israeli air force.*

**animals** Capitalize the name of a specific animal, and use Roman numerals to show sequence: *Bowser, Whirlaway II.* For breed names, follow the spelling and capitalization in *Webster's New World Dictionary.* For breeds not listed in the dictionary, capitalize words derived from proper nouns, use lowercase elsewhere: *basset hound, Boston terrier.*

**army** Capitalize when referring to U.S. forces: *the U.S. Army, the Army, Army regulations.* Use lowercase for the forces of other nations: *the French army.*

**Bible** Capitalize, without quotation marks, when referring to the Scriptures of the Old Testament or the New Testament. Capitalize related terms such as *the Gospels,*

*Gospel of St. Mark, the Scriptures, the Holy Scriptures.* Lowercase *biblical* in all uses. Lowercase *bible* as a nonreligious term: *My dictionary is my bible.*

**brand names** Capitalize them. Brand names normally should be used only if they are essential to a story. Sometimes the use of a brand name may not be essential but is acceptable because it lends an air of reality to a story: *He fished a Camel from his shirt pocket* may be preferable to the less specific *cigarette.*

**building** Capitalize the proper names of buildings, including the word *building* if it is an integral part of the proper name: *the Empire State Building.*

**bureau** Capitalize when part of the formal name for an organization or agency: *the Bureau of Labor Statistics, the Newspaper Advertising Bureau.* Lowercase when used alone or to designate a corporate subdivision: *the Washington bureau of The Associated Press.*

**cabinet** Capitalize references to a specific body of advisers heading executive departments for a president, king, governor, etc.: *The president-elect said he has not made his Cabinet selections.* The capital letter distinguishes the word from the common noun meaning "cupboard," which is lowercase.

**Cabinet titles** Capitalize the full title when used before a name, lowercase in other uses: *Secretary of State Warren Christopher,* but *Hazel O'Leary, secretary of energy.*

**century** Lowercase, spelling out numbers less than 10: *the first century, the 20th century.* For proper names, follow the organization's practice: *20th Century-Fox, Twentieth Century Fund, Twentieth Century Limited.*

**chairman, chairwoman** Capitalize as a formal title before a name: *company Chairman Henry Ford, committee Chairwoman Margaret Chase Smith.* Do not capitalize as a casual, temporary position: *meeting chairman Robert Jones.* Do not use *chairperson* unless it is an organization's formal title for an office.

**chief** Capitalize as a formal title before a name: He spoke to *Police Chief William Bratton. He spoke to Chief William Bratton of the New York police.* Lowercase when it is not a formal title: *union chief Walter Reuther.*

**church** Capitalize as part of the formal name of a building, a congregation or a denomination; lowercase in other uses: *St. Mary's Church, the Roman Catholic Church, the Catholic and Episcopal churches, a Roman Catholic church, a church.* Lowercase when the church is used in an institutional sense: *He believes in separation of church and state. The pope said the church opposes abortion.*

**city council** Capitalize when part of a proper noun: *the Boston City Council.* Retain capitalization if the reference is to a specific council but the context does not require the city name:

*BOSTON (AP)–The City Council...*

Lowercase in other uses: *the council, the Boston and New York city councils, a city council.*

**committee** Capitalize when part of a formal name: *the House Appropriations Committee.* Do not capitalize in shortened versions of long committee names: *the Special Senate Select Committee to Investigate Improper Labor-Management Practices,* for example, became the *rackets committee.*

**congress** Capitalize *U.S. Congress* and *Congress* when referring to the U.S. Senate and House of Representatives. Although Congress sometimes is used as a substitute for the House, it properly is reserved for reference to both the Senate and House.

**constitution** Capitalize references to the U.S. Constitution, with or without the *U.S.* modifier: *The president said he supports the Constitution.* When referring to constitutions of other nations or of states, capitalize only with the name of a nation or a state: *the French Constitution, the Massachusetts Constitution, the nation's constitution, the state constitution, the constitution.* Lowercase in other uses: *the organization's constitution.* Lowercase *constitutional* in all uses.

**courthouse** Capitalize with the name of a jurisdiction: *the Cook County Courthouse, the U.S. Courthouse.* Lowercase in other uses: *the county courthouse, the courthouse, the federal courthouse.*

**Court House** (two words) is used in the proper names of some communities: *Appomattox Court House, Va.*

**court names** Capitalize the full proper names of courts at all levels. Retain capitalization if *U.S.* or a state name is dropped: *the U.S. Supreme Court, the Supreme Court; the Massachusetts Superior Court, the state Superior Court, the Superior Court, Superior Court.* For courts identified by a numeral: *3rd District Court, 8th U.S. Circuit Court of Appeals.*

**directions and regions** In general, lowercase *north, south, northeast, northern,* etc., when they indicate compass direction; capitalize these words when they designate regions.

**federal** Use a capital letter for the architectural style and for corporate or governmental bodies that use the word as part of their formal names: *Federal Express, the Federal Trade Commission.* Lowercase when used as an adjective to distinguish something from state, county, city, town or private entities: *federal assistance, federal court, the federal government, a federal judge.* Also: *federal District Court* (but *U.S. District Court* is preferred) and *federal Judge Ann Aldrich* (but *U.S. District Judge Ann Aldrich* is preferred).

**federal court** Always lowercase. The preferred form for first reference is to use the proper name of the court. Do not create nonexistent entities such as *Manhattan Federal Court.* Instead, use *a federal court in Manhattan.*

**food** Most food names are lowercase: *apples, cheese, peanut butter.* Capitalize brand names and trademarks: *Roquefort cheese, Tabasco sauce.* Most proper nouns or adjectives are capitalized when they occur in a food name: *Boston brown bread, Russian dressing, Swiss cheese, Waldorf salad.* Lowercase is used, however, when the food does not depend on the proper noun or adjective for its meaning: *french fries, graham crackers, manhattan cocktail.*

**former** Always lowercase. But retain capitalization for a formal title used immediately before a name: *former President Bush.*

**geographic names** Capitalize common nouns when they form an integral part of a proper name, but lowercase them when they stand alone: *Pennsylvania Avenue, the avenue; the Philippine Islands, the islands; the Mississippi River, the river.* Lowercase common nouns that are not part of a specific proper name: *the Pacific islands, the Swiss mountains, Chekiang province.*

**government** Always lowercase: *the federal government, the state government, the U.S. government.*

**governmental bodies** Follow these guidelines:

FULL NAME: Capitalize the full proper names of governmental agencies, departments and offices: *the U.S. Department of State, the Georgia Department of Human Resources, the Boston City Council, the Chicago Fire Department.*

WITHOUT JURISDICTION: Retain capitalization in referring to a specific body if the dateline or context makes the name of the nation, state, county, city, etc., unnecessary: *the Department of State* (in a story from Washington), *the Department of Human Resources* or *the state Department of Human Resources* (in a story from Georgia), *the City Council* (in a story from Boston), *the Fire Department* or *the city Fire Department* (in a story from Chicago). Lowercase further condensations of the name: *the department, the council,* etc.

FLIP-FLOPPED NAMES: Retain capital letters for the name of a governmental body if its formal name is flopped to delete the word *of: the State Department, the Human Resources Department.*

GENERIC EQUIVALENTS: If a generic term has become the equivalent of a proper name in popular use, treat it as a proper name: *Walpole State Prison,* for example, even though the proper name is *the Massachusetts Correctional Institution-Walpole.*

PLURALS, NONSPECIFIC REFERENCES: All words that are capitalized when part of a proper name should be lowercased when they are used in the plural or do not refer to a specific existing body. Some examples: *All states except Nebraska have a state senate. The town does not have a fire department. The bill requires city councils to provide matching funds. The president will address the lower houses of the New York and New Jersey legislatures.*

**heavenly bodies** Capitalize the proper names of planets, stars, constellations, etc.: *Mars, Arcturus, the Big Dipper, Aries.* For comets, capitalize only the proper noun element of the name: *Halley's comet.* Lowercase *sun* and *moon,* but if their Greek names are used, capitalize them: *Helios* and *Luna.*

**historical periods and events** Capitalize the names of widely recognized epochs in anthropology, archaeology, geology and history: *the Bronze Age, the Dark Ages, the Middle Ages, the Pliocene Epoch.* Capitalize widely recognized popular names for periods and events: *the Atomic Age, the Boston Tea Party, the Civil War, the Exodus* (of the Israelites from Egypt), *the Great Depression, Prohibition.* Lowercase *century: the 18th century.* Capitalize only the proper nouns or adjectives in general descriptions of a period: *ancient Greece, classical Rome, the Victorian era, the fall of Rome.*

**holidays and holy days** Capitalize them: *New Year's Eve, New Year's Day, Groundhog Day, Easter, Hanukkah,* etc.

**house of representatives** Capitalize when referring to a specific governmental body: *the U.S. House of Representatives, the Massachusetts House of Representatives.* Capitalize shortened references that delete the words *of Representatives: the U.S. House, the Massachusetts House.* Retain capitalization if *U.S.* or the name of a state is dropped but the reference is to a specific body:

*BOSTON (AP)—The House has adjourned for the year.*

Lowercase plural uses: *the Massachusetts and Rhode Island houses.* Apply the same principles to similar legislative bodies such as *the Virginia House of Delegates.*

**judge** Capitalize before a name when it is the formal title for an individual who presides in a court of law. Do not continue to use the title in subsequent references. Do not use *court* as part of the title unless confusion would result without it.

No *Court* in the Title: *U.S. District Judge Ann Aldrich, District Judge Ann Aldrich, federal Judge Ann Aldrich, Judge Ann Aldrich, U.S. Circuit Judge Homer Thornberry, appellate Judge John Blair.*

*Court* Needed in the Title: *Juvenile Court Judge Angela Jones, Criminal Court Judge John Jones, Superior Court Judge Robert Harrison. state Supreme Court Judge William Cushing.*

When the formal title *chief judge* is relevant, put the court name after the judge's name: *Chief Judge Ann Aldrich of the U.S. District Court in Washington, D.C.; Chief Judge Sam J. Ervin III of the 4th U.S. Circuit Court of Appeals.* Do not pile up long court names before the name of a judge. Make it *Judge John Smith of Allegheny County Common Pleas Court.* Not: *Allegheny County Common Pleas Court Judge John Smith.* Lowercase *judge* as an occupational designation in phrases such as *beauty contest judge.*

**legislature** Capitalize when preceded by the name of a state: *the Kansas Legislature.* Retain capitalization when the state name is dropped but the reference is specifically to that state's legislature:

*TOPEKA, Kan. (AP)—Both houses of the Legislature adjourned today.*

Capitalize *legislature* in subsequent specific references and in such constructions as *the 100th Legislature, the state Legislature.* Lowercase *legislature* when used generically: *No legislature has approved the amendment.* Use *legislature* in lowercase for all plural references: *The Arkansas and Colorado legislatures are considering the amendment.*

**magazine names** Capitalize the name but do not place it in quotes. Lowercase *magazine* unless it is part of the publication's formal title: *Harper's Magazine, Newsweek magazine, Time magazine.* Check the masthead if in doubt.

**monuments** Capitalize the popular names of monuments and similar public attractions: *Lincoln Memorial, Statue of Liberty, Washington Monument, Leaning Tower of Pisa,* etc.

**mountains** Capitalize as part of a proper name: *Appalachian Mountains, Ozark Mountains, Rocky Mountains.* Or simply: *the Appalachians, the Ozarks, the Rockies.*

**nationalities and races** Capitalize the proper names of nationalities, peoples, races, tribes, etc.: *Arab, Arabic, African, African-American, American, Asian, Caucasian, Cherokee, Chinese* (both singular and plural), *Eskimo* (plural *Eskimos), French Canadian, Gypsy (Gypsies), Japanese* (singular and plural) *Jew, Jewish, Latin, Negro (Negroes), Nordic, Oriental, Sioux, Swede,* etc. Lowercase *black* (noun or adjective) *white, red, mulatto,* etc. Lowercase derogatory terms such as *honky* and *nigger.* Use them only in direct quotes when essential to the story.

**navy** Capitalize when referring to U.S. forces: *the U.S. Navy, the Navy, Navy policy.* Lowercase when referring to the naval forces of other nations: *the British navy.*

**newspaper names** Capitalize *the* in a newspaper's name if that is the way the publication prefers to be known. Lowercase *the* before newspaper names if a story mentions several papers, some of which use *the* as part of the name and some of which do not.

**organizations and institutions** Capitalize the full names of organizations and institutions: *the American Medical Association; First Presbyterian Church; General*

*Motors Corp.; Harvard University; Harvard University Medical School; the Procrastinators Club; the Society of Professional Journalists, Sigma Delta Chi.* Retain capitalization if *Co., Corp.* or a similar word is deleted from full proper name: *General Motors.*

FLIP-FLOPPED NAMES: Retain capital letters when commonly accepted practice flops a name to delete the word *of: College of the Holy Cross, Holy Cross College; Harvard School of Dental Medicine, Harvard Dental School.* Do not, however, flop formal names that are known to the public with the word *of: Massachusetts Institute of Technology,* for example, not *Massachusetts Technology Institute.*

**planets** Capitalize the proper names of planets: *Jupiter, Mars, Mercury, Neptune, Pluto, Saturn, Uranus, Venus.* Capitalize *earth* when used as the proper name of our planet: *The astronauts returned to Earth.* Lowercase nouns and adjectives derived from the proper names of planets and other heavenly bodies: *martian, jovian, lunar, solar, venusian.*

**plants** In general, lowercase the names of plants, but capitalize proper nouns or adjectives that occur in a name. Some examples: *tree, fir, white fir, Douglas fir, Dutch elm, Scotch pine, clover, white clover, white Dutch clover.*

**police department** In communities where this is the formal name, capitalize *police department* with or without the name of the community: *the Los Angeles Police Department, the Police Department.* If a police agency has some other formal name such as *Division of Police,* use that name if it is the way the department is known to the public. If the story uses *police department* as a generic term for such an agency, put *police department* in lowercase. If a police agency with an unusual formal name is known to the public as a police department, treat *police department* as the name, capitalizing it with or without the name of the community. Use the formal name only if there is a special reason in the story. If the proper name cannot be determined for some reason, such as the need to write about a police agency from a distance, treat *police department* as the proper name, capitalizing it with or without the name of the community. Lowercase police department in plural uses: *the Los Angeles and San Francisco police departments.* Lowercase *the department* whenever it stands alone.

**political parties and philosophies** Capitalize both the name of the party and the word *party* if it is customarily used as part of the organization's proper name: *the Democratic Party, the Republican Party.* Capitalize *Communist, Conservative, Democrat, Liberal, Republican, Socialist,* etc., when they refer to the activities of a specific party or to individuals who are members of it. Lowercase these words when they refer to political philosophy. Lowercase the name of a philosophy in noun and adjective forms unless it is the derivative of a proper name: *communism, communist; fascism, fascist.* But: *Marxism, Marxist; Nazism, Nazi.*

**pontiff** Not a formal title. Always lowercase.

**pope** Capitalize when used as a formal title before a name; lowercase in all other uses: *Pope John Paul II spoke to the crowd. At the close of his address, the pope gave his blessing.*

**presidency** Always lowercase.

**president** Capitalize *president* only as a formal title before one or more names: *President Clinton, Presidents Carter and Reagan.* Lowercase in all other uses: *The president said today. He is running for president. Lincoln was president during the Civil War.*

**religious references** The basic guidelines:

**DIETIES:** Capitalize the proper names of monotheistic deities: *God, Allah, the Father, the Son, Jesus Christ, the Son of God, the Redeemer, the Holy Spirit,* etc. Lowercase pronouns referring to the deity: *he, him, his, thee, thou, who, whose, thy,* etc. Lowercase *gods* in referring to the deities of polytheistic religions. Capitalize the proper names of pagan and mythological gods and goddesses: *Neptune, Thor, Venus,* etc. Lowercase such words as *god-awful, goddamn, godlike, godliness, godsend.*

**LIFE OF CHRIST:** Capitalize the names of major events in the life of Jesus Christ in references that do not use his name: *The doctrines of the Last Supper, the Crucifixion, the Resurrection* and *the Ascension* are central to Christian belief. But use lowercase when the words are used with his name: *The ascension of Jesus into heaven took place 40 days after his resurrection from the dead.* Apply the principle also to events in the life of his mother: *He cited the doctrine of the Immaculate Conception and the Assumption.* But: *She referred to the assumption of Mary into heaven.*

**RITES:** Capitalize proper names for rites that commemorate the Last Supper or signify a belief in Christ's presence: *the Lord's Supper, Holy Communion, Holy Eucharist.* Lowercase the names of other sacraments. Capitalize *Benediction* and the *Mass.* But: *a high Mass, a low Mass, a requiem Mass.*

**OTHER WORDS:** Lowercase *heaven, hell, devil, angel, cherub, an apostle, a priest,* etc. Capitalize *Hades* and *Satan.*

**seasons** Lowercase *spring, summer, fall, winter* and derivatives such as *springtime* unless part of a formal name: *Dartmouth Winter Carnival, Winter Olympics, Summer Olympics.*

**senate** Capitalize all specific references to governmental legislative bodies, regardless of whether the name of the nation or state is used: *the U.S. Senate, the Senate; the Virginia Senate, the state Senate, the Senate.* Lowercase plural uses: *the Virginia and North Carolina senates.* The same principles apply to foreign bodies. Lowercase references to nongovernmental bodies: *The student senate at Yale.*

**sentences** Capitalize the first word of every sentence, including quoted statements and direct questions: *Patrick Henry said, "I know not what course others may take, but as for me, give me liberty or give me death."* Capitalize the first word of a quoted statement if it constitutes a sentence, even if it was part of a larger sentence in the original: *Patrick Henry said, "Give me liberty or give me death."* Capitalize the first word of every direct question, even if it falls within a sentence and is not enclosed in quotation marks: *The story answers the question, Where does true happiness really lie?*

**Social Security** Capitalize all references to the U.S. system. Lowercase generic uses such as: *Is there a social security program in Sweden?*

**state** Lowercase in all *state of* constructions: *the state of Maine, the states of Maine and Vermont.* Do not capitalize *state* when used simply as an adjective to specify a level of jurisdiction: *state Rep. William Smith, the state Transportation Department, state funds.* Apply the same principle to phrases such as *the city of Chicago, the town of Auburn,* etc.

**statehouse** Capitalize all references to a specific statehouse, with or without the name of the state: *The Massachusetts Statehouse is in Boston. The governor will visit the Statehouse today.* Lowercase plural uses: *the Massachusetts and Rhode Island statehouses.*

**subcommittee** Lowercase when used with the name of a legislative body's full committee: *a Ways and Means subcommittee.* Capitalize when a subcommittee has a proper name of its own: *the Senate Permanent Subcommittee on Investigations.*

**titles** In general, confine capitalization to formal titles used directly before an individual's name. Lowercase and spell out titles when they are not used with an individual's name: *The president issued a statement. The pope gave his blessing.* Lowercase and spell out titles in constructions that set them off from a name by commas: *The vice president, Walter Mondale, declined to run again. John Paul II, the current pope, does not plan to retire.*

ABBREVIATED TITLES: The following formal titles are capitalized and abbreviated as shown when used before a name outside quotations: *Dr., Gov., Lt. Gov., Rep., Sen.* and certain military ranks. Spell out all except *Dr.* when they are used in quotations. All other formal titles are spelled out in all uses.

ACADEMIC TITLES: Capitalize and spell out formal titles such as *professor, dean, president, chancellor, chairman,* etc., when they precede a name. Lowercase elsewhere. Lowercase modifiers such as *history* in *history Professor Oscar Handlin* or *department* in *department Chairman Jerome Wiesner.*

FORMAL TITLES: Capitalize formal titles when they are used immediately before one or more names: *Pope John Paul II, President Washington, Vice Presidents John Jones and William Smith.*

LEGISLATIVE TITLES: Use *Rep., Reps., Sen.* and *Sens.* as formal titles before one or more names in regular text. Spell out and capitalize these titles before one or more names in a direct quotation. Spell out and lowercase *representative* and *senator* in other uses. Spell out other legislative titles in all uses. Capitalize formal titles such as *assemblyman, assemblywoman, city councilor, delegate,* etc., when they are used before a name. Lowercase in other uses. Add *U.S.* or *state* before a title only if necessary to avoid confusion: *U.S. Sen. Christopher Bond spoke with state Sen. Hugh Carter.*

First Reference Practice: The use of a title such as *Rep.* or *Sen.* in first reference is normal in most stories. It is not mandatory, however, provided an individual's title is given later in the story. Deletion of the title on first reference is frequently appropriate, for example, when an individual has become well-known: *Edward Kennedy endorsed President Clinton's budget plan today. The Massachusetts senator said he believes the budget is sound.*

Second Reference: Do not use a legislative title before a name on second reference unless part of a direct quotation. *Congressman, Congresswoman, Rep.* and *U.S. Rep.* are the preferred first-reference forms when a formal title is used before the name of a U.S. House member. The words *congressman* or *congresswoman,* in lowercase, may be used in subsequent references that do not use an individual's name, just as *senator* is used in references to members of the Senate. *Congressman and Congresswoman* should appear as capitalized formal titles before a name only in direct quotation.

Organizational Titles: Capitalize titles for formal, organizational offices within a legislative body when they are used before a name: *Majority Leader Dick Armey, Minority Leader Richard Gephardt.*

MILITARY TITLES: Capitalize a military rank when used as a formal title before an individual's name. Spell out and lowercase a title when it is substituted for a name: *Gen. John J. Pershing arrived today. An aide said the general would review the troops.*

ROYAL TITLES: Capitalize *king, queen,* etc., when used directly before a name.

**trademark** A trademark is a brand, symbol, word, etc., used by a manufacturer or dealer and protected by law to prevent a competitor from using it: *Astro Turf,* for a type of artificial grass, for example. In general, use a generic equivalent unless the trademark name is essential to the story. When a trademark is used, capitalize it.

**union names** The formal names of unions may be condensed to conventionally accepted short forms that capitalize characteristic words from the full name followed by *union* in lowercase.

## ABBREVIATIONS AND ACRONYMS

A few universally recognized abbreviations are required in some circumstances. Some others are acceptable depending on the context. But in general, avoid alphabet soup.

The same principle applies to acronyms—pronounceable words formed from the initial letters in a series of words: *ALCOA, NATO, radar, scuba,* etc.

Guidance on how to use some common abbreviations or acronyms is provided in entries alphabetized according to the sequence of letters in the word or phrase. You will find many more in the AP stylebook. Some general principles follow.

### Before a Name

Abbreviate the following titles when used before a full name outside direct quotations: *Dr., Gov., Lt. Gov., Mr., Mrs., Ms., Rep., the Rev., Sen.* and certain military designations. Spell out all except *Dr., Mr., Mrs.* and *Ms.* when they are used before a name in direct quotations.

### After a Name

Abbreviate *junior* or *senior* after an individual's name. Abbreviate *company, corporation, incorporated* and *limited* when used after the name of a corporate entity. In some cases, an academic degree may be abbreviated after an individual's name.

### With Dates or Numerals

Use the abbreviations *A.D., B.C., a.m., p.m., No.,* and abbreviate certain months when used with the day of the month.

### In Numbered Addresses

Abbreviate *avenue, boulevard* and *street* in numbered addresses: *He lives on Pennsylvania Avenue. He lives at 1600 Pennsylvania Ave.*

### States and Nations

The names of certain states, the United States and the former Union of Soviet Socialist Republics (but not of other nations) are abbreviated with periods in some circumstances.

## Acceptable but Not Required

Some organizations and government agencies are widely recognized by their initials: *CIA, FBI, GOP.* If the entry for such an organization notes that an abbreviation is acceptable in all references or on second reference, that does not mean that its use should be automatic. Let the context determine, for example, whether to use *Federal Bureau of Investigation* or *FBI.*

## Avoid Awkward Constructions

Do not follow an organization's full name with an abbreviation or acronym in parentheses or set off by dashes. If an abbreviation or acronym would not be clear on second reference without this arrangement, do not use it. Names not commonly before the public should not be reduced to acronyms solely to save a few words.

## Other Key Points of Abbreviations and Acronyms

**academic degrees** If mention of degrees is necessary to establish someone's credentials, the preferred form is to avoid an abbreviation and use instead a phrase such as *Alma Jones, who has a doctorate in psychology.* Use an apostrophe in *bachelor's degree, a master's,* etc. Use such abbreviations as *B.A., M.A., LL.D.* and *Ph.D.* only when the need to identify many individuals by degree on first reference would make the preferred form cumbersome. Use these abbreviations only after a full name, never after just a last name. When used after a name, an academic abbreviation is set off by commas: *Daniel Moynihan, Ph.D., spoke.* Do not precede a name with a courtesy title for an academic degree and follow it with the abbreviation for the degree in the same reference.

**addresses** Use the abbreviations *Ave., Blvd.* and *St.* only with a numbered address: *1600 Pennsylvania Ave.* Spell them out and capitalize when part of a formal street name without a number: *Pennsylvania Avenue.* Lowercase and spell out when used alone or with more than one street name: *Massachusetts and Pennsylvania avenues.* All similar words *(alley, drive, road, terrace,* etc.) always are spelled out. Capitalize them when part of a formal name without a number; lowercase when used alone or with two or more names. Always use figures for an address number: *9 Morningside Circle.* Spell out and capitalize *First* through *Ninth* when used as street names; use figures with two letters for *10th* and above: *7 Fifth Ave., 100 21st St.*

**aircraft names** Use a hyphen when changing from letters to figures; no hyphen when adding a letter after figures.

**AM** Acceptable in all references for the amplitude modulation system of radio transmission.

**a.m., p.m.** Lowercase, with periods. Avoid the redundant *10 a.m. this morning.*

**armed services** Do not use the abbreviations *U.S.A., USAF* and *USN.*

**assistant** Do not abbreviate. Capitalize only when part of a formal title before a name: *Assistant Secretary of State Thomas M. Tracy.* Wherever practical, however, an appositional construction should be used: *Thomas M. Tracy, assistant secretary of state.*

**association** Do not abbreviate. Capitalize as part of a proper name: *American Medical Association.*

**attorney general, attorneys general** Never abbreviate. Capitalize only when used as a title before a name: *Attorney General Richard Thornburgh.*

**Bible** Do not abbreviate individual books of the Bible. Citations listing the number of chapter(s) and verse(s) use this form: *Matthew 3:16, Luke 21:1:1–13, 1 Peter 2:1.*

**brothers** Abbreviate as *Bros.* in formal company names: *Warner Bros.* For possessives: *Warner Bros.' profits.*

**Christmas** Never abbreviate *Christmas* to *Xmas* or any other form.

**CIA** Acceptable in all references for *Central Intelligence Agency.*

**c.o.d.** Acceptable in all references for *cash on delivery* or *collect on delivery.* (The use of lowercase is an exception to the first listing in *Webster's New World.)*

**company, companies** Use *Co.* or *Cos.* when a business uses either word at the end of its proper name: *Ford Motor Co., American Broadcasting Cos.* But: *Aluminum Company of America.* If *company* or *companies* appears alone in second reference, spell the word out. The forms for possessives: *Ford Motor Co.'s profits, American Broadcasting Cos.' profits.*

**corporation** Abbreviate as *Corp.* when a company or government agency uses the word at the end of its name: *Gulf Oil Corp., the Federal Deposit Insurance Corp.* Spell out *corporation* when it occurs elsewhere in a name: *the Corporation for Public Broadcasting.* Spell out and lowercase *corporation* whenever it stands alone. The form for possessives: *Gulf Oil Corp.'s profits.*

**courtesy titles** [Note: Many newspapers consider the AP style (UPI style varies) on this subject to be sexist and have chosen to treat all women as men are treated—by first and last name on first reference and last name only on second reference. For the same reason, other newspapers use courtesy titles for both men and women on second reference. Despite those increasingly common deviations, the AP policy is quoted here; it is still a common policy at U.S. newspapers.]

In general, do not use the courtesy titles *Miss, Mr., Mrs.* or *Ms.* of first and last names of the person: *Hillary Rodham Clinton, Jimmy Carter.* Do not use *Mr.* in any reference unless it is combined with *Mrs.; Mr. and Mrs. John Smith.* On sports wires, do not use courtesy titles in any reference unless needed to distinguish among people of the same last name. On news wires, use courtesy titles for women on second reference, following the woman's preference. If the woman says she does not want a courtesy title, refer to her on second reference by last name only. Some guidelines:

MARRIED WOMEN: The preferred form on first reference is to identify a woman by her own first name and her husband's last name: *Susan Smith.* Use *Mrs.* on the first reference only if a woman requests that her husband's first name be used or her own first cannot be determined: *Mrs. John Smith.* On second reference, use *Mrs.* unless a woman initially identified by her own first name prefers *Ms.: Carla Hills, Mrs. Hills, Ms. Hills* or no title: *Carla Hills, Hills.* If a married woman is known by her maiden last name, precede it by *Miss* on second reference unless she prefers *Ms.: Jane Fonda, Miss Fonda, Ms. Fonda* or no title: *Jane Fonda, Fonda.*

UNMARRIED WOMEN: For women who have never been married, use *Miss, Ms.* or no title on second reference according to the woman's preferences. For divorced women and widows, the normal practice is to use *Mrs.* or no title, if she prefers, on second reference. But, if a woman returns to the use of her maiden name, use *Miss, Ms.* or no title if she prefers it.

MARITAL STATUS: If a woman prefers *Ms.* or no title, do not include her marital status in a story unless it is clearly pertinent.

**detective** Do not abbreviate.

**district attorney** Do not abbreviate.

**doctor** Use *Dr.* in first reference as a formal title before the name of an individual who holds a doctor of medicine degree: *Dr. Jonas Salk.* The form *Dr.*, or *Drs.* in plural construction, applies to all first-reference uses before a name, including direct quotations. If appropriate in the context, *Dr.* also may be used on first reference before the names of individuals who hold other types of doctoral degrees. However, because the public frequently identifies *Dr.* only with physicians, care should be taken to ensure that the individual's specialty is stated in first or second reference. The only exception would be a story in which the context left no doubt that the person was a dentist, psychologist, chemist, historian, etc. In some instances it also is necessary to specify that an individual identified as *Dr.* is a physician. One frequent case is a story reporting on joint research by physicians, biologists, etc. Do not use *Dr.* before the names of individuals who hold only honorary doctorates. Do not continue the use of *Dr.* in subsequent references.

**ERA** Acceptable in all references to baseball's *earned run average.* Acceptable on second reference for *Equal Rights Amendment.*

**FBI** Acceptable in all references for *Federal Bureau of Investigation.*

**FM** Acceptable in all references for the frequency modulation system of radio transmission.

**ICBM, ICBMs** Acceptable on first reference for intercontinental ballistic missile(s), but the term should be defined in the body of a story. Avoid the redundant *ICBM missiles.*

**incorporated** Abbreviate and capitalize as *Inc.* when used as part of a corporate name. It usually is not needed, but when it is used, do not set off with commas: *J.C. Penney Co. Inc. announced . . .*

**IQ** Acceptable in all references for *intelligence quotient.*

**junior, senior** Abbreviate as *Jr.* and *Sr.* only with full names of persons or animals. Do not precede by a comma: *Joseph P. Kennedy Jr.* The notation *II* or *2nd* may be used if it is the individual's preference. Note, however, that *II* and *2nd* are not necessarily the equivalent of junior—either may be used by a grandson or nephew. If necessary to distinguish between father and son in second reference, use *the elder Smith* or *the younger Smith.*

**mount** Spell out in all uses, including the names of communities and of mountains: *Mount Clements, Mich.; Mount Everest.*

**mph** Acceptable in all references for *miles per hour* or *miles an hour.*

**No.** Use as the abbreviation for *number* in conjunction with a figure to indicate position or rank: *No. 1 man, No. 3 choice.* Do not use in street addresses, with this exception: *No. 10 Downing St.,* the residence of Britain's prime minister. Do not use in names of schools: *Public School 19.*

**point** Do not abbreviate. Capitalize as part of a proper name: *Point Pleasant.*

**saint** Abbreviate as *St.* in the names of saints, cities and other places: *St. Jude; St. Paul, Minn.; St. John's, Newfoundland; St. Lawrence Seaway.*

**state names** Follow these guidelines:

STANDING ALONE: Spell out the names of the 50 U.S. states when they stand alone in textual material. Any state name may be condensed, however, to fit typographical requirements for tabular material.

EIGHT NOT ABBREVIATED: The names of eight states are never abbreviated in datelines or text: *Alaska, Hawaii, Idaho, Iowa, Maine, Ohio, Texas* and *Utah.*

ABBREVIATIONS REQUIRED: Use the state abbreviations listed here in the following contexts:

- In conjunction with the name of a city, town, village or military base in most datelines.
- In conjunction with the name of a city, county, town, village or military base in text. See examples in the punctuation section that follows.
- In short-form listings of party affiliation: D-Ala., R-Mont.

| | | | | |
|---|---|---|---|---|
| Ala. | Ill. | Miss. | N.C. | Vt. |
| Ariz. | Ind. | Mo. | N.D. | Va. |
| Ark. | Kan. | Mont. | Okla. | Wash. |
| Calif. | Ky. | Neb. | Ore. | W. Va. |
| Colo. | La. | Nev. | Pa. | Wis. |
| Conn. | Md. | N.H. | R.I. | Wyo. |
| Del. | Mass. | N.J. | S.C. | |
| Fla. | Mich. | N.M. | S.D. | |
| Ga. | Minn. | N.Y. | Tenn. | |

**TV** Acceptable as an adjective or in such constructions as *cable TV.* But do not normally use as a noun unless part of a quotation.

**UFO, UFOs** Acceptable in all references for *unidentified flying object(s).*

**U.N.** Used as an adjective, but not as a noun, for *United Nations.*

**U.S.** Used as an adjective, but not as a noun, for *United States.*

## PUNCTUATION AND HYPHENATION

Think of punctuation and hyphenation as a courtesy to your readers, designed to help them understand a story. Inevitably, a mandate of this scope involves gray areas. For this reason, the punctuation entries in the stylebooks refer to guidelines rather than rules. Guidelines should not be treated casually, however.

**all-** Use a hyphen.

| | |
|---|---|
| all-around (not all-round) | all-out |
| all-clear | all-star |

**ampersand (&)** Use the ampersand when it is part of a company's formal name: *Baltimore & Ohio Railroad, Newport News Shipbuilding & Dry Dock Co.* The ampersand should not otherwise be used in place of *and.*

**anti-** Hyphenate all except the following words, which have specific meanings of their own:

| | | |
|---|---|---|
| antibiotic | antiknock | antiphony |
| antibody | antimatter | antiseptic |
| anticlimax | antimony | antiserum |

| | | |
|---|---|---|
| antidote | antiparticle* | antithesis |
| antifreeze | antipasto | antitrust |
| antigen | antiperspirant | antitoxin |
| antihistamine | antiphon | antitussive |

* And similar terms in physics such as antiproton.

This approach has been adopted in the interests of readability and easily remembered consistency.

**apostrophe (')** Follow these guidelines:

POSSESSIVES: See the possessives entry.

OMITTED LETTERS: *I've, it's, don't, rock 'n' roll. 'Tis the season to be jolly. He is a ne'er-do-well.*

OMITTED FIGURES: *The class of '62. The Spirit of '76. The '20s.*

PLURALS OF A SINGLE LETTER: *Mind your p's and q's. He learned the three R's and brought home a report card with four A's and two B's. The Oakland A's won the pennant.*

DO NOT USE: For plurals of numerals or multiple-letter combinations.

**by** In general, no hyphen. Some examples:

| | |
|---|---|
| byline | byproduct |
| bypass | bystreet |

*By-election* is an exception.

**co-** Retain the hyphen when forming nouns, adjectives and verbs that indicate occupation or status:

| | | |
|---|---|---|
| co-author | co-owner | co-signer |
| co-chairman | co-partner | co-star |
| co-defendant | co-pilot | co-worker |
| co-host | co-respondent (in a divorce suit) | |

(Several are exceptions to *Webster's New World* in the interests of consistency.)

Use no hyphen in other combinations:

| | | |
|---|---|---|
| coed | coexist | cooperative |
| coeducation | coexistence | coordinate |
| coequal | cooperate | coordination |

*Cooperate, coordinate* and related words are exceptions to the rule that a hyphen is used if a prefix ends in a vowel and the word that follows begins with the same vowel.

**colon** The most frequent use of a colon is at the end of a sentence to introduce lists, tabulations, texts, etc.

Capitalize the first word after a colon only if it is a proper noun or the start of a complete sentence: *He promised this: The company will make good all the losses.* But: *There were three considerations: expense, time and feasibility.*

**INTRODUCING QUOTATIONS:** Use a comma to introduce a direct quotation of one sentence that remains within a paragraph. Use a colon to introduce longer quotations within a paragraph and to end all paragraphs that introduce a paragraph of quoted material.

**PLACEMENT WITH QUOTATION MARKS:** Colons go outside quotation marks unless they are part of the quotation itself.

**comma** The following guidelines treat some of the most frequent questions about the use of commas. Additional guidelines on specialized uses are provided in separate entries. For more detailed guidance, consult, "The Comma" and "Misused and Unnecessary Commas" in the Guide to Punctuation section in the back of *Webster's New World Dictionary.*

**IN A SERIES:** Use commas to separate elements in a series, but do not put a comma before the conjunction in a simple series: *The flag is red, white and blue. She would nominate Tom, Dick or Jane.* Put a comma before the conjunction in a series, however, if an integral element of the series requires a conjunction: *I had orange juice, toast, and ham and eggs for breakfast.* Use a comma also before the concluding conjunction in a complex series of phrases: *The main points to consider are whether the athletes are skillful enough to compete, whether they have the stamina to endure the training, and whether they have the proper mental attitude.*

**WITH EQUAL ADJECTIVES:** Use commas to separate a series of adjectives equal in rank. If the commas could be replaced by the word *and* without changing the sense, the adjectives are equal: *a thoughtful, precise manner; a dark, dangerous street.* Use no comma when the last adjective before a noun outranks its predecessors because it is an integral element of a noun phrase, which is the equivalent of a single noun: *a cheap fur coat* (the noun phrase is *fur coat); the old oaken bucket; a new, blue spring bonnet.*

**WITH INTRODUCTORY CLAUSES AND PHRASES:** A comma normally is used to separate an introductory clause or phrase from a main clause: *When he had tired of the mad pace of New York, he moved to Dubuque.* The comma may be omitted after short introductory phrases if no ambiguity would result: *During the night he heard many noises.* But use the comma if its omission would slow comprehension: *On the street below, the curious gathered.*

**WITH CONJUNCTIONS:** When a conjunction such as *and, but* or *for* links two clauses that could stand alone as separate sentences, use a comma before the conjunction in most cases: *She was glad she had looked, for a man was approaching the house.* As a rule of thumb, use a comma if the subject of each clause is expressly stated: *We are visiting Washington, and we also plan a side trip to Williamsburg. We visited Washington, and our senator greeted us personally.* But no comma when the subject of the two clauses is the same and is not repeated in the second: *We are visiting Washing-*

*ton and plan to see the White House.* The comma may be dropped if two clauses with expressly stated subjects are short. In general, however, favor use of a comma unless a particular literary effect is desired or it would distort the sense of a sentence.

**INTRODUCING DIRECT QUOTES:** Use a comma to introduce a complete, one-sentence quotation within a paragraph: *Wallace said, "She spent six months in Argentina and came back speaking English with a Spanish accent."* But use a colon to introduce quotations of more than one sentence. Do not use a comma at the start of an indirect or partial quotation: *He said his victory put him "firmly on the road to a first-ballot nomination."*

**BEFORE ATTRIBUTION:** Use a comma instead of a period at the end of a quote that is followed by attribution: *"Rub my shoulder," Miss Cawley suggested.* Do not use a comma, however, if the quoted statement ends with a question mark or exclamation point: *"Why should I?" he asked.*

**WITH HOMETOWNS AND AGES:** Use a comma to set off an individual's hometown when it is placed in apposition to a name: *Mary Richards, Minneapolis, and Maude Findlay, Tuckahoe, N.Y., were there.* However, the use of the word *of* without a comma between the individual's name and the city name generally is preferable: *Mary Richards of Minneapolis and Maude Findlay of Tuckahoe, N.Y., were there.* If an individual's age is used, set it off by commas: *Maude Findlay, 48, Tuckahoe, N.Y., was present.* The use of the word *of* eliminates the need for a comma after the hometown if a state name is not needed: *Mary Richards, 36, of Minneapolis and Maude Findlay, 48, of Tuckahoe, N. Y., attended the party.*

**IN LARGE FIGURES:** Use a comma for most figures higher than 999. The major exceptions are: street addresses *(1234 Main St.)*, broadcast frequencies *(1460 kilohertz)*, room numbers, serial numbers, telephone numbers and years *(1976)*.

**PLACEMENT WITH QUOTES:** Commas always go inside quotation marks.

**dash** Follow these guidelines:

**ABRUPT CHANGE:** Use dashes to denote an abrupt change in thought in a sentence or an emphatic pause: *We will fly to Paris in June—if I get a raise. Smith offered a plan—it was unprecedented—to raise revenues.*

**SERIES WITHIN A PHRASE:** When a phrase that otherwise would be set off by commas contains a series of words that must be separated by commas, use dashes to set off the full phrase: *He listed the qualities—intelligence, charm, beauty, independence—that he liked in women.*

**ATTRIBUTION:** Use a dash before an author's or composer's name at the end of a quotation: *"Who steals my purse steals trash."—Shakespeare.*

**IN DATELINES:**

*NEW YORK (AP)–The city is broke.*

IN LISTS: Dashes should be used to introduce individual sections of a list. [Note: Some newspapers and magazines use bullets instead.] Capitalize the first word following the dash. Use periods, not semicolons, at the end of each section. Example:

*Jones gave the following reasons:*
*—He never ordered the package.*
*—If he did, it didn't come.*
*—If it did, he sent it back.*

WITH SPACES: Put a space on both sides of a dash in all uses except the start of a paragraph and sports agate summaries.

**ellipsis (. . .)** In general, treat an ellipsis as a three-letter word, constructed with three periods and two spaces, as shown here. Use an ellipsis to indicate the deletion of one or more words in condensing quotes, texts and documents. Be especially careful to avoid deletions that would distort the meaning.

**ex-** Use no hyphen for words that use *ex-* in the sense of *out of:*

excommunicate     expropriate

Hyphenate when using *ex* in the sense of former:

ex-convict     ex-president

Do not capitalize *ex-* when attached to a formal title before a name: *ex-President Carter.* The prefix modifies the entire term: *ex-New Jersey Gov. Thomas Kean;* not *ex-Gov. Thomas Kean.*

Usually *former* is better.

**exclamation point (!)** Follow these guidelines:

EMPHATIC EXPRESSIONS: Use the mark to express a high degree of surprise, incredulity or other strong emotion.

AVOID OVERUSE: Use a comma after mild interjections. End mildly exclamatory sentences with a period.

PLACEMENT WITHIN QUOTES: Place the mark inside quotation marks when it is part of the quoted material: *"How wonderful!" he exclaimed. "Never!" she shouted.* Place the mark outside quotation marks when it is not part of the quoted material: *I hated reading Spenser's "Faerie Queene"!*

**extra-** Do not use a hyphen when *extra-* means *outside of* unless the prefix is followed by a word beginning with *a* or a capitalized word:

extralegal     extraterrestrial
extramarital     extraterritorial

But

extra-alimentary     extra-Britannic

Follow *extra-* with a hyphen when it is part of a compound modifier describing a condition beyond the usual size, extent or degree:

| | |
|---|---|
| extra-base hit | extra-dry drink |
| extra-large book | extra-mild taste |

**fore-** In general, no hyphen. Some examples:

| | |
|---|---|
| forebrain | foregoing |
| forefather | foretooth |

There are three nautical exceptions, based on long-standing practice:

| | |
|---|---|
| fore-topgallant | fore-topsail |
| fore-topmast | |

**full-** Hyphenate when used to form compound modifiers:

| | |
|---|---|
| full-dress | full-page |
| full-fledged | full-scale |
| full-length | |

**great-** Hyphenate *great-grandfather, great-great-grandmother,* etc. Use *great grandfather* only if the intended meaning is that the grandfather was a great man.

**hyphen** Hyphens are joiners. Use them to avoid ambiguity or to form a single idea from two or more words. Some guidelines:

AVOID AMBIGUITY: Use a hyphen whenever ambiguity would result if it were omitted: *The president will speak to small-business men. (Businessmen* normally is one word. But *The president will speak to small businessmen* is unclear.)

COMPOUND MODIFIERS: When a compound modifier—two or more words that express a single concept—precedes a noun, use hyphens to link all the words in the compound except the adverb *very* and all adverbs that end in *ly: a first-quarter touchdown, a bluish-green dress, a full-time job, a well-known man, a better-qualified woman, a know-it-all attitude, an easily remembered rule.* Many combinations that are hyphenated before a noun are not hyphenated when they occur after a noun: *The team scored in the first quarter. The dress, a bluish green, was attractive on her. She works full time. His attitude suggested that he knew it all.* But when a modifier that would be hyphenated before a noun occurs instead after a form of the verb *to be,* the hyphen usually must be retained to avoid confusion: *The man is well-known. The woman is quick-witted. The children are soft-spoken. The play is second-rate.* The principle of using a hyphen to avoid confusion explains why no hyphen is required with *very* and *ly* words. Readers can expect them to modify the word that follows. But if a combination such as *little-known man* were not hyphenated, readers could logically expect *little* to be followed by a noun, as in *little man.* Instead, readers encountering *little known* would have to back up mentally and make the compound connection on their own.

TWO-THOUGHT COMPOUNDS: *serio-comic, socio-economic.*

COMPOUND PROPER NOUNS AND ADJECTIVES: Use a hyphen to designate dual heritage: *Italian-American, Mexican-American.*

No hyphen, however, for *French Canadian* or *Latin American.*

**AVOID DUPLICATED VOWELS, TRIPLED CONSONANTS:** Examples:

anti-intellectual    shell-like
pre-empt

**WITH NUMERALS:** Use a hyphen to separate figures in betting odds, ratios, scores, some fractions and some election returns. See examples in entries under these headings. When large numbers must be spelled out, use a hyphen to connect a word ending in *y* to another word: *twenty-one, fifty-five,* etc.

**SUSPENSIVE HYPHENATION:** The form: *He received a 10- to 20-year sentence in prison.*

**in-** No hyphen when it means "not":

inaccurate    insufferable

Often solid in other cases:

inbound    infighting
indoor    inpatient (n., adj.)
infield

A few combinations take a hyphen, however:

in-depth    in-house
in-group    in-law

Follow *Webster's New World* when in doubt.

**-in** Precede with a hyphen:

break-in    walk-in
cave-in    write-in

**parentheses** In general, use parentheses around logos, as in datelines, but otherwise be sparing with them. Parentheses are jarring to the reader. The temptation to use parentheses is a clue that a sentence is becoming contorted. Try to write it another way. If a sentence must contain incidental material, then commas or two dashes are frequently more effective. Use these alternatives whenever possible. There are occasions, however, when parentheses are the only effective means of inserting necessary background or reference information.

**periods** Follow these guidelines:

**END OF DECLARATIVE SENTENCE:** *The stylebook is finished.*

**END OF A MILDLY IMPERATIVE SENTENCE:** *Shut the door.* Use an exclamation point if greater emphasis is desired: *Be careful!*

**END OF SOME RHETORICAL QUESTIONS:** A period is preferable if a statement is more a suggestion than a question: *Why don't we go.*

**END OF AN INDIRECT QUESTION:** *He asked what the score was.*

INITIALS: *John F. Kennedy, T.S. Eliot.* (No space between T. and S., to prevent them from being placed on two lines in typesetting.) Abbreviations using only the initials of a name do not take periods: *JFK, LBJ.*

ENUMERATIONS: After numbers or letters in enumerating elements of a summary: *1. Wash the car. 2. Clean the basement.* Or: *A. Punctuate properly. B. Write simply.*

**possessives** Follow these guidelines:

PLURAL NOUNS NOT ENDING IN S: Add *'s: the alumni's contributions, women's rights.*

PLURAL NOUNS ENDING IN S: Add only an apostrophe: *the churches' needs, the girls' toys, the horses' food, the ships' wake, states' rights, the VIPs' entrance.*

NOUNS PLURAL IN FORM, SINGULAR IN MEANING: Add only an apostrophe: *mathematics' rules, measles' effects.* (But see Inanimate Objects below.) Apply the same principle when a plural word occurs in the formal name of a singular entity: *General Motors' profits, the United States' wealth.*

NOUNS THE SAME IN SINGULAR AND PLURAL: Treat them the same as plurals, even if the meaning is singular: *one corps' location, the two deer's tracks, the lone moose's antlers.*

SINGULAR NOUNS NOT ENDING IN S: Add *'s: the church's needs, the girl's toys, the horse's food, the ship's route, the VIP's seat.* Some style guides say that singular nouns ending in *s* sounds such as *ce, x,* and *z* may take either the apostrophe alone or *'s.* See Special Expressions below, but otherwise, for consistency and ease in remembering a rule, always use *'s* if the word does not end in the letter *s: Butz's policies, the fox's den, the justice's verdict, Marx's theories, the prince's life, Xerox's profits.*

SINGULAR PROPER NAMES ENDING IN S: Use only an apostrophe: *Achilles' heel, Agnes' book, Ceres' rites, Descartes' theories, Dickens' novels, Euripides' dramas, Hercules' labors, Jesus' life, Jules' seat, Kansas' schools, Moses' law, Socrates' life, Tennessee Williams' plays, Xerxes' armies.*

SPECIAL EXPRESSIONS: The following exceptions to the general rule for words not ending in *s* apply to words that end in an *s* sound and are followed by a word that begins with *s: for appearance's sake, for conscience's sake.*

PRONOUNS: Personal, interrogative and relative pronouns have separate forms for the possessive. None involves an apostrophe: *mine, ours, yours, his, hers, its, theirs, whose.* Caution: If you are using an apostrophe with a pronoun, always double-check to be sure that the meaning calls for a contraction: *you're, it's, there's, who's.* Follow the rules listed above in forming the possessives of other pronouns: *another's idea, other's plans, someone's guess.*

COMPOUND WORDS: Applying the rules above, add an apostrophe or *'s* to the word closest to the object possessed: *the major general's decision, the major generals' decisions, the attorney general's request, the attorneys generals' request.* Also: *anyone else's attitude, John Adams Jr.'s father, Benjamin Franklin of Pennsylvania's motion.* Whenever practical, however, recast the phrase to avoid ambiguity: *the motion by Benjamin Franklin of Pennsylvania.*

JOINT POSSESSION, INDIVIDUAL POSSESSION: Use a possessive form after only the last word if ownership is joint: *Fred and Sylvia's apartment, Fred and Sylvia's stocks.* Use a possessive form after both words if the objects are individually owned: *Fred's and Sylvia's books.*

DESCRIPTIVE PHRASES: Do not add an apostrophe to a word ending in *s* when it is used primarily in a descriptive sense: *citizens band radio, a Cincinnati Reds infielder, a teachers college, a Teamsters request, a writers guide.* Memory Aid: The apostrophe usually is not used if *for* or *by* rather than *of* would be appropriate in the longer form: *a radio band for citizens, a college for teachers, a guide for writers, a request by the Teamsters.* An *'s* is required, however, when a term involves a plural word that does not end in *s: a children's hospital, a people's republic, the Young Women's Christian Association.*

DESCRIPTIVE NAMES: Some governmental, corporate and institutional organizations with a descriptive word in their names use an apostrophe; some do not. Follow the user's practice: *Actors Equity, Diners Club, the Ladies' Home Journal, the National Governors' Conference, the Veterans Administration.*

QUASI POSSESSIVES: Follow the rules above in composing the possessive form of words that occur in such phrases as *a day's pay, two weeks' vacation, three days' work, your money's worth.* Frequently, however, a hyphenated form is clearer: *a two-week vacation, a three-day job.*

DOUBLE POSSESSIVE: Two conditions must apply for a double possessive—a phrase such as *a friend of John's*—to occur: 1. The word after *of* must refer to an animate object, and 2. The word before *of* must involve only a portion of the animate object's possessions. Otherwise, do not use the possessive form on the word after *of: The friends of John Adams mourned his death.* (All the friends were involved.) *He is a friend of the college.* (Not *college's,* because *college* is inanimate.) Memory Aid: This construction occurs most often, and quite naturally, with the possessive forms of personal pronouns: *She is a friend of mine.*

INANIMATE OBJECTS: There is no blanket rule against creating a possessive form for an inanimate object, particularly if the object is treated in a personified sense. See some of the earlier examples, and note these: *death's call, the wind's murmur.* In general, however, avoid excessive personalization of inanimate objects, and give preference to an *of* construction when it fits the makeup of the sentence. For example, the earlier references to mathematics' rules and measles' effects would better be phrased: *the rules of mathematics, the effects of measles.*

**post-** Follow *Webster's New World.* Hyphenate if not listed there. Some words without a hyphen:

| | |
|---|---|
| postdate | postnuptial |
| postdoctoral | postoperative |
| postelection | postscript |
| postgraduate | postwar |

Some words that use a hyphen:

| | |
|---|---|
| post-bellum | post-mortem |

**prefixes** See separate listings for commonly used prefixes. Three rules are constant, although they yield some exceptions to first-listed spellings in *Webster's New World Dictionary:*

- Except for *cooperate* and *coordinate,* use a hyphen if the prefix ends in a vowel.
- Use a hyphen if the word that follows is capitalized.
- Use a hyphen to join doubled prefixes: *sub-subparagraph.*

**pro-** Use a hyphen when coining words that denote support for something. Some examples:

| | |
|---|---|
| pro-business | pro-life |
| pro-labor | pro-war |

No hyphen when *pro* is used in other senses:

| | |
|---|---|
| produce | pronoun |
| profile | |

**question mark** Follow these guidelines:

END OF A DIRECT QUESTION: *Who started the riot? Did she ask who started the riot?* (The sentence as a whole is a direct question despite the indirect question at the end.) *You started the riot?* (A question in the form of a declarative statement.)

INTERPOLATED QUESTION: *You told me—Did I hear you correctly?—that you started the riot.*

MULTIPLE QUESTIONS: Use a single question mark at the end of the full sentence: *Did you hear him say, "What right have you to ask about the riot?" Did he plan the riot, employ assistants and give the signal to begin?* Or, to cause full stops and throw emphasis on each element, break into separate sentences: *Did he plan the riot? Employ assistants? Give the signal to begin?*

CAUTION: Do not use question marks to indicate the end of an indirect question: *She asked who started the riot. To ask why the riot started is unnecessary. I want to know what the cause of the riot was. How foolish it is to ask what caused the riot.*

QUESTION AND ANSWER FORMAT: Do not use quotation marks. Paragraph each speaker's words:

*Q. Where did you keep it?*

*A. In a little tin box.*

PLACEMENT WITH QUOTATION MARKS: Inside or outside, depending on the meaning: *Who wrote "Gone With the Wind"? He asked, "How long will it take?"*

MISCELLANEOUS: The question mark supersedes the comma that normally is used when supplying attribution for a quotation: *"Who is there?" she asked.*

**quotation marks** The basic guidelines for open-quote marks (") and close-quote marks ("):

FOR DIRECT QUOTATIONS: To surround the exact words of a speaker or writer when reported in a story:

*"I have no intention of staying," she replied.*
*"I do not object," he said, "to the tenor of the report."*
*Franklin said, "A penny saved is a penny earned."*
*A speaker said the practice is "too conservative for inflationary times."*

RUNNING QUOTATIONS: If a full paragraph of quoted material is followed by a paragraph that continues the quotation, do not put close-quote marks at the end of the first paragraph. Do, however, put open-quote marks at the start of the second paragraph. Continue in this way for any succeeding paragraphs, using close-quote marks only at the end of the quoted material. If a paragraph does not start with quotation marks but ends with a quotation that is continued in the next paragraph, do not use close-quote marks at the end of the introductory paragraph if the quoted material constitutes a full sentence. Use close-quote marks, however, if the quoted material does not constitute a full sentence.

DIALOGUE OR CONVERSATION: Each person's words, no matter how brief, are placed in a separate paragraph, with quotation marks at the beginning and the end of each person's speech:

*"Will you go?"*
*"Yes."*
*"When?"*
*"Thursday."*

NOT IN Q/A: Quotation marks are not required in formats that identify questions and answers by Q. and A.

NOT IN TEXTS: Quotation marks are not required in full texts, condensed texts or textual excerpts.

IRONY: Put quotation marks around a word or words used in an ironic sense: *The "debate" turned into a free-for-all.*

UNFAMILIAR TERMS: A word or words being introduced to readers may be placed in quotation marks on first reference: *Broadcast frequencies are measured in "kilohertz."* Do not put subsequent references to kilohertz in quotation marks.

AVOID UNNECESSARY FRAGMENTS: Do not use quotation marks to report a few ordinary words that a speaker or writer has used:

Wrong: *The senator said he would "go home to Michigan" if he lost the election.*

Right: *The senator said he would go home to Michigan if he lost the election.*

PARTIAL QUOTES: When a partial quote is used, do not put quotation marks around words that the speaker could not have used. Suppose the individual said, "I am horrified at your slovenly manners."

Wrong: *She said she "was horrified at their slovenly manners."*

Right: *She said she was horrified at their "slovenly manners."*

Better When Practical: Use the full quote.

QUOTES WITHIN QUOTES: Alternate between double quotation marks *("or")* and single marks *('or'): She said, "I quote from his letter, 'I agree with Kipling that "the female of the species is more deadly than the male," but the phenomenon is not an unchangeable law of nature,' a remark he did not explain."* Use three marks together if two quoted elements end at same time: *She said, "He told me, 'I love you.' "*

PLACEMENT WITH OTHER PUNCTUATION: Follow these long-established printers' rules: The period and the comma always go within the quotation marks. The dash, the semicolon, the question mark and the exclamation point go within the quotation marks when they apply to the quoted matter only. They go outside when they apply to the whole sentence.

**re-** The rules in prefixes apply. The following examples of exceptions to first-listed spellings in *Webster's New World* are based on the general rule that a hyphen is used if a prefix ends in a vowel and the word that follows begins with the same vowel:

| | |
|---|---|
| re-elect | re-enlist |
| re-election | re-enter |
| re-emerge | re-entry |
| re-employ | re-equip |
| re-enact | re-establish |
| re-engage | re-examine |

For many other words, the sense is the governing factor:

| | |
|---|---|
| recover (regain) | re-sign (sign again) |
| re-cover (cover again) | reform (improve) |
| resign (quit) | re-form (form again) |

Otherwise, follow *Webster's New World.* Use a hyphen for words not listed there unless the hyphen would distort the sense.

**semicolon** In general, use the semicolon to indicate a greater separation of thought and information than a comma can convey but less than the separation that a period implies.

**suffixes** See separate listings in the AP stylebook for commonly used suffixes. Follow *Webster's New World Dictionary* for words not in this appendix. If a word combination is not listed in *Webster's New World,* use two words for the verb form; hyphenate any noun or adjective forms.

**suspensive hyphenation** The form: The 5- and 6-year-olds attend morning classes.

## NUMERALS

A numeral is a figure, letter, word or group of words expressing a number. Roman numerals use the letters *I, V, X, L, C, D* and *M.* Use Roman numerals for wars and to show personal sequence for animals and people: *World War II, Native Dancer II, King George VI, Pope John Paul II.* Arabic numerals use the figure *1, 2, 3, 4, 5, 6, 7, 8, 9* and *0.* Use Arabic forms unless Roman numerals are specifically required. The figures *1, 2, 10, 101,* etc., and the corresponding words—*one, two, ten, one hundred one,* etc.—are called cardinal numbers. The term ordinal number applies to *1st, 2nd, 10th, 101st, first, second, tenth, one hundred first,* etc. The following guidelines cover the essentials of using numerals.

### Large Numbers

When large numbers must be spelled out, use a hyphen to connect a word ending in *y* to another word; do not use commas between other separate words that are part of one number: *twenty; thirty; twenty-one; thirty-one; one hundred forty-three; one thousand one hundred fifty-five; one million two hundred seventy-six thousand five hundred eighty-seven.*

### Sentence Start

Spell out a numeral at the beginning of a sentence. If necessary, recast the sentence. There is one exception—a numeral that identifies a calendar year.

Wrong: *993 freshmen entered the college last year.*
Right: *Last year 993 freshmen entered the college.*
Right: *1976 was a very good year.*

### Casual Uses

Spell out casual expressions: *A thousand times no! Thanks a million. He walked a quarter of a mile.*

### Proper Names

Use words or numerals according to an organization's practice: *20th Century-Fox, Twentieth Century Fund, Big Ten.*

## Figures or Words

For ordinals: Spell out *first* through *ninth* when they indicate sequence in time or location—*first base, the First Amendment, he was first in line.* Starting with *10th,* use figures. Use *1st, 2nd, 3rd, 4th,* etc., when the sequence has been assigned in forming names. The principal examples are geographic, military and political designations such as *1st Ward, 7th Fleet* and *1st Sgt.*

## Other Uses

For uses not covered by these listings: Spell out whole numbers below 10, use figures for 10 and above. Typical examples: *The woman has three sons and two daughters. She has a fleet of 10 station wagons and two buses.*

## In a Series

Apply the appropriate guidelines: *They had 10 dogs, six cats and 97 hamsters. They had four four-room houses, 10 three-room houses and 12 10-room houses.*

## Other Key Points of Using Numerals

**addresses** Always use figures for an address number: *9 Morningside Circle.* Spell out and capitalize *First* through *Ninth* when used as street names; use figures with two letters for *10th* and above: 7 *Fifth Ave., 100 21st St.*

**ages** Always use figures. When the context does not require years or years old, the figure is presumed to be years.

**aircraft names** Use a hyphen when changing from letters to figures; no hyphen when adding a letter after figures. Some examples for aircraft often in the news: *B-1, BAC-111, C-5A, DC-10, FH-227, F-14, Phantom II, F-86 Sabre, L-1011, Mig-21, TU-144, 727-100c, 747, 747B, 757, 767, VC-10.*

**amendments to the Constitution** Use *First Amendment, 10th Amendment,* etc. Colloquial references to the Fifth Amendment's protection against self-incrimination are best avoided, but where appropriate: *He took the Fifth seven times.*

**Arabic numerals** The numerical figures *1, 2, 3, 4, 5, 6, 7, 8, 9, 10.* In general, use Arabic forms unless denoting the sequence of wars or establishing a personal sequence for people and animals.

**betting odds** Use figures and a hyphen: *The odds were 5-4. He won despite 3-2 odds against him.* The word *to* seldom is necessary, but when it appears it should be hyphenated in all constructions: *3-to-2 odds, odds of 3-to-2, the odds were 3-to-2.*

**Celsius** Use this term rather than centigrade for the temperature scale that is part of the metric system. When giving a Celsius temperature, use these forms: *40* *degrees Celsius* or *40 C* (note the space and no period after the capital C) if degrees and Celsius are clear from the context.

**cents** Spell out the word *cents* and lowercase, using numerals for amounts less than a dollar: *5 cents, 12 cents.* Use the $ sign and decimal system for larger amounts: *$1.01, $2.50.* Numerals alone, with or without a decimal point as appropriate, may be used in tabular matter.

**congressional districts** Use figures and capitalize *district* when joined with a figure: *the 1st Congressional District, the 1st District.* Lowercase *district* whenever it stands alone.

**court decisions** Use figures and a hyphen: *The Supreme Court ruled 5-4, a 5-4 decision.* The word *to* is not needed, but use hyphens if it appears in quoted matter: *"the court ruled 5-to-4, the 5-to-4 decision."*

**court names** For courts identified by a numeral: *2nd District Court, 8th U.S. Circuit Court of Appeals.*

**dates** Always use Arabic figures, without *st, nd, rd* or *th.*

**decades** Use Arabic figures to indicate decades of history. Use an apostrophe to indicate numerals that are left out; show plural by adding the letter *s: the 1890s, the '90s, the Gay '90s, the 1920s, the mid-1930s.*

**decimal units** Use a period and numerals to indicate decimal amounts. Decimalization should not exceed two places in textual material unless there are special circumstances.

**dimensions** Use figures and spell out inches, feet, yards, etc., to indicate depth, height, length and width. Hyphenate adjectival forms before nouns. Use an apostrophe to indicate feet and quote marks to indicate inches *(5'6")* only in very technical contexts.

**distances** Use figures for 10 and above, spell out one through nine: *He walked four miles.*

**district** Use a figure and capitalize district when forming a proper name: *the 2nd District.*

**dollars** Use figures and the $ sign in all except casual references or amounts without a figure: *The book cost $4. Dad, please give me a dollar. Dollars are flowing overseas.* For specified amounts, the word takes a singular verb: *He said $500,000 is what they want.* For amounts of more than $1 million, use the $ and numerals up to two decimal places. Do not link the numerals and the word by a hyphen: *She is worth $4.35 million. She is worth exactly $4,351,242. He proposed a $300 billion budget.* The form for amounts less than $1 million: *$4, $25, $500, $1,000, $650,000.*

**election returns** Use figures, with commas every three digits starting at the right and counting left. Use the word *to* (not a hyphen) in separating different totals listed together: *Bill Clinton defeated George Bush 43,682,625 to 38,117,331. Ross Perot got 19,217,213 votes* (these are the actual final figures). Use the word *votes* if there is any possibility that the figures could be confused with a ratio: *Bush defeated Dukakis 16 votes to 3 votes in Dixville Notch.* Do not attempt to create adjectival forms such as *the 48,881,221-41,805,422 vote.*

**fractions** Spell out amounts less than 1 in stories, using hyphens between the words: *two-thirds, four-fifths, seven-sixteenths,* etc. Use figures for precise amounts larger than 1, converting to decimals whenever practical. Fractions are preferred, however, in stories about stocks. When using fractional characters, remember that most newspaper type fonts can set only $\frac{1}{8}$, $\frac{1}{4}$, $\frac{3}{8}$, $\frac{1}{2}$, $\frac{5}{8}$, $\frac{3}{4}$ and $\frac{7}{8}$ as one unit; use $1\frac{1}{2}$, $2\frac{5}{8}$, etc. with no space between the figure and the fraction. Other fractions require a hyphen and individual figures, with a space between the whole number and the fraction: *1 3-16, 2 1-3, 5 9-10.*

**highway designations** Use these forms, as appropriate in the context, for highways identified by number: *U.S. Highway 1, U.S. Route 1, U.S. 1, Route Q, Illinois 34, Illinois Route 34, state Route 34, Route 34, Interstate Highway 495.* On second reference only for Interstate: *I-495.*

**mile** Use figures for amounts under 10 in dimensions, formulas and speeds: *The farm measures 5 miles by 4 miles. The car slowed to 7 mph. The new model gets 4 miles more per gallon.* Spell out below 10 in distances: *He drove four miles.*

**millions, billions** Use figures with million or billion in all except casual uses: *I'd like to make a billion dollars.* But: *The nation has 1 million citizens. I need $7 billion.* Do not go beyond two decimals: *7.51 million persons, $2.56 billion, 7,542,500 persons, $2,565,750,000.* Decimals are preferred where practical: *1.5 million.* Not *$1\frac{1}{2}$ million.* Do not mix millions and billions in the same figure: *2.6 billion.* Not *2 billion, 600 million.* Do not drop the word *million* or *billion* in the first figure of a range: *He is worth from $2 million to $4 million.* Not *$2 to $4 million,* unless you really mean $2. Note that a hyphen is not used to join the figures and the word *million* or *billion,* even in this type of phrase: *The president submitted a $3 billion budget.*

**minus sign** Use a hyphen, not a dash, but use the word *minus* if there is any danger of confusion. Use a word, not a minus sign, to indicate temperatures below zero: *minus 10* or *5 below zero.*

**No.** Use as the abbreviation for number in conjunction with a figure to indicate position or rank: *No. 1 person, No. 3 choice.* Do not use in street addresses, with this exception: *No. 10 Downing St.,* the residence of Britain's prime minister. Do not use in the names of schools: *Public School 19.*

**page numbers** Use figures and capitalize *page* when used with a figure. When a letter is appended to the figure, capitalize it but do not use a hyphen: *Page 2, Page 10, Page 20A.* One exception: *It's a Page One story.*

**percentages** Use figures: *1 percent, 2.5 percent* (use decimals not fractions), *10 percent.* For amounts less than 1 percent, precede the decimal with a zero: *The cost of living rose 0.6 percent.* Repeat percent with each individual figure: *He said 10 percent to 30 percent of the electorate may not vote.*

**political divisions** Use Arabic figures and capitalize the accompanying word when used with the figure: *1st Ward, 10th Ward, 3rd Precinct, 22nd Precinct, the ward, the precinct.*

**proportions** Always use figures: *2 parts powder to 6 parts water.*

**ratios** Use figures and a hyphen: *the ratio was 2-to-1, a ratio of 2-to-1, a 2-1 ratio.* As illustrated, the word *to* should be omitted when the numbers precede the word *ratio.* Always use the word *ratio* or a phrase such as *a 2-1 majority* to avoid confusion with actual figures.

**scores** Use figures exclusively, placing a hyphen between the totals of the winning and losing teams: *The Reds defeated the Red Sox 4-3, the Giants scored a 12-6 football victory over the Cardinals, the golfer had a 5 on the first hole but finished with a 2-under-par score.* Use a comma in this format: *Boston 6, Baltimore 5.*

**sizes** Use figures: *a size 9 dress, size 40 long, 10½ B shoes, a 34½ sleeve.*

**speeds** Use figures. *The car slowed to 7 mph, winds of 5 to 10 mph, winds of 7 to 9 knots, 10-knot winds.* Avoid extensively hyphenated constructions such as *5-mile-per-hour winds.*

**telephone numbers** Use figures. The forms: *(212) 262-4000, 262-4000.* If extension numbers are given: *Ext. 2, Ext. 364, Ext. 4071.*

**temperatures** Use figures for all except zero. Use a word, not a minus sign, to indicate temperatures below zero.

**times** Use figures except for noon and midnight. Use a colon to separate hours from minutes: *11 a.m., 1 p.m., 3:30 p.m.* Avoid such redundancies as *10 a.m. this morning, 10 p.m. tonight* or *10 p.m. Monday night.* Use *10 a.m. today, 10 p.m. Monday,* etc., as required by the norms in time element. The construction *4 o'clock* is acceptable, but time listings with *a.m.* or *p.m.* are preferred.

**weights** Use figures: *The baby weighed 9 pounds, 7 ounces. She had a 9-pound, 7-ounce boy.*

**years** Use figures, without commas: *1985.* Use an *s* without an apostrophe to indicate spans of decades or centuries: *the 1890s, the 1800s.* Years are the lone exception to the general rule in numerals that a figure is not used to start a sentence: *1984 was a very good year.*

## GRAMMAR, SPELLING AND WORD USAGE

This section lists common problems of grammatical usage, word selection and spelling.

**a, an** Use the article *a* before consonant sounds: *a historic event, a one-year term* (sounds as if it begins with the letter *w*), *a united stand* (sounds like *you*). Use the article *an* before vowel sounds: *an energy crisis, an honorable man* (the *h* is silent), *an MBA record* (sounds as if it begins with the letter *e*), *an 1890s celebration.*

**accept, except** *Accept* means to receive. *Except* means to exclude.

**adverse, averse** *Adverse* means unfavorable: *She predicted adverse weather. Averse* means reluctant, opposed: *He is averse to change.*

**affect, effect** *Affect,* as a verb, means to influence: *The game will affect the standings. Affect,* as a noun, is best avoided. It occasionally is used in psychology to describe an emotion, but there is no need for it in everyday language. *Effect,* as a verb, means to cause: *He will effect many changes in the company. Effect,* as a noun, means result: *The effect was overwhelming. She miscalculated the effect of her actions. It was a law of little effect.*

**aid, aide** *Aid* is assistance. An *aide* is a person who serves as an assistant.

**ain't** A dialectical or substandard contraction. Use it only in quoted matter or special contexts.

**allude, refer** To *allude* to something is to speak of it without specifically mentioning it. To *refer* is to mention it directly.

**allusion, illusion** *Allusion* means an indirect reference: *The allusion was to his opponent's record. Illusion* means an unreal or false impression: *The scenic director created the illusion of choppy seas.*

**among, between** The maxim that *between* introduces two items and *among* introduces more than two covers most questions about how to use these words: *The funds were divided among Susan, Robert and William.* However, *between* is the correct word when expressing the relationships of three or more items considered one pair at a time: *Negotiations on a debate format are under way between the network and the Clinton, Bush and Perot committees.* As with all prepositions, any pronouns that follow these words must be in the objective case: *among us, between him and her, between you and me.*

**anticipate, expect** *Anticipate* means to expect and prepare for something; *expect* does not include the notion of preparation: *They expect a record crowd. They have anticipated it by adding more seats to the auditorium.*

**anybody, any body, anyone, any one** One word for an indefinite reference: *Anyone can do that.* Two words when the emphasis is on singling out one element of a group: *Any one of them may speak up.*

**apposition** A decision on whether to put commas around a word, phrase or clause used in apposition depends on whether it is essential to the meaning of the sentence (no commas) or not essential (use commas).

**because, since** Use *because* to denote a specific cause-effect relationship: *He went because he was told. Since* is acceptable in causal sense when the first event in a sequence led logically to the second but was not its direct cause: *He went to the game, since he had been given the tickets.*

**blond, blonde** Use *blond* as a noun for males and as the adjective for all applications: *She has blond hair.* Use *blonde* as a noun for females.

**boy** Applicable until 18th birthday is reached. Use *man* or *young man* afterward.

**brunet, brunette** Use *brunet* as a noun for males and as the adjective for both sexes. Use *brunette* as a noun for females,

**burglary, larceny, robbery, theft** Legal definitions of burglary vary, but in general a *burglary* involves entering a building (not necessarily by breaking in) and remaining unlawfully with the intention of committing a crime. *Larceny* is the legal term for the wrongful taking of property. Its nonlegal equivalents are stealing or theft. *Robbery* in the legal sense involves the use of violence or threat in committing larceny. In a wider sense it means to plunder or rifle, and may thus be used even if a person was not present: *Her house was robbed while she was away. Theft* describes a larceny that did not involve threat, violence or plundering. Usage Note: You rob a person, bank, house, etc., but you steal the money or the jewels.

**collective nouns** Nouns that denote a unit take singular verbs and pronouns: *class, committee, crowd, family group, herd, jury, orchestra, team.* Some usage examples: *The committee is meeting to set its agenda. The jury reached its verdict. A herd of cattle was sold.*

**PLURAL FORM:** Some words that are plural in form become collective nouns and take singular verbs when the group or quantity is regarded as a unit.

Right: *A thousand bushels is a good yield.* (A unit.)
Right: A *thousand bushels were created.* (Individual items.)
Right: *The data is sound.* (A unit.)
Right: *The data have been carefully collected* (Individual items.)

**compose, comprise, constitute** *Compose* means to create or put together. It commonly is used in both the active and passive voices: *She composed a song. The United States is composed of 50 states. The zoo is composed of many animals. Comprise* means to contain, to include all or embrace. It is best used only in the active voice, followed by a direct object: *The United States comprises 50 states. The jury comprises five men and seven women. The zoo comprises many animals. Constitute,* in the sense of form or make up, may be the best word if neither *compose* nor *comprise* seems to fit: *Fifty states constitute the United States. Five men and seven women constitute the jury. A collection of animals can constitute a zoo.* Use *include* when what follows is only part of the total: *The price includes breakfast. The zoo includes lions and tigers.*

**contractions** Contractions reflect informal speech and writing. *Webster's New World Dictionary* includes many entries for contractions: *aren't* for *are not,* for example. Avoid excessive use of contractions. Contractions listed in the dictionary are acceptable, however, in informal contexts or circumstances in which they reflect the way a phrase commonly appears in speech or writing.

**contrasted to, contrasted with** Use *contrasted to* when the intent is to assert, without the need for elaboration, that two items have opposite characteristics: *He contrasted the appearance of the house today to its ramshackle look last year.* Use *contrasted with* when juxtaposing two or more items to illustrate similarities and/or differences: *He contrasted the Republican platform with the Democratic platform.*

**dangling modifiers** Avoid modifiers that do not refer clearly and logically to some word in the sentence.

Dangling: *Taking our seats, the game started.* (Taking does not refer to the subject, game, nor to any other word in the sentence.)

Correct: *Taking our seats, we watched the opening of the game.* (Taking refers to we, the subject of the sentence.)

**either** Use it to mean *one or the other,* not *both.*

Right: *She said to use either door.*

Wrong: *There were lions on either side of the door.*

Right: *There were lions on each side of the door. There were lions on both sides of the door.*

**either . . . or, neither . . . nor** The nouns that follow these words do not constitute a compound subject; they are alternate subjects and require a verb that agrees with the nearer subject: *Neither they nor he is going. Neither he nor they are going.*

**essential clauses, nonessential clauses** These terms are used instead of *restrictive clause* and *nonrestrictive clause* to make the distinction between the two easier to remember. Both types of clauses provide additional information about a word or phrase in the sentence. The difference between them is that the essential clause cannot be eliminated without changing the meaning of the sentence—it so "restricts" the meaning of the word or phrase that its absence would lead to a substantially different interpretation of what the author meant. The nonessential clauses, however, can be eliminated without altering the basic meaning of the sentence—it does not "restrict" the meaning so significantly that its absence would radically alter the author's thought.

PUNCTUATION: An essential clause must not be set off from the rest of a sentence by commas. A nonessential clause must be set off by commas. The presence or absence of commas provides the reader with critical information about the writer's intended meaning. Note the following examples: *Reporters who do not read the stylebook should not criticize their editors.* (The writer is saying that only one class of reporters, those who do not read the stylebook, should not criticize their editors. If the *who... stylebook* phrase were deleted, the meaning of the sentence would be changed substantially.) *Reporters, who do not read the stylebook, should not criticize their editors.* (The writer is saying that all reporters should not criticize their editors. If the *who... stylebook* phrase were deleted, this meaning would not be changed.)

USE OF WHO, THAT, WHICH: When an essential or nonessential clause refers to a human being or an animal with a name, it should be introduced by *who* or *whom.* (see the Who, Whom entry.) Do not use commas if the clause is essential to the meaning; use them if it is not. *That* is the preferred pronoun to introduce essential clauses that refer to an inanimate object or an animal without a name. *Which* is the only acceptable pronoun to introduce a nonessential clause that refers to an inanimate object or an animal without a name. The pronoun *which* occasionally may be substituted for *that* in the introduction of an essential clause that refers to an inanimate object or an animal without a name. In general, this use of *which* should appear only when *that* is used as a conjunction to introduce another clause in the same sentence: *He said Monday that the part of the army which suffered severe casualties needs reinforcement.*

**essential phrase, nonessential phrase** These terms are used instead of *restrictive phrase* and *nonrestrictive phrase* to make the distinction between the two easier to remember. The underlying concept is the one that also applies to clauses: An *essential phrase* is a word or group of words critical to the reader's understanding of what the author had in mind. A *nonessential phrase* provides more information about something. Although the information may be helpful to the reader's comprehension, the reader would not be misled if the information were not there.

PUNCTUATION: Do not set off an essential phrase from the rest of a sentence by commas: *We saw the award-winning movie "One Flew Over the Cuckoo's Nest."* (No comma, because many movies have won awards, and without the name of the movie the reader would not know which movie was meant.) *They ate dinner with their daughter Julie.* (Because they have more than one daughter, the inclusion of Julie's name is critical if the reader is to know which daughter is meant.) Set off nonessential phrases by commas: *We saw the 1976 winner in the Academy Award competition for best movie, "One Flew Over the Cuckoo's Nest."* (Only one movie won the award. The name is informative, but even without the name no other movie could be meant.) *They ate dinner with their daughter Julie and her husband, David.* (Julie has only one husband. If the phrase read *and her husband David,* it would suggest that she had more than one husband.) *The company chairman, Henry Ford II, spoke.* (In the context, only one person could be meant.) *Indian corn, or maize, was harvested.* (*Maize* provides the reader with the name of the corn, but its absence would not change the meaning of the sentence.)

**DESCRIPTIVE WORDS:** Do not confuse punctuation rules for nonessential clauses with the correct punctuation when a nonessential word is used as a descriptive adjective. The distinguishing clue often is the lack of an article or pronoun:

Right: *Julie and husband David went shopping. Julie and her husband, David, went shopping.*

Right: *Company Chairman Henry Ford II made the announcement. The company chairman, Henry Ford II, made the announcement.*

**every one, everyone** Two words when it means each individual item: *Every one of the clues was worthless.* One word when used as a pronoun meaning all persons: *Everyone wants his life to be happy.* (Note that *everyone* takes singular verbs and pronouns.)

**farther, further** *Farther* refers to physical distance: *He walked farther into the woods. Further* refers to an extension of time or degree: *She will look further into the mystery.*

**fewer, less** In general, use *fewer* for individual items, *less* for bulk or quantity.

**flaunt, flout** To *flaunt* is to make an ostentatious or defiant display: *She flaunted her beauty.* To *flout* is to show contempt for: *She flouts the law.*

**flier, flyer** *Flier* is the preferred term for an aviator or a handbill. *Flyer* is the proper name of some trains and buses: *The Western Flyer.*

**good, well** *Good* is an adjective that means something is as it should be or is better than average. When used as an adjective, *well* means suitable, proper, healthy. When used as an adverb, *well* means in a satisfactory manner or skillfully. *Good* should not be used as an adverb. It does not lose its status as an adjective in a sentence such as *I feel good.* Such a statement is the idiomatic equivalent of *I am in good health.* An alternative, *I feel well,* could be interpreted as meaning that your sense of touch was good.

**hopefully** It means in a hopeful manner. Do not use it to mean *it is hoped, let us hope* or *we hope.*

Right: *It is hoped that we will complete our work in June.*

Right: *We hope that we will complete our work in June.*

Wrong as a saying to express the thought in the previous two sentences: *Hopefully, we will complete our work in June.*

**imply, infer** Writers or speakers imply in the words they use. A listener or reader *infers* something from the words.

**in, into** *In* indicates location: *He was in the room. Into* indicates *motion. She walked into the room.*

**lay, lie** The action word is *lay.* It takes a direct object. *Laid* is the form for its past tense and its past participle. Its present participle is *laying. Lie* indicates a state of reclining along a horizontal plane. It does not take a direct object. Its past tense is *lay.* Its past participle is *lain.* Its present participle is *lying.* When *lie* means to make an untrue statement, the verb forms are *lie, lied, lying.*

**like, as** Use *like* as a preposition to compare nouns and pronouns. It requires an object: *Jim blocks like a pro.* The conjunction *as* is the correct word to introduce clauses: *Jim blocks the linebacker as he should.*

**majority, plurality** *Majority* means more than half of an amount. *Plurality* means more than the next highest number.

**marshal, Marshall, marshaled, marshaling** *Marshal* is the spelling for both the verb and the noun: *Marilyn will marshal her forces. Erwin Rommel was a field marshal. Marshall* is used in proper names: *George C. Marshall, John Marshall, the Marshall Islands.*

**obscenities, profanities, vulgarities** Do not use them in stories unless they are part of direct quotations and there is a compelling reason for them. Confine the offending language, in quotation marks, to a separate paragraph that can be deleted easily. In reporting profanity that normally would use the words *damn* or *god,* lowercase *god* and use the following forms: *damn, damn it, goddamn it.* Do not, however, change the offending words to euphemisms. Do not, for example, change *damn it* to *darn it.* If a full quote that contains profanity, obscenity or vulgarity cannot be dropped but there is no compelling reason for the offensive language, replace letters of an offensive word with a hyphen. The word *damn,* for example, would become *d—* or —.

**off of** The *of* is unnecessary: *He fell off the bed.* Not *He fell off of the bed.*

**on** Do not use *on* before a date or day of the week when its absence would not lead to confusion: *The meeting will be held Monday. He will be inaugurated Jan. 20.* Use *on* to avoid an awkward juxtaposition of a date and a proper name: *John met Mary on Monday. He told Clinton on Thursday that the bill was doomed.* Use *on* also to avoid any suggestion that a date is the object of a transitive verb: *The House killed on Tuesday a bid to raise taxes. The Senate postponed on Wednesday its consideration of a bill to reduce import duties.*

**over** It is interchangeable with *more than.* Usually, however, *over* refers to spatial relationships: *The plane flew over the city. More than* is used with figures: *More than 40,000 fans were in the stadium.*

**people, persons** Use *people* when speaking of a large or uncounted number of individuals: *Thousands of people attended the fair. Some rich people pay few taxes. What will people say?* Do not use *persons* in this sense. *Persons* usually is used when speaking of a relatively small number of people who can be counted, but people also may be used:

Right: *There were 20 persons in the room.*

Right: *There were 20 people in the room.*

*People* is also a collective noun that takes a plural verb and is used to refer to a single race or nation: *The American people are united.* In this sense, the plural is *peoples: The peoples of Africa speak many languages.*

**principal, principle** *Principal* is a noun and adjective meaning someone or something first in rank, authority, importance or degree: *She is the school principal. He was the principal player in the trade. Money is the principal problem. Principle* is a noun that means a fundamental truth, law, doctrine or motivating force: *They fought for the principle of self-determination.*

**prior to** *Before* is less stilted for most uses. *Prior to* is appropriate, however, when a notion of requirement is involved: *The fee must be paid prior to the examination.*

**reign, rein** The leather strap for a horse is a *rein,* hence figuratively: *seize the reins, give free rein to, put a check rein on. Reign* is the period a ruler is on the throne: *The king began his reign.*

**should, would** Use *should* to express an obligation: *We should help the needy.* Use *would* to express a customary action: *In the summer we would spend hours by the seashore.* Use *would* also in constructing a conditional past tense, but be careful:

Wrong: *If Soderholm would not have had an injured foot, Thompson would not have been in the lineup.*

Right: *If Soderholm had not had an injured foot, Thompson would not have been in the lineup.*

**spelling** The basic rule when in doubt is to consult the stylebooks followed by, if necessary, a dictionary. Memory Aid: Noah Webster developed the following rule of thumb for the frequently vexing question of whether to double a final consonant in forming the present participle and past tense of a verb: If the stress in pronunciation is on the first syllable, do not double the consonant: *cancel, canceling, canceled.* If the stress in pronunciation is on the second syllable, double the consonant: *control, controlling, controlled; refer, referring, referred.* If the word is only one syllable, double a consonant unless confusion would result: *jut, jutted, jutting.* An exception, to avoid confusion with *buss,* is *bus, bused, busing.* Here is a list of commonly misspelled words:

| | | |
|---|---|---|
| accommodate | council | likable |
| adviser | counsel | machine gun |
| Asian flu | drought | percent |
| ax | drunken | percentage |
| baby-sit | embarrass | restaurant |
| baby sitter | employee | restaurateur |
| baby-sitting | eyewitness | rock 'n' roll |
| cannot | firefighter | skillful |
| cave-in (v.) | fulfill | subpoena |
| cave-in (n., adj.) | goodbye | teen-age (adj.) |
| chauffeur | hanged | teen-ager (n.) |
| cigarette | harass | under way |
| clue | hitchhiker | vacuum |
| commitment | homemade | weird |
| consensus | impostor | whiskey |
| consul | judgment | X-ray (n., v. and adj.) |
| copter | kidnapping | |

**subjunctive mood** Use the subjunctive mood of a verb for contrary-to-fact conditions and expressions of doubts, wishes or regrets: *If I were a rich man, I wouldn't have to work hard. I doubt that more money would be the answer. I wish it were possible to take back my words.* Sentences that express a contingency or hypothesis may use either the subjunctive or the indicative mood depending on the context. In general, use the subjunctive if there is little likelihood that a contingency might come true: *If I were to marry a millionaire, I wouldn't have to worry about money. If the bill should overcome the opposition against it, it would provide extensive tax relief.* But: *If I marry a millionaire, I won't have to worry about money. If the bill passes as expected, it will provide an immediate tax cut.*

**that (conjunction)** Use the conjunction *that* to introduce a dependent clause if the sentence sounds or looks awkward without it. There are no hard-and-fast rules, but in general: *That* usually may be omitted when a dependent clause immediately follows a form of the verb *to say: The president said he had signed the bill. That* should be used when a time element intervenes between the verb and the dependent clause: *The president said Monday that he had signed the bill. That* usually is necessary after some verbs. They include: *advocate, assert, contend, declare, estimate, make clear, point out, propose* and *state. That* is required before subordinate clauses beginning with conjunctions such as *after, although, because, before, in addition to, until* and *while: Haldeman said that after he learned of Nixon's intention to resign, he sought pardons for all connected with Watergate.* When in doubt, include *that.* Omission can hurt. Inclusion never does.

**that, which, who, whom (pronouns)** Use *who* and *whom* in referring to persons and to animals with a name: *John Jones is the man who helped me.* See the Who, Whom entry. Use *that* and *which* in referring to inanimate objects and to animals without a name. See the Essential Clauses, Nonessential Clauses entry for guidelines on using *that* and *which* to introduce phrases and clauses.

**under way** Two words in virtually all uses: *The project is under way. The naval maneuvers are under way.* One word only when used as an adjective before a noun in a nautical sense: *an underway flotilla.*

**verbs** In general, avoid awkward constructions that split infinitive forms of a verb *(to leave, to help,* etc.) or compound forms *(had left, are found out,* etc.).

Awkward: *She was ordered to immediately leave on an assignment.*

Preferred: *She was ordered to leave immediately on an assignment.*

Awkward: *There stood the wagon that we had early last autumn left by the barn.*

Preferred: *There stood the wagon that we had left by the barn early last autumn.*

Occasionally, however, a split is not awkward and is necessary to convey the meaning: *He wanted to really help his mother. Those who lie are often found out. How has your health been? The budget was tentatively approved.*

**who, whom** Use *who* and *whom* for references to human beings and to animals with a name. Use *that* and *which* for inanimate objects and animals without a name. *Who* is the word when someone is the subject of a sentence, clause or phrase: *The woman who rented the room left the window open. Whom do you wish to see?* See the Essential Clauses, Nonessential Clauses entry for guidelines on how to punctuate clauses introduced by *who, whom, that* and *which.*

**who's, whose** *Who's* is a contraction for *who is,* not a possessive: *Who's there? Whose* is the possessive: *I do not know whose coat it is.*

**widow, widower** In obituaries: A man is survived by his wife, or leaves his wife. A woman is survived by her husband, or leaves her husband. Guard against the redundant *widow of the late.* Use *wife of the late* or *widow of.*

# APPENDIX II

# COPY-EDITING AND PROOFREADING SYMBOLS

Today's editors do most of their work on computers, so the editing of paper manuscripts, or *copy* (see Figure II–1), is increasingly rare. A few small newspapers and magazines, however, still use manual methods of production, and many book publishers still do so. In such cases, knowing the long-established copy-editing symbols is essential. Even at large newspapers and magazines, editors pass printouts of articles to publishers, attorneys and others, and those individuals use copy-editing symbols to make suggested changes. Therefore, it's a good idea to know the standard symbols and how to use them, even if you usually edit with a computer.

Copy-editing symbols (see Figure II–2), which are universally understood in the business, are used to make changes in paper manuscripts or on printouts of stories that reside in computers. In modern newsrooms, those changes usually are

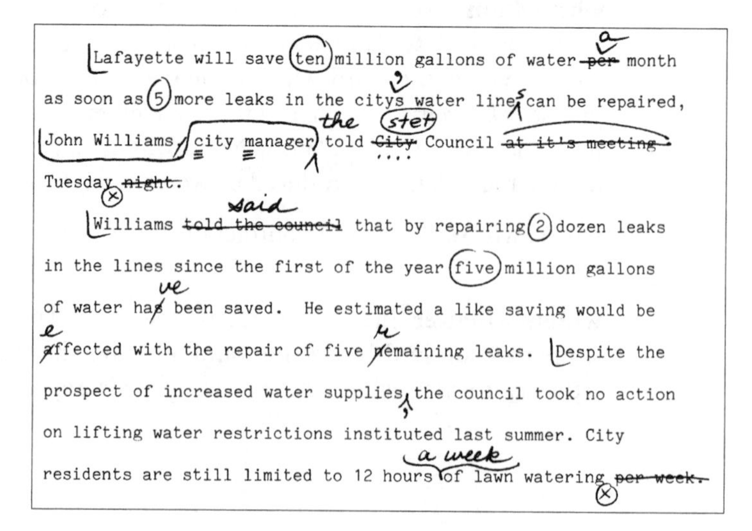
Lafayette will save ten million gallons of water per month as soon as 5 more leaks in the citys water line can be repaired, John Williams city manager told City Council at it's meeting Tuesday night.

Williams told the council that by repairing 2 dozen leaks in the lines since the first of the year five million gallons of water has been saved. He estimated a like saving would be affected with the repair of five remaining leaks. Despite the prospect of increased water supplies the council took no action on lifting water restrictions instituted last summer. City residents are still limited to 12 hours of lawn watering per week.

***Figure II-1** An example of edited copy.*

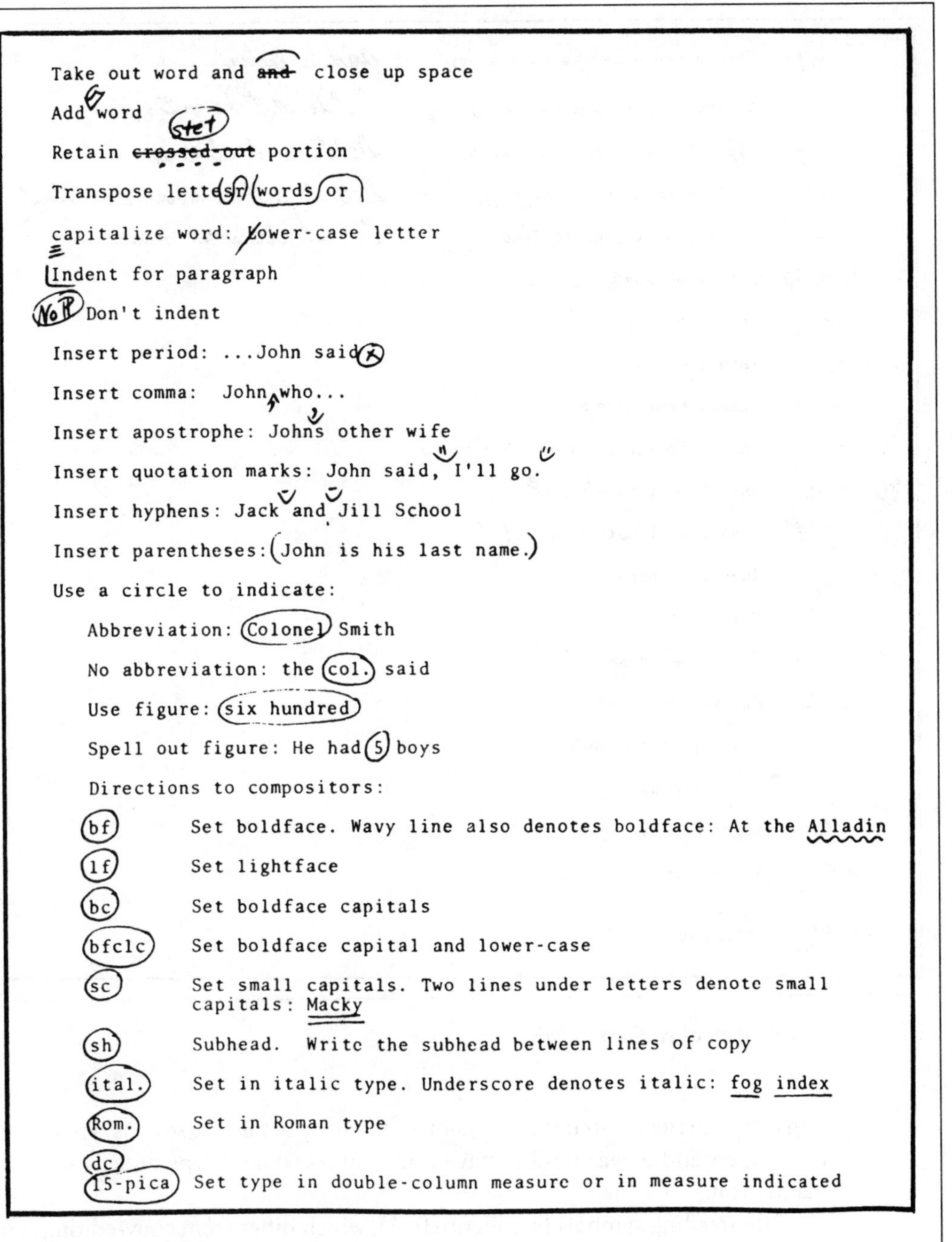
Take out word and and close up space

Add word

Retain crossed-out portion

Transpose letters/words or

capitalize word: lower-case letter

Indent for paragraph

Don't indent

Insert period: ...John said

Insert comma: John who...

Insert apostrophe: Johns other wife

Insert quotation marks: John said, I'll go.

Insert hyphens: Jack and Jill School

Insert parentheses: John is his last name.

Use a circle to indicate:

Abbreviation: Colonel Smith

No abbreviation: the col. said

Use figure: six hundred

Spell out figure: He had 5 boys

Directions to compositors:

| | |
|---|---|
| bf | Set boldface. Wavy line also denotes boldface: At the Alladin |
| lf | Set lightface |
| bc | Set boldface capitals |
| bfclc | Set boldface capital and lower-case |
| sc | Set small capitals. Two lines under letters denote small capitals: Macky |
| sh | Subhead. Write the subhead between lines of copy |
| ital. | Set in italic type. Underscore denotes italic: fog index |
| Rom. | Set in Roman type |
| dc 15-pica | Set type in double-column measure or in measure indicated |

*Figure II-2 Copy-editing symbols.*

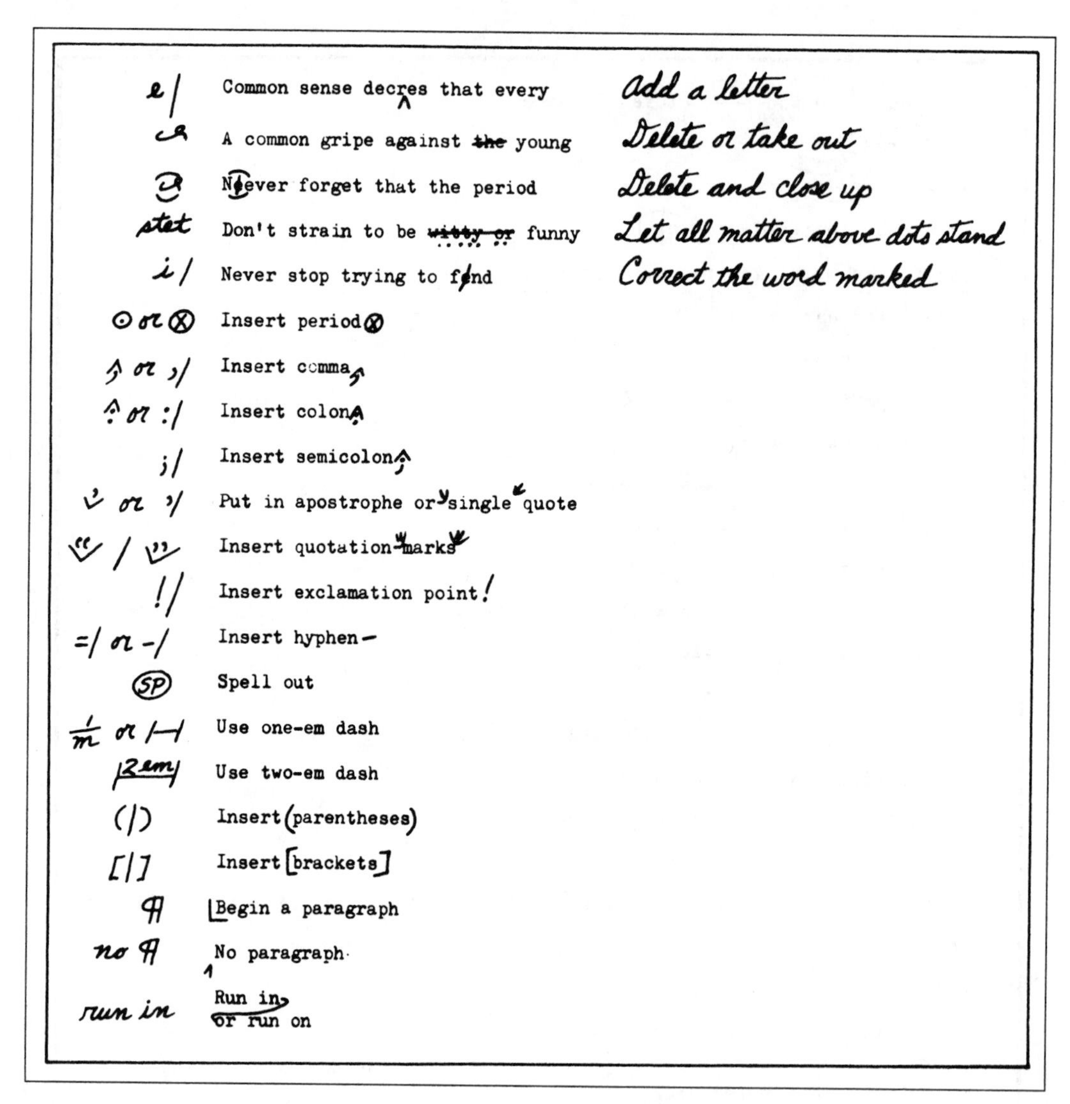


*Figure II-3 Standard proofreading symbols.*

transferred to the computerized version of the story before the story is typeset. Few newspapers and magazines still have compositors who set type directly from the manuscript.

Proofreading symbols (see Figure II–3), which differ from copy-editing symbols somewhat, are used to make corrections in type once it has been typeset. In the newspaper and magazine businesses, editors often use proofreading symbols to mark minor changes on *page proofs* before those pages are sent to the platemaking department. Newspaper editors also use page proofs to make minor changes in pages between editions. Newspapers and magazines use the informal method of marking proofs, while book publishers use a more formalized version (see Figures II–4 and 5).

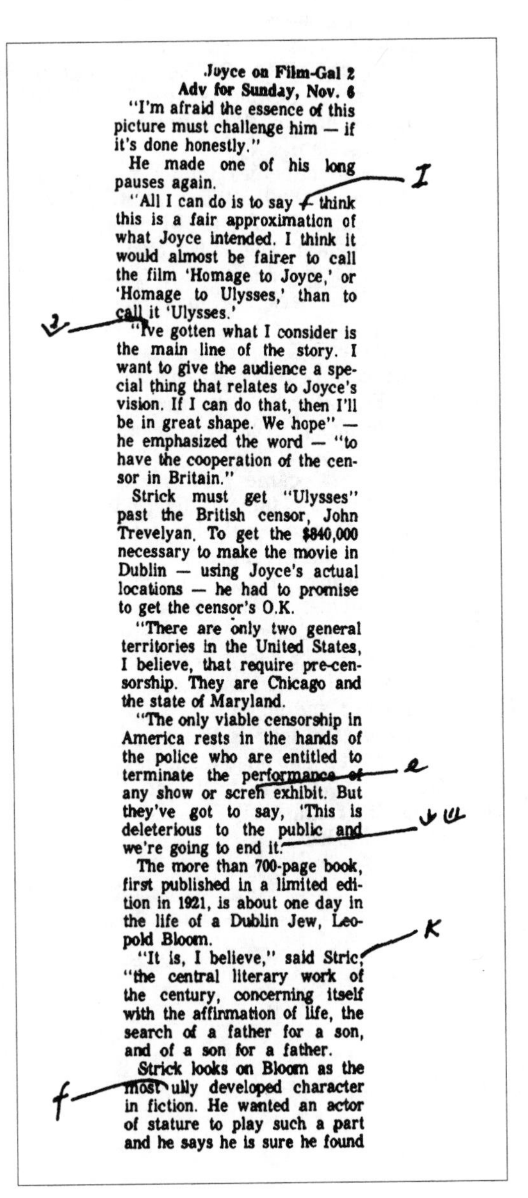

Joyce on Film-Gal 2
Adv for Sunday, Nov. 6

"I'm afraid the essence of this picture must challenge him — if it's done honestly."

He made one of his long pauses again.

"All I can do is to say think this is a fair approximation of what Joyce intended. I think it would almost be fairer to call the film 'Homage to Joyce,' or 'Homage to Ulysses,' than to call it 'Ulysses.'

"Ive gotten what I consider is the main line of the story. I want to give the audience a special thing that relates to Joyce's vision. If I can do that, then I'll be in great shape. We hope" — he emphasized the word — "to have the cooperation of the censor in Britain."

Strick must get "Ulysses" past the British censor, John Trevelyan. To get the $840,000 necessary to make the movie in Dublin — using Joyce's actual locations — he had to promise to get the censor's O.K.

"There are only two general territories in the United States, I believe, that require pre-censorship. They are Chicago and the state of Maryland.

"The only viable censorship in America rests in the hands of the police who are entitled to terminate the performance of any show or scren exhibit. But they've got to say, 'This is deleterious to the public and we're going to end it.

The more than 700-page book, first published in a limited edition in 1921, is about one day in the life of a Dublin Jew, Leopold Bloom.

"It is, I believe," said Stric, "the central literary work of the century, concerning itself with the affirmation of life, the search of a father for a son, and of a son for a father.

Strick looks on Bloom as the most ully developed character in fiction. He wanted an actor of stature to play such a part and he says he is sure he found



*Figure II–4 Informal or newspaper method of marking proofs.*

*Figure II–5 Formal method of marking proofs.*

Arlette Schmitt, a brown-eyed blonde from Nice, France, speaks four languages and can tell a housewife how to use each item in the store.

Arlette and store manager Bal Raj DKOGRA, OF New De India, agree on one point: The shoppers are mostly American — not foreign.

"A woman comes up to me and says, 'What's this stuff? Dogra relates. "I tell her it's Egyptian jam. She says, 'okay, I'll try it,' and dumps it in her cart. Next week she may buy three jars."

Two years ago, Dogra says, items were new to most customers. But now he says they've become picky, even over brand names.

Dogra, 29, came here seven years ago as a student. Married to an American girl, his food favorites still are Indian.

He's proud of an "instant curry dinner from Bombay. Just heat and serve it." There are packets of curry rice, herb rice and rice flamenco, too — and water and boil.

Dogra says food is bought as soon as it's put on the shelves. "You can't believe how fast it goes. It's fantastic."

**MORE MORE**

Dogra/Delhi

Although he is married

Add

# APPENDIX III

# DATABASE SEARCHES

Each day thousands of news stories and magazine articles are written, and many of them contain material that is invaluable to reporters researching stories. Those articles also can be invaluable for editors who need to check facts. Until recently, it was impractical for editors to consult such resources if they were not located in the newspaper library. Going to the public library to check a fact was out of the question because of the time involved. Now, however, many media operations have installed computer terminals that provide access to literally thousands of databases with material on almost every conceivable subject.

Among the most useful databases to journalists are the comprehensive files of Dialog, a Palo Alto, Calif., company that has files on everything from national statistics (American Statistics Index) to magazine article listings (Magazine Index).

Also of interest to the fast-paced world of daily journalism are services such as Dow Jones News Retrieval. This service allows the user to read a story exactly as was published in *The Wall Street Journal.*

Let's say you are searching for an offbeat story you remember reading in the *Journal* about a haunted house in Louisiana. You go to your computer terminal and sign on to the Dow Jones News Retrieval free-text search service. The service asks you to describe key words that might help locate the article. You enter *Louisiana* and find that the service has 441 listings, or articles, containing that word. That won't do. It would take far too long to search all those articles for the one in question, and you are being charged for each minute connected to the service. You must narrow the search, so you decide to ask the computer for the number of stories that contain the words *Louisiana* and *ghosts.* Two stories fit the description. You give the computer a simple command to display the first of the two stories on the screen. The result is shown in Figure III–1.

That's it. The search has turned up the story you remembered. But the search words you used were unique, and the computer indicated there were two stories that matched the description. Is it possible there was a follow-up you missed? To determine that, you decide to look at the other story (see Figure III–2).

```
            DOCUMENT=        1 OF        2    PAGE =      1 OF      16
AN        841031-0161.
HL          As Spooky Places Go,
            An Inn in the Bayous
            Goes a Bit Too Far

            ---
            'Most Haunted House' Teems
            With Spirits, Legend Says;
            How Our Man Faces Them
            ---
            By Bryan Burrough
                                       - MORE -


            DOCUMENT=        1 OF        2    PAGE =      2 OF      16
            Staff Reporter of The Wall Street Journal
DD        10/31/84
SO        WALL STREET JOURNAL (J)
TX           ST. FRANCISVILLE, La. -- They never extinguish all the
          lights at the Myrtles, a 188-year-old mansion outside this
          sleepy bayou town. Total darkness, legend says, brings out
          ghoulish things that go bump in the night.
             At first glance, the house doesn't look haunted. Huge oaks
          dripping Spanish moss cover the plantation, and out back a
          gazebo sits on an island in a tiny pond. The white two-story
                                       - MORE -


            DOCUMENT=        1 OF        2    PAGE =      3 OF      16
          mansion is attractively encircled by a veranda lined with
          colonial-blue wrought iron.
             But those who know about the Myrtles whisper stories of
          phantoms and voodoo curses, of two long-dead girls who peer
          into the mansion's windows at night, of picture frames that
          fly from the wall and of a ghostly one-eared slave woman who
          glides by the canopied beds at night, chilling the bones of
          those she encounters.
             Unexplained sights and sounds have been observed by
          everyone from psychics to local officials, and since the
                                       - MORE -


            DOCUMENT=        1 OF        2    PAGE =      4 OF      16
          Myrtles was converted into a six-bedroom inn four years ago,
          the owner says, many tourists have been so spooked in the
          night that they have vowed never to return. As for me, in the
          words of the "Ghostbusters" movie theme song, "I ain't 'fraid
          of no ghosts." I agree to spend a weekend there, in the
          building that the book "Houses of Horror" by Richard Winer
          calls America's most haunted house. . . .
```

*Figure III–1 Dow Jones News Retrieval displays a story located by searching for key words.*

```
          DOCUMENT=          2 OF          2     PAGE =         1 OF        16
AN        840614-0143.
HL          LEISURE & ARTS:
           Mississippi Mermaids: World's Fair Takes a Dive
           ---
           By Manuela Hoelterhoff
DD        06/14/84
SO        WALL STREET JOURNAL (J)
TX          New Orleans -- The last time the Crescent City had a huge
          exposition was 100 years ago, and the event is not happily
          remembered. The director absconded with part of the state
                                       - MORE -


          DOCUMENT=          2 OF          2     PAGE =         2 OF        16
          treasury, attendance was lower than projected and reviews
          were mixed, despite such unusual exhibitions as 200,000
          insect species from Maryland and a reception hosted by
          costumed pigs.
            Maybe a combo of smart porkers playing jazz would help out
          the current World's Fair, which opened on May 12 for a
          six-month run. Despite a decade of planning and a cost of
          $350 million, America, never mind the world, hasn't been
          rushing down to this 84-acre extravaganza deposited along the
          Mississippi River. Attendance figures have been well below
                                       - MORE -
```

*Figure III–II The text search found this story as well.*

```
  DOCUMENT=          2 OF          2     PAGE =        15 OF        16
  After Nov. 11, this show will be returning to Paris via a
stop in Washington, and the expo site will be redeveloped
into yet another waterside shrine of gentrification via Rouse
Co. and a second investment group. Mr. Gehry's striking
theater will come down, and at most a small chunk of
Wonderwall will be preserved. I cannot think of a city less
in need of additional ethnic harbor-view food shops than New
Orleans, which overflows with gumbo and already has a market
by the river. Still, I guess even this dull prospect will be
better than what happened to the fairgrounds in Knoxville --
                             - MORE -


  DOCUMENT=          2 OF          2     PAGE =        16 OF        16
decrepit and abandoned, frequented only by joggers and
ghosts.
```

*Figure III–III The second story contains the same key words as the first.*

It's obvious why the search revealed *Louisiana* for the second story, but it isn't about a haunted house. The last paragraph, however, reveals why the search gave two choices (see Figure III–3). The writer happened to write a story from Louisiana that included the word *ghosts.*

The two stories found in this search indicate not only how useful such databases can be but also the limitations of search techniques.

On one hand, you found the article you sought, but you wasted time looking at one that produced nothing of use. Such limitations apply to searching libraries' card catalogs, too.

Database search techniques may not be perfect, but the results give the copy editor far more flexibility in checking facts than ever was possible before. Thus, the ability of the editor to check facts is vastly expanded. Widespread use of database searches at publications and broadcast stations should make the news reports of the future more accurate than ever.

Here are some of the major databases of interest to journalists. There are thousands, and this is merely a sample:

America Online
8619 Westwood Center Drive
Vienna, Va. 22182-2285
(800) 827-6364

AT&T Services
25 First St.
Cambridge, Mass. 02141
(617) 252-5477

CompuServe
P.O. Box 20212
5000 Arlington Centre Blvd.
Columbus, Ohio 43220
(800) 848-8199

Dialog
3460 Hillview Ave., Dept. 79
Palo Alto, Calif. 94304
(800) 227-1927

Dow Jones News Retrieval
P.O. Box 300
Princeton, N.J. 08540
(800) 345-8500

The Microsoft Network
One Microsoft Way
Redmond, Wash. 98052-6399
(800) 386-5550

Prodigy Services
445 Hamilton Ave.
White Plains, N.Y. 10601
(800) 776-3449

Vu/Text Information Services
325 Chestnut St.
Philadelphia, Pa. 19106
(800) 258-8080

# APPENDIX IV

# GLOSSARY

**Accordion folds** parallel folds used to determine an imposition pattern on a sheet of paper.

**Ad** short for advertisement.

**Ad alley** section in mechanical department where ads are assembled.

**Add** material to be added to news story, usually with a number and slug: *add 1-fire.*

**Add end** addendum to a story after it has been apparently closed.

**Ad lib** unscripted comment made before a microphone.

**Ad-side** advertising department as distinguished from editorial department.

**ADV** abbreviation for advance. A story intended for later use.

**Advance** story sent out in advance of the scheduled publication date.

**Agate** name of a type size, e.g., agate type, or type that is $5\frac{1}{2}$ points high. Advertising is also sold on the basis of agate lines, or 14 agate lines to an inch.

**Agenda setter** journalists often are referred to in this way because their traditional role has been to set the agenda for public action through the emphasis of public affairs in the news.

**Air** white space.

**Align** to place adjacent to an even baseline on a horizontal plane.

**Alphanumeric** pertaining to a character set that contains letters, numerals and usually other characters.

**A.M.** morning edition.

**Analog** data in the form of varying physical characteristics such as voltage, pressure, speed, etc. A computer term.

**Angle** special aspect of a story; a slant.

**Antique** rough-surfaced paper resembling old handmade papers.

**AP** short for The Associated Press, a major news agency.

**Art** newspaper or magazine illustrations excluding photos.

**Attribution** source of the material in a story.

**Audio tape** tape on which sound has been transcribed.

**Audiotext** delivery of news verbally over telephone lines, a process increasingly being automated with digital technology.

**Autofunctions** copy symbols or commands placed on copy. Used to instruct the computer on type sizes, column widths, etc.

**Auxiliary storage** any peripheral devices (tape, disk, etc.) upon which computer data may be stored.

**Back timing** timing of a broadcast script so that it ends at the proper time.

**Back up** printing the reverse side of a sheet.

**Bad break** bad phrasing of a headline; bad wrapping of headline type; bad arrangement of type in columns that gives the reader the impression a paragraph ending is the end of the story.

**Bagasse** residue of sugar cane stalks used experimentally as a substitute for wood pulp in making newsprint.

**Bank** the lower portion of a headline (deck).

**Banner** usually a headline stretching across all columns of a newspaper.

**Barker** reversed kicker in which the kicker is in larger type than the lines below it. Also called a *hammer.*

**Bastard type** type that varies from the standard column width.

**Baud** speed rate of transmission. News personnel speak in terms of 66 or 1,200 words a minute. Computer personnel talk of 1,200 bits (or 150 characters) a second.

**Binary** a base 2 numbering system using the digits 0 and 1. Widely used in computer systems whose binary code may represent numbers, letters and punctuation marks. Performs "in" and "out" or "on" and "off" function.

**Binder** inside page streamer; head that binds together two or more related stories.

**Bit** one-eighth of a byte in computer terminology. Combinations of eight basic bits yield up to 256 characters.

**Black and white** reproduction in one color (black).

**Black Letter** text or Old English style of type.

**Blanket head** headline over several columns of type or over type and illustration.

**Bleed** running an illustration off the page.

**Blooper** any embarrassing error in print or broadcast.

**Body** main story or text; body type is the size of type used for the contents.

**Boil** to trim or reduce wordage of a story.

**Boilerplate** syndicated material.

**Boldface** type that is blacker than normal typeface; also *black face.* Abbreviated *bf.*

**Book** a basic category of printing paper.

**Book number** number assigned to each item in a wire service report.

**Box** unit of type enclosed by a border.

**Brace** type of makeup, usually with a banner headline and the story in the right-hand column. A bracket.

**Break** point at which a story turns from one column to another. Also, an exclusive story.

**Break over** story that jumps from one page to another.

**Brightener** short, amusing item.

**Broadsheet** term used to describe a full-size newspaper page as opposed to tabloid.

**Broadside** large sheet printed on only one side.

**Broken heads** headlines with lines of different widths.

**Budget (or BJT, called "News Digest" by AP)** listing of the major stories expected to be delivered by the wire service.

**Bug** type ornament; a logotype; a star or other element that designates makeovers.

**Bulldog** an early edition of a newspaper.

**Bulletin** last-minute story of significance; a wire service designation of a story of major importance, usually followed by bulletin matter. Abbreviated *bun.*

**Bullets** large periods used for decoration, usually at the beginning of paragraphs.

**Bumper** two elements placed side by side or one immediately beneath the other. A bumped headline is also called a *tombstone.*

**Bureau code letters** each wire service uses its own code letters to designate a bureau.

**Byline** credit given to the author.

**Byte** in computer terms, one alphanumeric character.

**Canned copy** copy released by press agents or syndicates.

**Canopy head** streamer headline from which two or more readout heads drop.

**Caps** short for capital or uppercase letters.

**Caption** display line over a picture or over the cutline. Also used as a synonym for a *cutline.*

**Cartridge** holder for audio tapes. Also a *staccato lead.*

**CD-ROM** a compact disc that can contain information for delivery on a computer.

**Centered** type placed in the middle of a line.

**Center spread** two facing pages made up as one in the center of a newspaper section; also called *double truck.*

**Character** an alphanumeric or special symbol.

**Character generator** a cathode ray tube or laser used to display characters in high-speed phototypesetters.

**Chase** metal frame in which forms are locked before printing or stereotyping in the old hot-type process.

**Chaser** fast, urgent replate.

**Cheesecake** slang for photographs emphasizing women's legs. Good newspapers don't use such pictures.

**Circuits** refers to wire services used. The A wire is the main trunk news circuit. Regional news trunk systems carry letter designations such as B, G and E wires. High-speed circuits have replaced many of these.

**Circumlocution** wordy, roundabout expressions. Also *redundancy.*

**Circus makeup** flamboyant makeup featuring a variety of typefaces and sizes.

**City room** main newsroom of a newspaper. The city editor presides over the city desk.

**Civic journalism** a brand of local journalism that involves activist media participation in focusing the public's attention on key issues.

**Clean copy** copy with a minimum of typographical or editing corrections. Clean proof is proof that requires few corrections.

**Clips** short for clippings of newspaper stories.

**Close-up** photo showing head or head and shoulders or an object seen at close range.

**Closing** time at which pages are closed. Also *ending.*

**Cloze** method of testing readability. Respondents are asked to fill in words in blank spaces.

**Col.** abbreviation for column.

**Cold type** reproduction of characters composed photographically.

**Color page** page on which color is used.

**Column inch** unit of space measurement; one column wide and one inch deep.

**Column rule** printing units that create vertical lines of separation on a page.

**Combination cut** illustration that includes both halftone and line work.

**Combo** short for combination; pictures of the same subject used as a single unit.

**Compose** type is set or composed in a composing room by a compositor.

**Composition** all typesetting.

**Computer** a device capable of accepting information, applying prescribed processes to the information, and supplying the results of these processes; a computer system usually consists of input and output devices, storage, arithmetic and logical units and a control unit.

**Computer program** a set of stored instructions that controls all hardware.

**Constant** element used regularly without change. Also called *standing* or *stet* material.

**Copy** words keyed by reporters or editors from which type is set.

**Copy fitting** editing copy to fit a required space.

**Copyreader** antiquated term for *copy editor,* one who edits copy and writes headlines.

**Core** the main memory of a cornputer in which data may be stored. Also called *core memory.*

**CPU** an abbreviation for central processing unit. That portion of the computer that handles input, output and memory.

**CQ** abbreviation for correct.

**CQN** abbreviation for correction.

**Credit line** line indicating source of a story.

**Crop** to eliminate unwanted portions of a photograph. Marks used to show the elimination are called *crop marks.*

**Crossline** headline composed of a single line.

**Cue** signal given to announcer; a line in a script indicating a change.

**Cursor** a mobile block of light the size of a single character on the computer screen indicating the position at which an editing change is to be made.

**Cut** illustration or engraving. Or a direction to trim or shorten a story.

**Cut-in** may refer to an *initial letter* beginning a paragraph or to a side head that is set into the opening lines of a paragraph.

**Cutline** explanatory material, usually placed beneath a picture. Also called *underline, legend* and *caption.*

**Cutoff** hairline that marks the point where the story moves from one column to another or to separate boxes and photos from text material or to separate a headline from other elements.

**Cx** short for correction. Indicates that corrections are to be made in type. Also called *fix.*

**Cycle** complete news report for either morning or afternoon newspapers.

**Cycle time** the length of time used by a computer for one operation, usually measured in microseconds (millionths) or nanoseconds (billionths).

**Dangler** short for *dangling participle* or similar grammatical error.

**Dash** short line separating parts of headlines or headline and story.

**Data** any kind of information that can be processed by a computer.

**Dateline** opening phrase of story showing origin, source and sometimes date of story.

**Dead** newspaper copy or type that is no longer usable.

**Deadline** the shutoff time for copy for an edition.

**Debug** eliminating problems in a new or revised computer program or system.

**Deck** section of a headline.

**Delete** take out. The proofreader uses a symbol for a delete mark.

**Deontological ethics** a form of ethics in which there is always an absolute right or wrong.

**Desk** standing alone, usually the copy desk. Also *city desk, sports desk,* etc.

**Digit** in computers, a character used to represent one of the non-negative integers smaller than the radix, e.g., in decimal notation, one of the characters 0 to 9.

**Digital** in computers, pertaining to data in the form of digits, in contrast with analog.

**Dingbat** typographic decoration.

**Disk drive** a unit somewhat like a record turntable that contains a computer storage disk pack and rotates it at high speed.

**Disk pack** a set of magnetic disks used by computers for storing data or text. Also called *disk cartridge.*

**Disk storage** a means of storing data on a magnetic disk, a technique similar to magnetic tape or record, so that data can be read into the computer, changed by the computer or erased.

**Display** term given to a type of advertising that distinguishes it from classified advertising. Display lines are those set in larger sizes than regular body type.

**Dissolve** in broadcasting, a smooth transition from one image to another.

**Dog watch** late shift of an afternoon paper or early shift of a morning paper. Also *lobster trick.*

**Double-spacer** term used on broadcast news wire to designate a story of unusual significance. Extra space is used between copy lines to alert the editor to the story.

**Double truck** two pages at the center of a section made up as a single unit.

**Downstyle** headline style using a minimum of capital letters.

**Dropout** a subsidiary headline. Also called *deck.*

**Dub** transfer of film or tape.

**Dummy** diagram outlining the makeup scheme. A rough dummy has little detail; a pasteup dummy is created by pasting page elements on a sheet of paper the actual size of the page.

**Dump** a routine that causes the computer to dump data from storage to another unit.

**Dupe** short for *duplicate* or *carbon copy.*

**Dutch wrap** breaking body type from one column to another not covered by the display line. Also called *dutch turn* or *raw wrap.*

**Ears** small box on one or both sides of the nameplate carrying brief announcements of weather, circulation, edition and the like.

**Edition** one of several press runs such as *city edition* and *home edition.*

**Em** measurement of type that is as wide as it is high. A *pica em* is 12 points wide. Some printers still refer to all picas as *ems.*

**En** one-half em. Mostly used to express space. If the type measures 10 points, an indentation of an en would equal 5 points.

**Endmark** symbol (such as # or 30) to indicate the close of the story. An *end dash* (sometimes called a *30 dash)* is used at the end of the story in type.

**Etching** process of removing nonprinting areas from a relief plate by acid.

**Extra** now rare, a special edition published to carry an important news break.

**Eyebrow** another name for a *kicker* head.

**Face** style or cut of type; the printing surface of type or of a plate.

**Family** as applied to type, all the type in any one design. Usually designated by a trade name.

**Fat head** headline too large for the space allowed for it.

**Fax** short for facsimile or transmission by wire of a picture.

**Fax newspaper** a brief news update transmitted to clients by facsimile machine.

**Feature** to give special prominence to a story or illustration. A story that stresses a human-interest angle.

**Feed** story or program electronically transmitted to other stations or broadcast to the public.

**Filler** short items, usually set in type in advance and used to fill out space in a column of type. Also called *briefs* or *shorts.*

**Fingernails** parentheses.

**Fix** to correct or a correction.

**Flag** title of paper appearing on Page One. Also *nameplate.*

**Flash** rare announcement by a wire service of extremely urgent news. Usually followed by a *bulletin.*

**Float** ruled sidebar that may go anywhere in a story. To center an element in space that is not large enough to fill.

**Flop** illustration reversed in engraving.

**Flush** even with the column margin. Type aligned on one side. Alignment may be on either left or right side.

**Focus groups** groups of news consumers brought together to discuss what they like and don't like about the media.

**Folio** lines showing the newspaper name, date and page number.

**Follow** related matter that follows main story. Abbreviated *folo.*

**Follow copy** set the story as set; disregard seeming errors. Abbreviated *fc.*

**Follow-up** second-day story.

**Format** physical form of a publication. Also, a series of alphas and/or numerics that cause a computer to function in certain special ways.

**Frame makeup** vertical makeup.

**Freaks** devices that depart from normal indented body and headline type, etc.

**Futures file** record kept by the city desk of future events.

**FYI** for your information.

**Gain** sound level.

**Galley** metal tray used to hold hot type.

**Galley proof** print of the assembled type, used in proofreading.

**Gatekeeper** one who decides whether to pass a news story along. The account of an event goes through many gatekeepers before it reaches the reader.

**Glossy** photograph with a hard, shiny finish.

**Gobbledygook** editor's slang for material characterized by jargon and circumlocution.

**Gothic** sans serif type. Also called *block letter.*

**Graf** short for paragraph.

**Gravure** process of photomechanical printing. Also *rotogravure* or *intaglio* (printing ink is transferred to paper from areas sunk below the surface) printing.

**Guideline** instructions on copy to direct a printer. Usually includes slug, edition, section, etc.

**Gutter** vertical space that separates one page from another on two facing pages. Also, long, unbroken space between two columns of type.

**Hairline** finest line available in printing. Often used for rules between columns.

**Halftone** photoengraving. A dot pattern gives the illusion of tones.

**Hammer** see *barker.*

**Handout** release from a public relations agency.

**Hanger** headline that descends from a banner. Also called *readout* or *deck.*

**Hanging indent** headline style in which the top line is set full measure and succeeding lines are indented from the left.

**Hard copy** original copy, distinguishing it from monitor copy or carbon copy. Also a glossy photographic print as contrasted to facsimile.

**Hard news** spot news or news or record as contrasted to features and background material.

**Hardware** the physical equipment that is part of the overall production system (computers, phototypesetting machines, etc.).

**Hard-wired** electronically connected, as in the case of one or more pieces of hardware connected to others.

**Head count** number of letters and spaces available for a headline.

**Headlinese** overworked short words in a headline, such as *cop, nab, hit, set.*

**Head shot** photo of person's head or head and shoulders. Also called *face shots* and *mug shots.*

**HFR** hold for release.

**Highlight** white or light portions of a photograph. Also the high point of a story.

**Hold for release** copy that is not to be used until a specified time.

**Holdout** portion held out of a story and placed in overset.

**Hot type** linecaster slugs as opposed to cold type or type set photographically.

**HTK** headline to come. Also HTC.

**HTML** HyperText Markup Language. Used to create documents for the World Wide Web.

**Hugger mugger** newspaper lead crammed with details.

**Hypertext** links within electronic documents that send the reader to other information about the subject.

**Hyphenless justification** a system of justifying lines by interword and interletter spaces, without breaking any words at the end of lines. Usually accomplished by computers.

**Imposition** process of placing type and illustrations in pages.

**Impression** any printing of ink on paper. Also appearance of the printed page. Also the number of times a press has completed a printing cycle.

**Index** newspaper's table of contents, usually found on Page One.

**Infomedium** a term sometimes used to describe a new medium of the future that would incorporate both text and moving pictures.

**Information Age** sociologists and others often describe the current era this way because computers have made available a vast amount of information.

**Information graphics** illustrations, charts, tables or maps used to convey information graphically.

**Information Superhighway** a term widely attributed to Vice President Al Gore to refer to computerized information delivery into the home.

**Initial (Initial cap)** first letter in a paragraph set in type larger than the body type.

**Input** bringing data or text from external sources into computer storage.

**Insert** addition to a story placed within the story.

**Interface** a device through which one piece of computer hardware or equipment communicates with another. The translation of output to input.

**Internet** a worldwide computer network.

**Intro** short for introduction. Opening copy to film or tape. Also, wire service term for lead of story.

**Inverted pyramid** news story structure in which the parts are placed in a descending order of their importance. Also, a headline in inverted pyramid shape.

**Issue** all copies produced by a newspaper in a day.

**Italics** slanted letter form. Shortened form is *itals.*

**Jargon** language of a profession, trade or group. Newspaper jargon.

**Jump** to continue a story from one page to another.

**Jump head** headline over the continued portion of a story.

**Jump lines** continuation lines: *continued on Page X.*

**Justify** spacing out a line of type to fill the column; spacing elements in a form so the form can be locked up.

**K** thousands of units of core or computer storage referring to storage for bytes, usually in multiples of 4.

**Kenaf** field crop used experimentally as a substitute for wood pulp in making newsprint.

**Kicker** overline over a headline. Also *eyebrow.*

**Kill** to discard copy, type, mats and so on.

**Label head** dull, lifeless headline. Sometimes a standing head such as **NEWS OF THE WORLD.**

**Layout** pattern of typographic arrangement. Similar to dummy.

**Lead** beginning of a story or the most important story of the day (pronounced "leed").

**Lead** piece of metal or spacing increment varying from ½ to 3 points placed between lines of type for spacing purposes (pronounced "led").

**Leaders** line of dots.

**Leading** the space between lines of type.

**Lead out** to justify a line of type.

**Legend** information under an illustration. Also *cutline.*

**Letterpress** technique of printing from raised letters. Ink is applied to the letters, paper is placed over the type and impression implied to the paper, resulting in printing on paper.

**Library** newspaper's collection of books, files and so on. Also called *morgue.*

**Ligature** two or more letters run together for superior typesetting—*fi, ffi.*

**Linecaster** any keyboarded machine that casts lines of type.

**Line cut** illustration without tones. Used for maps, charts and so on.

**Line gauge** pica rule or a ruler marked off in pica segments.

**Logotype** single matrix of type containing two or more letters commonly used together: AP, UPI. Also a combination of the nameplate and other matter to identify a section. Also an advertising signature. Commonly abbreviated *logo.*

**Lowercase** small letter as distinguished from a capital letter. Abbreviated *lc.*

**Machine language** a language used directly by a computer; a set of instructions a computer can recognize and execute.

**Macro instruction** a single instruction that is expanded to a predetermined sequence of instructions during the assembly of a computer program.

**Magnetic storage** any device using magnetic materials as a medium for computer storage of data: magnetic disk, film, tape, drum, core, etc.

**Makeover** to change page content or layout.

**Makeup** design of a newspaper page. Assembling elements in a page.

**Marker** proof or tearsheet used to show where inserts are to go after story has been sent to the composing room or other instructions for guidance of printers and makeup editors.

**Masthead** informational material about a newspaper, usually placed on the editorial page.

**Mat** short for matrix or mold for making a stereotype plate in letterpress printing. The mat of a page is made of papier-mâché. In typesetting machines, the matrix is made of brass.

**Measure** length of a line of type.

**Modem** a piece of hardware that facilitates connections between two other pieces of hardware in a system.

**Monitor copy** tearsheet copy or copy produced electronically. A monitor is also a television or radio receiver. To monitor is to watch or time a radio or television program.

**Montage** succession of pictures assembled to create an overall effect. Usually a single photograph using several negatives. This technique is usually discouraged.

**Morgue** newspaper reference library or repository for clippings.

**Mortise** cutaway section of a photo into which type is inserted. Usually considered bad practice.

**Mug shot** same as *photo close-up* or *face shot.*

**Multimedia** the use of various media—text, graphics, video and audio—to deliver information.

**Multiplexor (MUX)** a device that allows the computer to handle input or output simultaneously from several different devices connected to it.

**Must** matter that someone in authority has ordered published.

**Nameplate** name of newspaper displayed on Page One. Also called *flag, title* or *line.*

**New media** media delivered via computer.

**News hole** space left for news and editorial matter after ads have been placed on pages.

**Newsprint** low-quality paper used to print newspapers.

**NL** slug on copy and notation on a marker indicating new lead. Also wire service designation for *night lead.*

**Obit** abbreviation for obituary.

**OCR** optical character reader, or "scanner," which converts typewritten material to electronic impulses for processing within a computer system. The process itself also is known as OCR, optical character recognition. Employs technology now considered outdated.

**Offset** method of printing differing from letterpress. A photograph is taken after the page has been assembled. The negatives are placed over a light-sensitive printing plate and light is exposed through the open spaces of the negative. The result is that the letters are hardened and the nonprinting surface is washed away. This method of printing involves inking the printed plate with a water solution and then with ink. The water resides only on the nonprinting surface, whereas the ink resides on the printing surface. The inked letters are then printed on a rubber blanket, which in turn prints (or offsets) on paper.

**On-line** a unit "on-line" is connected to units operating independently of the system.

**Op. ed.** page opposite the editorial page.

**Optional** matter that may be used without regard to the time element. Also called *time copy, grape* and *AOT* (any old time).

**Out-cue** cue telling a news director or engineer that a film or tape is near the end.

**Outlined cut** halftone with background cut out. Also *silhouette.*

**Output** data coming from the computer system in either printed, typeset or paper tape form.

**Overset** type in excess of amount needed.

**Pad** to make a story longer with excess words.

**Page proof** proof of an entire page.

**Pagination** makeup of a newspaper page on a computer to allow output of the composed page. Also the organization of pages within the newspaper.

**Parameter** symbol in computer programming indicating a constant such as a figure.

**Parens** short for parentheses.

**Photocomposition** type composed photographically.

**Photolithography** printing process such as offset in which the impression is transferred from a plate to a rubber roller and to paper.

**Pica** linear measure in 12 points. A *pica em* is a standard measure, but only 12-point type can be a *pica em.*

**Pix** short for pictures. The singular may be *pic* or *pix.*

**Pixel** small picture element to denote gray or color values for computer storage.

**Plate** stereotyped or lithographic page ready for the press.

**Play** prominence given a story, its display. Also the principal story.

**Point** unit of printing measurement, approximately 1/72 inch. Actually 0.01384 inches. Also any punctuation mark.

**Pos.** positive film image.

**Precede** material such as a bulletin or an editor's note appearing at the top of a story.

**Predate** edition delivered before its announced date. Usually a Sunday edition delivered to outlying areas.

**Printer** machine that produces copy by telegraphic impulses. A Teletype machine. Also a person who prepares composition for imprinting operations.

**Printout** visual copy produced by a computer, usually for proofreading. Same as *master copy* or *tear sheet.*

**Process color** method of printing that duplicates a full-color original copy.

**Proof** print of type used for proofreading.

**Proofreader** one who corrects proofs.

**Prop** an object used during a newscast to give credence to an item.

**Public information utilities** commercial computer services marketed to home users.

**Public journalism** see *civic journalism.*

**Pullout** a special section within a paper, designed to be removed from the main portion.

**Purge** a method by which data or text is removed from computer storage by erasure from the disk.

**Put to bed** to lock up forms for an edition.

**Pyramid** arrangement of ads in half-pyramid form from top right to lower left.

**Quad** short for quadrant, a blank printing unit for spacing.

**Quadrant** layout pattern in which the page is divided into fourths.

**Query** brief message outlining a story. Also a question put to a news source.

**Queue** directing a story file to another operation, such as a line printer or photocomposition machine. The electronic equivalent of an in-basket or holding area.

**Quote** short for quotation marks.

**Race** classification of type, such as *roman, text, script.*

**Railroad** to rush copy to the printer before it is edited; to rush type to press without proofreading. Also a term for a headline type.

**Read in** secondary head leading in to the main head.

**Readout** secondary head accompanying a main head.

**Rear projection** projection of a film, photo, map or graph placed on a screen behind the newscaster.

**Register** alignment of printing plates to get true color reproduction.

**Release copy** copy to be held until a specified release time. Same as advance copy.

**Repeat** a rerun of a story for a wire service member or client.

**Replate** to make a page over after an edition has gone to press.

**Retouch** to alter a photograph by painting or airbrushing.

**Reverse plate** reversing the color values so that white letters are on a black background.

**Revised proof** second proof after corrections have been made.

**Ribbon** another name for a *banner* or *streamer* headline.

**Rim** outer edge of a copy desk. Copy editors once were known as *rimmen* or *rim women.*

**Rip and read** derogatory expression applied to radio newspersons who simply read the latest summary from the radio wire without careful editing.

**Rising initial** initial capital letter that aligns with the body type at the baseline.

**Rivers** streaks of white space within typeset columns caused by excessive word spacing or letterspacing.

**ROP** run of the paper. Stories or art that do not demand upfront position. Ads that may appear anywhere in the paper. Color printed in a newspaper without the use of special presses.

**Rotogravure** means of printing from recessed letters. One of the major printing techniques (along with letterpress and offset). Used mostly in catalogs, magazines and fine color work.

**Rough** may be applied to a dummy that gives little or no detail or to an uncorrected, unjustified proof.

**Roundup** compilation of stories.

**Rules** any line that is printed. Lines are cast in type metal form. Hairline rules are often used in newspaper work. The underscore of the preceding is a type rule.

**Run** reporter's beat.

**Runaround** method of setting type around a picture.

**Run-in** to incorporate sentences or lists into one paragraph.

**Running story** story handled in takes or small segments. Each take is sent to the composing room as soon as it is edited.

**Runover** portion of a story that continues from page to the next. Also a *jump story.*

**Sans serif** typeface without serifs.

**sc** proofreader's mark meaning *see copy.*

**Schedule** list of available stories and pictures; desk's record of stories edited.

**Scoop** to get an exclusive story. Also a *beat.*

**Screen** to view film or videotape.

**Script** in broadcast news, the arrangement of news, together with an opening and closing and leads to commercials.

**Scroll** a means of moving story text forward or backward so it can be displayed on a computer screen.

**Second front page** first page of the second section. Also called a *split page.*

**Section page** first page of a pullout section.

**Serifs** the fine cross strokes at the top and bottom of most styles of letters.

**Set solid** lines of type without extra spacing between lines.

**Shirt tail** slang for follow story.

**Short** brief item of filler.

**Sidebar** brief story with a special angle that goes with the main story.

**Signature** group of pages on one sheet. Also an advertiser's name displayed in an ad.

**Silhouette** form of halftone with the background removed. Same as *outline cut.*

**Situation ethics** a form of ethics in which journalists take a stand based on the facts of the situation.

**Skyline** headline across top of page over nameplate.

**Slant** angle of a story. A story written a certain way.

**Slug** label identifying a story. Same as *guideline* or *catchline.* Also a piece of metal used for spacing. Also used to designate linecaster slugs.

**Software** the programs that make computer hardware work.

**Sound-on-film** film carrying its own sound track. Abbreviated *SOF.*

**Sound under** audio level where background sounds may be heard.

**Space out** direction to the printer to add space between lines until the story fills the space allotted for it.

**Split** term used to designate a break in a wire service circuit to permit the filing of other material, such as regional news.

**Split page** first page of the second section of a newspaper.

**Split run** making a change in part of a press run of the same edition.

**Spot news** news obtained firsthand; fresh news.

**Spread** story prominently displayed, often over several columns and with art.

**Squib** short news item or filler.

**Stand-alone** a picture without an accompanying story.

**Standing box** type box kept on hand for repeated use. Likewise with standing head.

**Standing type** similar to standing boxes and head; type kept standing for future use.

**Standupper** television report at the scene with the camera on the reporter.

**Steplines** headline with top line flush left, second line centered and third line flush right.

**Stereotype** process of casting a plate from a papier-mâché mold.

**Stet** let it stand. Disregard correction.

**Stinger** another term for *kicker* or *eyebrow.*

**Stock** paper used for any printing job.

**Straight matter** copy set in one size of type for the main reading matter of a page. Also called body type.

**Streamer** another name for a *banner* or a *ribbon* headline.

**String** clippings of stories, usually from a correspondent.

**Stringer** correspondent paid on space rate. In television news, a free-lance camera operator.

**Strip-in** to insert one illustrative element into another.

**Sub** short for substitute. *Sub bomber* means a new story for a story slugged *bomber.*

**Subhead** one- or two-line head used within the body of a story in type.

**Summary** may be a news index or a news roundup. A summary lead gives the gist of the facts in the story.

**Supplemental service** syndicated service in addition to major wire service.

**System** a combination of computer programs and hardware designed to perform specific tasks.

**TAB** indicates tabular matter in wire service copy.

**Tabloid** newspaper format, usually four or five columns wide and approximately 14 inches deep.

**Take** small part of a running story. Also the part of a story given to a compositor.

**Tear sheet** sheet or part of a sheet used for corrections. Also copy produced by a computerized copy follower.

**Tease** news announcement before the station break with details to follow the break.

**Teleological ethics** a form of ethics in which the consequences of an act, not the act itself, make something ethical or unethical.

**Teletext** home information retrieval system using television signal for transmission of data.

**Teletype** automatic printer used to send and receive wire copy.

**Terminal** a point in a system or communication network at which data can either enter or leave.

**Thirty dash** end mark.

**Thumbnail** half-column portrait.

**Tie-back** part of a story providing background material.

**Tight paper** paper containing so much advertising there is limited space for news.

**Time copy** copy that may be used anytime. Also called *grape, plug copy* and so on.

**Tombstone** to place headlines of the same type side by side. Such adjacent heads are called *bumped heads.*

**Turn story** same as jump story (continues from last column on one page to first column on the next page).

**Typo** short for typographical error.

**Undated** story without a dateline (but usually a credit line) summarizing related events from different origins.

**Underline** same as cutline.

**Update** to bring a story up to date or to give it a timely angle.

**UPI** short for United Press International, a major news agency.

**Upstyle** headline style in which each word is capitalized.

**VDT** an acronym for video display terminal, a device that looks like a typewriter with a small television set attached. Stories may be composed or corrected on these minicomputer-based units, which increasingly are being replaced by networked personal computers.

**Videotape** tape that projects pictures.

**Videotex** home information retrieval systems using television-type displays.

**Vignette** halftone with a fading background. Also a feature story or sketch.

**Visible** tape perforations arranged so that they spell words or symbols.

**VTR** short for videotape recording.

**wf** short for wrong font or type of a different size or style from that used in text.

**Wicket** kicker-like element placed to one side of a headline.

**Widow** one or two words appearing at the end of a paragraph and on the last line. It is unsightly because of the excessive white space appearing after the widow, particularly at the top of a column.

**Wirephoto** AP system of transmitting pictures by wire.

**Wooden head** one that is dull and lifeless.

**World Wide Web** the user-friendly portion of the Internet capable of displaying photos and graphics as well as text.

**Wrap around** ending the top line of a headline with a preposition, conjunction or the like, or splitting words that are properly a unit. Also, setting type to wrap around a picture.

**Wrapup** complete story. Wire services use a wrapup to contain in one story all elements of the same story sent previously.

# INDEX

*Note:* Italicized words and phrases (with the exception of titles and foreign words) indicate discussions in the Appendixes.